Geodynamics Series

Published under the aegis of the AGU Books Board.

ISSN — 0277-6669. ISBN — 0-87590-528-5

Figures, tables, and short excerpts may be reprinted in scientific books and journals if the source is properly cited.

Authorization to photocopy items for internal or personal use, or the internal or personal use of specific clients, is granted by the American Geophysical Union for libraries and other users registered with the Copyright Clearance Center (CCC) Transactional Reporting Service, provided that the base fee of $01.50 per copy plus $0.50 per page is paid directly to CCC, 222 Rosewood Dr., Danvers, MA 01923.0277-6669/98/$01.50+0.5. This consent does not extend to other kinds of copying, such as copying for creating new collective works or for resale. The reproduction of multiple copies and the use of full articles or the use of extracts, including figures and tables, for commercial purposes requires permission from the American Geophysical Union.

Copyright 1998
Published by the American Geophysical Union
2000 Florida Avenue, N.W.
Washington, D.C. 20009

Printed in the United States of America.

Structure and Evolution of the Australian Continent

Jean Braun
Jim Dooley
Bruce Goleby
Rob van der Hilst
Chris Klootwijk
Editors

Geodynamics Series   Volume 26

American Geophysical Union
Washington, D.C.

Geodynamics Series

1. Dynamics of Plate Interiors
   A. W. Bally, P. L. Bender, T. R. McGetchin, and R. I. Walcott (Editors)

2. Paleoreconstruction of the Continents
   M. W. McElhinny and D. A. Valencio (Editors)

3. Zagros, Hindu Kush, Himalaya, Geodynamic Evolution
   H. K. Gupta and F. M. Delany (Editors)

4. Anelasticity in the Earth
   F. D. Stacey, M. S. Patterson, and A. Nicholas (Editors)

5. Evolution of the Earth
   R. J. O'Connell and W. S. Fyfe (Editors)

6. Dynamics of Passive Margins
   R. A. Scrutton (Editor)

7. Alpine-Mediterranean Geodynamics
   H. Berckhemer and K. Hsü (Editors)

8. Continental and Oceanic Rifts
   G. Pálmason, P. Mohr, K. Burke, R. W. Girdler, R. J. Bridwell, and G. E. Sigvaldason (Editors)

9. Geodynamics of the Eastern Pacific Region, Caribbean, and Scotia Arcs
   Rámon Cabré, S. J. (Editor)

10. Profiles of Orogenic Belts
    N. Rast and F. M. Delaney (Editors)

11. Geodynamics of the Western Pacific-Indonesian Region
    Thomas W. C. Hilde and Seiya Uyeda (Editors)

12. Plate Reconstruction From Paleozoic Paleomagnetism
    R. Van der Voo, C. R. Scotese, and N. Bonhommet (Editors)

13. Reflection Seismology: A Global Perspective
    Muawia Barazangi and Larry Brown (Editors)

14. Reflection Seismology: The Continental Crust
    Muawia Barazangi and Larry Brown (Editors)

15. Mesozoic and Cenozoic Oceans
    Kenneth J. Hsü (Editor)

16. Composition, Structure, and Dynamics of the Lithosphere-Asthenosphere System
    K. Fuchs and C. Froidevaux (Editors)

17. Proterozoic Lithospheric Evolution
    A. Kröner (Editor)

18. Circum-Pacific Orogenic Belts and Evolution of the Pacific Ocean Basin
    J. W. H. Monger and J. Francheteau (Editors)

19. Terrane Accretion and Orogenic Belts
    Evan C. Leitch and Erwin Scheibner (Editors)

20. Recent Plate Movements and Deformation
    K. Kasahara (Editor)

21. Geology of the USSR: A Plate-Tectonic Synthesis
    L. P. Zonenshain, M. I. Kuzmin, and L. M. Natapov   B. M. Page (Editor)

22. Continental Lithosphere: Deep Seismic Reflections
    R. Meissner, L. Brown, H. Dürbaum, W. Franke, K. Fuchs, F. Seifert (Editors)

23. Contributions of Space Geodesy to Geodynamics: Crustal Dynamics
    D. E. Smith, D. L. Turcotte (Editors)

24. Contributions of Space Geodesy to Geodynamics: Earth Dynamics
    D. E. Smith, D. L. Turcotte (Editors)

25. Contributions of Space Geodesy to Geodynamics: Technology
    D. E. Smith, D. L. Turcotte (Editors)

CONTENTS

Preface
J. Braun, J. C. Dooley, B. R. Goleby, R. D. van der Hilst, and C. T. Klootwijk iv

Secular Variation in the Composition of Subcontinental Lithospheric Mantle: Geophysical and Geodynamic Implications
W. L. Griffin, S. Y. O'Reilly, C. G. Ryan, O. Gaul, and D. A. Ionov 1

Hypotheses Relevant to Crustal Growth
A. L. Hales 27

Upper Mantle Structure beneath Australia from Portable Array Deployments
R. D. van der Hilst, B. L. N. Kennett, and T. Shibutani 39

Mapping of Geophysical Domains in the Australian Continental Crust Using Gravity and Magnetic Anomalies
P. Wellman 59

Complex Anisotropy in the Australian Lithosphere from Shear-wave Splitting in Broad-band SKS Records
G. Clitheroe and R.. D. van der Hilst 73

A Brief Review of Differences in Lithosphere Seismic Properties Under Western and Eastern Australia Stimulated by Seismograms from the Marryat Creek Earthquakes of 1986
B. J. Drummond 79

Lithospheric Structure in Southeast Australia: a Model Based on Gravity, Geoid and Mechanical Analyses
Y. Zhang, E. Scheibner, B. E. Hobbs, A. Ord, B. J. Drummond, and S. J. D. Cox 89

The Mount Isa Geodynamic Transect - Crustal Implications
B. R. Goleby, T. MacCready, B. J. Drummond, and A. G. Goncharov 109

Intra-Crustal "Seismic Isostasy" in the Baltic Shield and Australian Precambrian Cratons from Deep Seismic Profiles and the Kola Superdeep Bore Hole Data
A. G. Goncharov, M. D. Lizinsky, C. D. N. Collins, K. A. Kalnin, T. N. Fomin, B. J. Drummond, B. R. Goleby, and L. N. Platonenkova 119

Contrasting Styles of Lithospheric Deformation Along the Northern Margin of the Amadeus Basin, Central Australia
J. Braun and R. Shaw 139

Extension in the Fitzroy Trough, Western Australia: an Example of Reactivation Tectonics
J. Braun and R. Shaw 157

Granite-Greenstone Zircon U-Pb Chronology of the Gum Creek Greenstone Belt, Southern Cross Province, Yilgarn Craton: Tectonic Implications
Q. Wang, J. Beeson, and I. H. Campbell 175

PREFACE

Recent geophysical, geochemical and geological studies have led to a much improved understanding of the structure and evolution of the Australian continent from its Archaean nuclei to its present-day morphology. This new wealth of information has raised additional questions on the continent's geological past, and has led to the formulation of new hypotheses on continental growth and dynamics which will serve to inspire further investigations. This volume compiles the most current geological and geophysical data pertaining to the formation and evolution of the Australian continent through geological time. Although the main focus of this monograph is the structure and evolution of the Australian continent, many of the observations and interpretations are discussed in a global framework and are relevant for studies of other continents.

The contributors describe our state of knowledge on the structure of the Australian continent, and our present understanding of how this structure evolved through geological time by accretion of the central Proterozoic crustal blocks and eastern Phanerozoic terranes to the Archaean nuclei of western Australia.

New data are presented from a wide range of disciplines including seismology, petrophysics, petrochemistry, geochronology, and potential field studies. The data include the seismic-velocity structure under most parts of the continent to depths of 1000 km from the inversion of seismic data from a portable array of broad-band seismometers (the SKIPPY experiment), deep reflection seismic soundings, observations of seismic anisotropy from shear-wave splitting, high-resolution gravity and magnetic anomaly maps, modal and trace element data from xenoliths, and high-resolution sensitive high resolution ion microprobe (SHRIMP) U-Pb zircon dating of Archaean rocks in the Yilgarn Block.

Hypotheses and models are formulated concerning the growth and re-structuring of the continent through a large number of tectonic events, such as the Proterozoic Isan Orogeny of northeastern Australia, the late Palaeozoic Alice Springs Orogeny of central and northwestern Australia, and the Mesozoic continental extension associated with the opening of the Tasman Sea along the southeastern margin of the continent.

This volume is based on papers presented at the 1996 Western Pacific Geophysics Meeting in Brisbane. The meeting was co-sponsored by the Specialist Group on Solid Earth Geophysics of the Geological Society of Australia and the American Geophysical Union. The editors wish to thank the many individuals who have contributed to this monograph, and especially the authors and reviewers, who worked diligently to permit its timely publication.

Finally, we hope that this Australian perspective will help other Earth scientists to improve our understanding of the evolution and dynamics of the Earth's continents.

Jean Braun
Jim Dooley
Bruce Goleby
Rob van der Hilst
Chris Klootwijk
Editors

Secular Variation in the Composition of Subcontinental Lithospheric Mantle: Geophysical and Geodynamic Implications

W. L. Griffin

National Key Centre for the Geochemical Evolution and Metallogeny of Continents, School of Earth Sciences, Macquarie University, Sydney, NSW 2109, Australia, and CSIRO Exploration and Mining, P.O. Box 136, North Ryde, NSW, Australia

Suzanne Y. O'Reilly

National Key Centre for the Geochemical Evolution and Metallogeny of Continents, School of Earth Sciences, Macquarie University, Sydney, NSW, Australia

C. G. Ryan

CSIRO Exploration and Mining, P.O. Box 136, North Ryde, NSW, Australia

O. Gaul and D. A. Ionov

National Key Centre for the Geochemical Evolution and Metallogeny of Continents, School of Earth Sciences, Macquarie University, Sydney, NSW, Australia

A synthesis of modal and trace element data for mantle-derived peridotites and the compositions of over 8000 mantle-derived Cr-pyrope garnets, documents a secular and apparently irreversible change in the chemical composition of newly created lithospheric mantle throughout the Earth's history. This change suggests an evolution in fundamental large-scale Earth processes; it has important implications for the interpretation of seismic tomography, and means that lithosphere erosion will have major tectonic consequences. The average composition of peridotitic garnet xenocrysts from volcanic rocks is strongly correlated with the tectonothermal age of the continental crust penetrated by the eruptions. Garnets derived from harzburgitic or lherzolitic rock types can be recognised by comparison with data from mantle-derived xenoliths, and used to estimate relative abundances of these rock types in individual mantle sections. Subcalcic harzburgites are found only in lithospheric mantle beneath Archean terrains; mildly subcalcic harzburgites are common beneath Archean terrains, less abundant beneath Proterozoic terrains, and essentially absent beneath terrains with tectonothermal ages less than 1 Ga. Garnets from lherzolites (clinopyroxene-bearing peridotites) show a decrease of mean Cr content and Zr/Y, and a rise in Y and Y/Ga, with decreasing crustal age. Modeling using empirical element distribution coefficients suggests that these changes reflect a rise in (cpx+gnt) and cpx/gnt, and a decrease in mg#, from Early Proterozoic time to the present. The Archean-Proterozoic boundary represents a major change in the processes that form continental lithospheric mantle; since 2.5 Ga there has been a pronounced, but more gradual, secular change in the nature of these processes. Actualistic models of lithosphere formation based on modern processes may be inadequate, even for Proterozoic time. The correlation between mantle type and crustal age indicates that the continental crust and the underlying lithospheric mantle are formed together, and generally stay coupled together for periods of eons. The stability and thickness of Archean lithospheric mantle is directly related to its low density, which in turn reflects both its high degree of depletion in basaltic components, and its low Mg/Si. These chemical characteristics produce high seismic velocities, and compositional factors may account for at least half of the velocity contrast between Archean and younger areas, seen in seismic tomography. The higher density and mantle heat flow of younger, less depleted mantle sections imposes severe limits on their thickness and ultimate stability, because the cooler upper parts of these sections will be negatively buoyant relative to the underlying asthenosphere.

1. INTRODUCTION

The aim of this paper is to examine changes in the composition of the subcontinental lithospheric mantle from Archean time to the present. The nature and scale of these variations represent fundamental information relating to the geodynamic evolution of Earth, and changes through time in the processes that have produced the continents. Understanding of the evolution of these processes impacts on our ideas about major aspects of geology, from the composition of the Earth, to the formation and localisation of large ore deposits. Knowledge of the composition of the mantle is essential also for realistic modeling of geophysical data, especially seismic and gravity.

Diamond exploration activities have made a major contribution to understanding the nature of the continental lithosphere. One fundamental observation from this body of work is summarised in Clifford's Rule (as expanded by Janse [1994]): kimberlites with economic concentrations of diamonds are restricted to cratons with crustal ages greater than or equal to 2.5 Ga, and diamondiferous lamproites are restricted to areas with crustal ages between 2.5 and 1.6 Ga. Combined with experimental data on diamond stability in the mantle, this observation has led to the concept of a thick cold "root" or "keel" beneath Archean cratons, and its corollary, that the lithospheric mantle beneath younger terrains should be thinner or hotter, or both, to explain the scarcity of diamonds in regions with Proterozoic and Phanerozoic crustal ages [Boyd and Gurney, 1986].

These ideas have been largely substantiated by the results of seismic tomography studies, which show regions of high seismic velocity extending to 150-300 km depth beneath some Archean cratons, but not beneath the younger parts of continents [Su et al., 1994]. These high-velocity volumes are interpreted as being cooler than the lower-velocity volumes, reflecting the generally lower surface heat flow measured in many Archean cratons [Morgan, 1995]. Archean model ages on inclusions in diamonds [Richardson et al., 1984] and Archean Re-Os depletion ages on mantle-derived xenoliths in kimberlites from Archean cratons [Pearson et al., 1995] strongly suggest that the lithospheric keels beneath these areas formed in Archean time, and have persisted to the present. The repeated intrusion of diamondiferous kimberlites in some areas (e.g. the Kaapvaal Craton of South Africa, from at least 1600 Ma to 80 Ma [Smith et al., 1995]) also testifies to the long-term stability of at least some Archean keels.

This general model of lithospheric keels beneath some Archean cratons raises the fundamental question of their origin and the reasons for their persistence through geologic time. Are the causes tectonic, thermal or compositional? This paper will briefly examine the evidence from mantle-derived xenoliths, and then expand on this using a large body of compositional data on mantle-derived garnets, to trace the evolution of subcontinental mantle through time.

2. DATA BASES AND DEFINITIONS

Two data bases are used in this paper. One consists of averaged data on garnet concentrates from volcanic rocks of widely different ages and tectonic settings (Table 1). The other is a compilation of modal and compositional data on garnet peridotite xenoliths from a variety of tectonic settings (Table 2). All of the garnets used in Table 1 are Cr-pyropes, judged on the basis of composition to be derived from ultramafic wall rocks during ascent of the host volcanic rock. Many of the samples are derived from diamond exploration activities; others have been collected or analysed specifically for the purposes of this research. The garnets have been analysed for major elements by electron microprobe, and for trace elements either by proton

Table 1. Garnet data

Locality	no. deposits	age of intrusion	Ca-harzb gnt %	low-Ca gnt %	diamonds	max. grade cts/tonne	T range selected	no. gnts total	no. in T range	lherz no. in T range	mean Zr/Y ± 1 std dev	mean Y/Ga ± 1 std dev	mean Cr2O3 wt % ± 1 std dev	mean Zr ppm	med. Zr ppm	mean Y ppm	med. Y ppm	mean TiO2 wt % ± 1 std dev	med. TiO2 ppm	mean Ti/Y	med. Ti/Y
Archons																					
Malo-Bot.	4	E. Paleozoic	13	2.5	abundant	5	600-1100	440	375	278	2.5±2	2±1.5	4.6±1.6	34	28	17	17	1700±1100	1600	60	56
Daldyn	9	E. Paleozoic	24	32	abundant	5	600-1100	826	692	389	3±3	1.1±0.7	5.2±1.9	29	20	11	9	2100±1900	1900	113	127
Alakit	9	E. Paleozoic	23	11	abundant	7	600-1100	440	356	280	3.6±3	1.4±0.9	5±2.2	40	26	15	15	2600±1900	2425	105	97
Upper Muna	3	E. Paleozoic	17	3.5	abundant	1	600-1100	295	75	75	3.9±3	1.3±0.6	4.8±1.9	55	45	13	14	3900±2300	4000	178	171
Liaoning	3	E. Paleozoic	11	7	abundant	2.5	600-1100	318	246	173	8.1±9	1.3±1	6.7±2.8	36	18	9	6	1200±1150	875	81	87
Shandong	4	E. Paleozoic	25	4.5	abundant	1	600-1200	314	118	56	7.9±9	1±0.7	5.6±2.8	47	32	10	8	3100±2300	2670	182	200
Kaapvaal >90MA	5	Cretaceous	10	7	abundant	1.5	600-1100	288	256	165	5.7±8	2.4±1.8	4.6±2.2	46	37	14	14	1800±1500	1320	78	57
Kaapvaal <90MA	9	Cretaceous	12	1.5	abundant	0.8	600-1100	318	239	185	7±8	2.4±2.4	3.8±1.4	52	39	11	10	1200±1300	825	66	49
Slave	18	Mesozoic	31	4.5	abundant	5	900-1150	529	529	481	4.5±4	1.3±0.8	6±2.4	36	34	11	10	2400±1200	2320	132	139
Guaniamo	4	Proterozoic	15	12	abundant	1.5	600-1100	179	172	113	3±2.5	1.3±1	5.7±2.7	26	21	11	10	2100±920	2160	116	129
Majhgawan	1	Proterozoic	16	1.5	common	0.2	800-1300	157	153	126	5.1±7	1.3±.8	5.7±1.6	40	29	11	11	1900±1600	1550	104	84
Arch/Prot																					
Saskatchewan	6	Cretaceous	10	0	comon	0.1	600-1100	221	150	138	3±2.5	2±1.8	5.2±1.8	35	24	14	14	2100±1900	1650	90	71
N. Michigan	8	Paleozoic	8.5	0	rare	0.05	600-1100	397	245	206	4.9±6	2.1±1.3	4.8±1.2	45	36	13	12	1750±1200	1470	80	73
State Line	7	Devonian	7.5	1	common	0.8	600-1100	492	340	284	5±5.5	1.7±1.4	5.5±2.3	38	25	11	10	1750±1900	1150	95	69
Finland	3	Paleozoic?	18	1	common	?	600-1200	204	162	119	3.4±3	1.5±1.8	5.7±1.8	32	32	11	12	3300±2700	2880	181	144
Protons																					
Merchimden	5	E. Paleozoic	0.5	none	none	0	600-1100	213	133	133	2±2	3±2	3.6±1.6	29	21	18	18	1300±600	1260	43	42
U. Molodo	3	Mesozoic	0	none	none	0	600-1100	100	75	75	1±1.5	2.9±1.4	3.2±1	12	9	19	18	1000±440	1015	32	34
Kuoika	4	Mesozoic	1	none	none	0	600-1100	410	309	309	1.9±2	3.2±3	3.5±1.2	23	18	16	15	960±550	970	36	39
Guizhou	1	Paleozoic	1	none	rare	0.15	800-1300	82	82	82	1.5±1	1.3±0.5	3.8±2	19	19	14	16	3000±1300	3080	130	115
Hunan	4	Mesozoic?	0	none	none	0	800-1300	153	147	147	4±3	1.2±1	6.6±2.2	19	19	8	5	1650±1500	1330	125	159
Hubei	4	Paleozoic	0.5	none	none	0	600-1100	157	135	114	1.4±2.5	3.3±1.4	2.7±1	13	12	18	19	1050±600	980	36	31
Arkansas	3	Cretaceous	10	none	common	0.1	600-1100	110	85	77	2±2.5	4.5±3.3	3.7±2.3	19	10	24	24	1650±1200	1380	41	34
S. Australia	8	Mesozoic	1.5	none	rare	0.1	600-1100	639	429	365	4.9±5.5	2.7±3	4.9±2.7	43	29	14	10	1500±1700	850	64	51
Ellendale	4	Cenozoic	5.5	none	common	0.15	600-1200	283	270	251	2.3±2.5	1.5±1.7	4.6±1.6	17	10	9	9	1700±1100	1640	113	109
Tectons																					
Vitim	1	Paleozoic	none	none	none	none	1000-1200	30	30	30	1.5±1	7±3	1.3±0.2	28	28	32	32	2100±800	1860	39	35
Taihang	6	Meso.-Tert.	none	none	none	none	600-1100	80	70	70	0.3±0.3	5.9±2.5	2±0.4	9	8	28	26	670±390	640	14	15
Teiling	3	Meso.-Tert.	none	none	none	none	600-1100	80	74	63	0.65±0.4	10±4	2.2±0.2	24	23	35	36	840±390	860	14	14
Nushan/Mingxi	2	Tertiary	none	none	none	none	1000-1300	103	103	103	0.88±0.25	10.4±8	1.2±0.6	32	35	43	37	2300±650	2200	32	36
Mt. Horeb	1	E. Paleozoic	none	none	none	none	800-1000	42	40	40	1±0.4	6.8±2.6	1.4±0.4	32	33	35	33	1850±500	1960	32	36
Malaita	1	Tertiary	none	none	none	none	800-1200	81	79	79	0.53±0.5	7±4	1.4±0.4	14	12	31	29	700±750	500	14	10
W. Cliffs	3	Mesozoic	none	none	none	none	600-1200	278	246	205	1±0.8	4.2±1.5	2.1±0.6	20	19	24	24	1800±850	1670	45	42
Renmark	1	Mesozoic	none	none	none	none	600-1100	53	53	53	1.3±1.2	4.4±1.7	2.1±1.5	19	17	23	26	1500±1700			
Jugiong	1	Tertiary	none	none	none	none	1000-1200	90	90	90	0.93±0.5	11±6	1.7±1.7	37	35	44	39	1850±450	1820	25	28
Colorado Plat.	4	Tertiary	2	none	none	none	600-1000	236	181	128	1.2±2	4.5±3.6	2.6±1	18	10	21	19	950±700	845	27	27

Table 2. Xenolith data

Location/reference	Sample no.	Mode oliv %	opx %	cpx %	gnt %	cpx/gnt %	cpx+gnt %	Gnt Composition Cr2O3 %	CaO %	Olivine %Fo	Bulk Rock Al2O3 %	CaO %	MgO %	mg#
ARCHONS														
Kaapvaal Craton	EJB 4	60.8	31.0	3.0	4.2	0.71	7.2	4.49	5.13	92.1	1.51	1.28	42.43	91.9
Cox et al., 1987	EJb 48	63.1	30.5	1.6	4.5	0.36	6.1	4.84	5.15	92.9	1.13	0.64	45.53	92.8
	mb3	60.7	31.1	2.5	5.7	0.44	8.2	4.11	4.84	92.5	1.65	0.92	43.84	92.5
	mb 7	52.4	39.3	1.5	5.1	0.29	6.6	4.75	5.24	92.9	1.43	0.95	44.00	92.8
	mb 12	63.7	29.5	2.0	4.4	0.45	6.4	4.80	5.04	92.3	1.47	1.03	42.95	92.6
	mb 13	66.5	27.2	0.7	4.4	0.16	5.1	4.45	5.00	92.8	1.56	0.74	43.39	92.9
Cox et al., 1973	lbm9	48.7	40.6	2.7	6.7	0.40	9.4	4.00	5.07	92.3	2.45	1.36	38.84	91.8
	11	47.5	47.0	0.4	5.0	0.08	5.4	5.83	5.24	92.3	1.41	0.75	41.12	92.2
	12	42.0	38.4	9.2	10.4	0.88	19.6	3.96	4.96	83.4	3.60	2.90	33.45	80.9
	17	49.1	44.4	0.1	6.1	0.02	6.2	3.74	4.80	92.6	2.45	0.89	38.33	92.9
	32	27.7	43.8	4.2	20.4	0.21	24.6	3.61	4.55	89.0	4.71	2.29	33.94	88.2
	33-c	13.9	20.7	49.0	16.4	2.99	65.4	2.13	4.72	87.5	5.02	9.10	24.89	86.6
	36-a	54.0	23.0	12.0	11.0	1.09	23.0	2.47	4.79	87.4	3.23	3.20	35.38	86.7
	37	42.2	31.1	13.0	13.7	0.95	26.7	2.33	4.68	87.5	4.10	6.67	27.33	87.1
	38	67.4	22.8	2.1	7.7	0.27	9.8	1.73	4.46	88.6	2.27	3.55	30.34	86.1
	bd 1355	54.0	34.0	4.0	7.0	0.57	11.0	4.21	5.22	92.5	1.50	1.03	41.80	92.8
Boyd et al., 1993	frb932	61.4	30.0	0	7.3	0	7.3	3.78	3.42	93.9	1.67	0.61	43.69	93.6
	frb978	64.7	28.3	0	5.1	0	5.1	4.93	3.94	93.5	1.23	0.76	43.73	93.3
	frb 1013	65.2	27.4	0	4.5	0	4.5	7.86	4.18	93.5	0.97	0.71	43.58	92.4
	phn 4254	63.6	32.1	0	3.9	0	3.9	6.16	4.00	93.5	0.99	0.32	43.77	93.7
	phn 5596	77.3	18.3	0	3.5	0	3.5	9.47	4.84	92.6	0.68	0.30	44.13	92.3
	frb 1402	69.8	25.2	0	3.7	0	3.7	4.51	3.92	93.2	0.89	0.29	43.70	92.5
	frb 1404	57.6	35.8	0	5.8	0	5.8	4.14	3.77	93.5	1.39	0.45	43.07	93.2
	frb 1409	57.6	35.0	0	6.8	0	6.8	4.23	3.96	93.7	1.57	0.59	43.04	93.4
	frb 1422	63.7	30.9	0	4.5	0	4.5	4.25	3.54	93.5	1.11	0.35	43.79	93.2
	frb 1447	65.3	28.9	0	4.7	0	4.7	4.39	3.77	93.4	1.13	0.32	43.75	93.1
	118	47.4	40.2	2.8	9.6	0.29	12.4	2.58	4.64	92.9	2.41	1.03	41.35	92.3
	175	58.4	29.7	7.4	4.5	1.66	11.9	2.30	4.60	92.2	1.32	1.70	42.61	92.2
	181	82.2	15.1	0.4	2.4	0.16	2.7	6.22	5.53	93.2	0.55	0.33	48.14	92.9
	197	57.1	20.1	8.8	14.0	0.63	22.8	2.17	3.87	91.7	3.47	2.14	40.13	90.1
	127	64.4	31.6	0	4.0	0	4.0	5.89	4.26	92.7	1.09	0.48	45.41	92.8
	168	63.1	31.9	0	5.0	0	5.0	5.92	4.21	92.7	1.15	0.51	44.80	92.3
	184	72.3	22.0	0	5.6	0	5.6	4.29	3.68	93.8	1.23	0.56	46.79	93.5
Carswell et al., 1984	PTH207	64.9	31.3	2.2	1.6	1.34	3.8	8.34	6.90	92.0	0.68	0.73	44.27	92.6
	PTH400	63.4	28.7	6.6	1.3	5.05	7.9	4.34	5.45	91.4	0.42	1.35	43.07	90.7
	PTH403	61.7	29.1	3.4	5.8	0.58	9.2	6.80	6.02	91.9	1.19	1.01	42.36	91.0
	PTH405	60.3	33.6	1.3	4.7	0.28	6.1	6.33	5.35	92.5	1.08	0.58	43.66	91.8
	PTH409	67.2	29.8	1.2	1.8	0.69	3.0	7.70	6.57	92.5	0.61	0.45	44.97	92.7
Danchin, 1979	118	47.4	40.2	2.8	9.6	0.29	12.4	2.58	4.64	80.7	2.41	1.03	41.35	92.3
	175	58.4	29.7	7.4	4.5	1.66	11.9	2.30	4.60	58.0	1.32	1.7	42.61	92.2
	181	82.2	15.1	0.4	2.4	0.16	2.7	6.22	5.53	74.8	0.55	0.33	48.14	92.9
	197	57.1	20.1	8.8	14.0	0.63	22.8	2.17	3.87	74.3	3.47	2.14	40.13	90.1
	168	63.1	31.9	0.0	5.0	0.00	5.0	5.92	4.21	80.1	1.15	0.51	44.80	92.3
	184	72.3	22.0	0.0	5.6	0.00	5.6	4.29	3.68	79.7	1.23	0.56	46.79	93.5
Lashaine	775	74.2	15.7	0	8.5	0	8.5	4.51	2.85	92.7	2.06	0.42	45.6	94.1
Reid et al., 1974	797	62.7	28.1	1.3	4.8	0.27	6.1	4.49	4.39	92.7	1.42	0.57	44.3	92.8
Rhodes & Dawson, 1974	796	72.6	18.0	0.9	6.7	0.13	7.6	3.53	4.63	92.2	1.67	0.59	45.0	92.2
	740	68.0	20.7	2.0	8.1	0.25	10.1	2.78	4.92	92.0	2.09	0.91	44.0	92.1
	794	69.2	21.0	2.3	5.8	0.40	8.1	3.59	5.07	92.0	1.53	0.87	44.2	92.2
	776a	77.5	10.7	5.0	6.5	0.77	11.5	4.21	4.44	91.7	1.53	1.26	44.1	91.2
	782	79.0	11.9	0.8	4.4	0.18	5.2	3.44	3.93	91.3	1.16	2.55	44.3	91.4
TECTONS														
The Thumb	112	82	11	4	2	2.00	6.0	5.81	5.94	86.8	1.14	1.28	41.4	85.6
Ehrenberg, 1982	ao82	61	16	16	9	1.78	25.0	2.40	5.20	89.7	2.60	3.26	39.0	88.2
	110	62	14	15	9	1.67	24.0	1.68	4.66	90.4	3.52	2.96	38.0	89.3
	128	64	11	14	11	1.27	25.0	3.06	5.01	91.1	3.40	2.63	39.8	90.0
	ro77	57	18	14	11	1.27	25.0				2.75	2.80	39.3	90.2
	qo77	72	22	2.0	3.0	0.67	5.0	2.87	6.03	91.2	1.07	0.96	44.3	90.5
	ho77	80	15	2.7	2.2	1.23	4.9	2.71	5.93	91.7	0.89	0.82	45.2	91.2
	io78	70	20	3.3	4.4	0.75	7.7				1.31	1.24	42.7	91.3

Notes: Blank entries = no data; * average for gnt peridotites (Nixon and Boyd, 1979) and Stern et al., (1989)

Table 2. Xenolith data continued

Location/ reference	Sample no.	Mode				cpx/gnt %	cpx+gnt %	Gnt Composition		Olivine %Fo	Bulk Rock			mg#
		oliv %	opx %	cpx %	gnt %			Cr2O3 %	CaO %		Al2O3 %	CaO %	MgO %	
TECTONS ctd														
The Thumb ctd	126	88	9	0.7	2.0	0.35	2.7	6.09	6.32	91.4	0.54	0.52	47.0	91.5
Ehrenberg, 1982	117	96	3	1.0	1.0	1.00	2.0				0.24	0.44	47.6	91.6
	no77	74	19	1.7	3.5	0.49	5.2	3.57	6.40	90.8	1.54	0.82	43.9	91.7
	156	79	19	1.0	1.0	1.00	2.0				0.97	0.60	45.8	91.7
	104	76	18	3.0	3.0	1.00	6.0				0.54	0.72	45.3	91.8
	120	81.0	16.0	1.0	2.0	0.50	3.0				0.53	0.50	47.2	91.9
	105	81	16	1.0	2.0	0.50	3.0	4.10	6.26	92.2	1.21	1.18	44.9	92.1
Malaita	phn 4002	66	10	15	4	3.75	19.0	1.01	5.02	89.5*	3.55	3.58	39.2	88.6
Neal, 1988	phn 4009	57	12	20	1	20.00	21.0	0.61	4.84	89.5*	4.93	4.78	36.5	88.6
	phn 4013	56	12	13	10	1.30	23.0	1.32	6.00	89.5*	4.99	4.47	36.5	88.3
	phn 4016	60	12	24	2	12.00	26.0	0.77	5.15	89.5*	3.48	5.23	37.8	89.2
	phn 4034	66	5	9	2	4.50	11.0	1.78	5.14	89.5*	3.52	3.78	38.9	89.0
	phn 4064	66	5	17	8	2.13	25.0	0.90	4.93	89.5*	4.59	3.92	38.4	88.3
	phn 4067	58	5	28	6	4.67	34.0	0.84	4.95	89.5*	4.76	6.08	35.6	88.8
	phn 4069	60	8	15	7	2.14	22.0	1.32	5.89	89.5*	4.51	4.64	37.2	88.3
	crn 209	66	8	16	2	8.00	18.0	4.90	5.54	89.5*	2.95	2.17	39.9	89.0
	crn 213	61	12	18	4	4.50	22.0	0.68	4.79	89.5*	5.11	4.01	37.8	88.7
Pali-Aike	TM 2	60	20	20	10	2	30	1.23	5.05	87*	4.0	3.0	37.7	88.7
Stern et al., 1989	TM1	55	15	15	15	1	30	1.28	4.96	87*	4.1	3.2	37.6	89.1
	BN4	50	20	15	15	1	30	1.20	4.96	87*	4.6	3.4	36.9	88.3
	BN35	45	25	15	15	1	30	1.16	4.97	87*	4.3	3.3	37.1	89.3
	LS4	65	20	10	5	2	15	1.69	5.13	87*	3.8	2.9	38.3	89.6
	LS101	60	20	10	10	1	20	1.75	5.11	87*	3.9	3.0	37.9	89.2
	LS33	55	20	15	10	2	25	1.90	5.02	87*	3.9	3.0	37.3	89.1
Vitim	313-1	60.8	13.7	12.1	13.4	0.90	25.5	1.18	4.82	90.1	4.37	3	38.50	89.6
Ionov et al., 1993	313-2	58.0	22.1	13.7	6.3	2.17	20.0	1.32	4.96	89.8	3.23	3.18	39.70	90.2
Ionov, unpubl.	313-3	61.1	12.7	14.3	11.9	1.20	26.2	1.14	4.9	90.1	4.03	3.35	38.65	89.7
	313-5	64.0	11.2	13.4	11.4	1.18	24.8	1.52	4.83	90.5	3.95	3	39.10	89.9
	313-6	63.7	12.1	12.9	11.3	1.14	24.2	1.59	5.11	90.4	3.88	3.04	38.95	89.3
	313-8	57.0	13.4	15.6	14	1.11	29.6	1.23	4.9	90.5	4.82	3.7	37.40	89.5
	313-37	60.9	16.8	15.5	6.5	2.38	22.0	1.18	5.18	90.2	3.27	3.45	39.20	90.1
	313-54	60.6	13.3	14.7	11.4	1.29	26.1	1.32	4.87	90.7	4.1	3.42	38.70	89.7
	313-104	57.6	16.8	14.1	11.1	1.27	25.2	1.06	4.94	89.7	4.32	3.43	38.52	89.5
	313-105	59.8	19.5	12.1	8.0	1.51	20.1	1.11	4.71	90.0	3.41	2.83	39.15	89.8
	313-106	63.2	15.4	13.1	8.4	1.56	21.5	1.01	4.85	89.8	3.29	3.04	39.72	89.5
	313-110	61.4	14.3	12.3	11.7	1.05	24.0	1.2	4.95	90.5	4.01	3.07	39.56	89.9
	313-240	66.7	14.1	11.3	7.9	1.43	19.2	1.16	4.99	90.0	3.19	2.69	41.06	89.9
	313-241	61.3	16.8	11.9	10.0	1.19	21.9	1.15	4.95	89.9	3.62	2.84	39.55	89.4
	314-74	67.9	16.5	10.4	3.9	2.67	14.3	1.16	5.05	90.9	2.84	2.35	41.82	90.4
	314-580	65.6	14.9	13.9	3.5	3.97	17.4	1.54	5.04	90.8	3.17	2.68	40.41	89.1
	313-113sg	70.2	8.6	13.3	7.6	1.75	20.9	0.99	4.80	90.2	3.05	2.93	41.20	89.6
E. China	m33	42.3	23.4	24.1	10.2	2.4	34.3	1.50	5.00	89.7	5.37	5.34	33.80	88.6
Qi et al., 1995	m38	52.5	25.3	15.2	7.0	2.2	22.2	1.47	5.02	89.6	3.91	3.38	36.80	88.8
	m31	57.6	20.2	15.2	7.0	2.2	22.2	1.51	4.99	89.7	3.54	3.23	37.90	89.1
	m6	60.3	19.3	12.5	7.3	1.7	19.8	1.29	4.99	89.4	3.66	2.93	38.50	89.2
	M8	64.0	21.0	12.0	2.0	6.0	14.0	1.51	5.18	90.0	2.15	2.6	40.36	89.9
	M30	55.0	22.0	16.0	8.0	2.0	24.0	1.36	5.01	89.6	3.8	3.5	37.90	89.2
	M22	44.0	32.0	14.0	10.0	1.4	24.0	1.37	5.07	89.7	4.61	3.35	36.10	89.0
	M34	59.0	29.0	8.3	4.0	2.1	12.3	1.97	5.37	90.5	2.81	2.15	39.63	90.4
	M35	53.0	26.0	10.0	11.0	0.9	21.0	1.43	5.11	89.8	4.43	2.78	37.48	89.1
	M7	52.0	27.0	13.0	8.3	1.6	21.3	1.40	4.78	90.0	3.94	3.32	36.40	89.3
	M32	61.0	20.0	10.0	7.5	1.3	17.5	1.86	5.34	90.5	3.35	2.75	39.20	90.1
	X77	53.0	28.0	15.0	4.5	3.3	19.5	2.26	5.62	89.9	3.26	3.25	37.30	89.0
	M3	63.0	13.1	14.1	9.7	1.5	23.8	1.27	5.07	89.7	2.93	3.03	39.15	88.1
	MD-4	70.2	13.4	5.9	10.5	0.6	16.4	2.22	5.45	90.7	2.77	2.08	43.35	92.2
Liu & Fan, 1990	ZN-12	40.8	32.3	14.1	12.8	1.1	26.9	1.15	5.65	88.8	5.16	3.28	35.28	89.2
Fan & Hooper, 1989	MQ-8	44.5	29.4	17.9	8.2	2.2	26.1	1.53	5.37	89.9	4.8	3.66	36.11	89.9

Notes: Blank entries = no data; * average for gnt peridotites (Nixon and Boyd, 1979) and Stern et al., (1989)

microprobe (as described by Ryan et al. [1990] or by laser-ablation ICPMS microprobe (as described by Norman et al., [1996]). The typical analytical uncertainty on the major-element data given here is 2-3%; the typical uncertainty on the trace-element data (for an individual analysis) is 5%, increasing to 10% for some of the lowest-level data.

The temperature of a given grain of Cr-pyrope garnet can be estimated by its Ni content, using the Ni-in-gnt thermometer of Griffin et al. [1989b] as recalibrated by Ryan et al. [1996]. This Nickel Temperature (T_{Ni}) records the temperature of equilibration between the garnet and olivine, which is the major reservoir of Ni in the mantle, and is relatively constant in composition (mean Ni = 2900±360 ppm; Ryan et al. [1996]). T_{Ni} is interpreted as representing the ambient temperature at the time at which the garnet grain, or the enclosing xenolith, was entrained in the host volcanic rock. The local geotherm within the mantle at the time of eruption of the volcanic rock can be determined from analysis of garnet ± chromite concentrates [Ryan et al., 1996], and the depth from which each garnet grain was derived can be estimated by referring its T_{Ni}) to this geotherm. This information gives insight into the vertical distribution of rock types and processes in each mantle section, as will be discussed below.

In many mantle samples, high-T garnets show a characteristic geochemical signatures, with high contents of Ti, Zr, Y, Ga and Na [Griffin and Ryan, 1995]. These signatures are similar to those found in garnets from high-T, typically sheared, lherzolite xenoliths, and ascribed to metasomatic alteration by infiltrating melts [Griffin et al., 1989a, 1996c; Smith and Boyd, 1987; Smith et al., 1991, 1993]. Zoning profiles in garnets from these xenoliths indicate that the metasomatism occurred immediately prior to eruption of the host volcanic rock, and probably was accompanied by extensive growth of both garnet and clinopyroxene [Griffin et al., 1989a, 1996c; Smith et al., 1993]. It is evident from these zoning studies that despite their chemically enriched nature, these high-T, sheared, metasomatised xenoliths do not represent any long-term or large-scale mantle reservoir, and are irrelevant to the problem of estimating the composition of mantle of different ages. We therefore have attempted to exclude such garnets from consideration, by selecting an appropriate range of T_{Ni}) from each mantle section. These T ranges, and the proportion of the total sample represented by them, are given in Table 1. Beneath most Archean and Proterozoic areas, the T ranges encompass most of the lithospheric mantle.

The mean or median values of element contents for each garnet suite used here are our best estimate of the average composition of the mantle garnets, excluding late-stage high-T metasomatism. We consider the eruption rates of the host volcanic rocks to be so high that interaction between the transporting magma and the garnet grains is negligible. This is borne out by the rarity of zoning in individual xenocrysts; the analyses used here represent the core of each xenocryst, to reduce further the probability of such interaction affecting the chemical signatures of the garnets. In most areas represented in Table 1, we have been able to combine data from several volcanic eruptions. This is important, because within a single volcanic field, individual eruptions may preferentially sample different stratigraphic levels of the mantle through which the magma has passed, leaving other levels under represented. Where several eruptive centres can be used, the estimate of "average composition" is likely to be more representative of the whole lithosphere.

The samples have been divided into three main groups on the basis of the "tectonothermal age" of the crust penetrated by the volcanic rock from which the sample is derived. This age represents the last major crust-forming event, either major magmatic activity (such as granite intrusion), or tectonic activity represented by high-grade metamorphism, continental accretion or rifting. A division of cratons into Archons (tectonothermal age ≥2.5 Ga), Protons (2.5-1.6 Ga) and Tectons (1.6-1.0 Ga) has been suggested by Janse (1994, and references therein) and the concept has proven useful in diamond-exploration work. However, this terminology, designed for application to cratons, does not encompass the Phanerozoic mobile belts, and our data, as will be seen below, suggest that the boundary at 1.6 Ga, used by Janse, does not correspond to a globally significant break. On the other hand, as recognised by Janse, the development of the linear continental-margin orogenic belts characteristic of the modern plate-tectonic regime appears to have begun ca 1 Ga ago. To provide a useful shorthand, we therefore have adopted a modified version of Janse's scheme, and divide our samples into those from Archons (>2.5 Ga), Protons (2.5-1 Ga) and Tectons (<1 Ga). One group of four samples is classified as Archon/Proton, in recognition that these terrains contain significant amounts of Archean crust, but also have undergone major reworking during Early-Mid Proterozoic (2.5-1.8 Ga) time. Notes on the tectonic situation of each region are given in Appendix 1.

The mantle rock type from which a peridotitic garnet was derived can be deduced, with some limitations, from the $CaO-Cr_2O_3$ relationship of the garnet (Figure 1). For the purposes of this paper, "harzburgite" is defined as an ultramafic rock that lacks clinopyroxene (cpx), rather than simply having less than 5% cpx (a typical petrographic classification criterion). This "thermodynamic" definition emphasises the important distinction between rocks in which the garnet is Ca-saturated through equilibrium with cpx, and those in which the Ca content of the garnet is not buffered by the mineral assemblage. Various divisions between these two fields have been suggested (Figure 1); in this report we adopt the one proposed by Gurney [1984], although Sobolev [1974] has pointed out that some garnets classified as harzburgitic by this method could be derived from lherzolites in which the cpx coexisting with garnet is

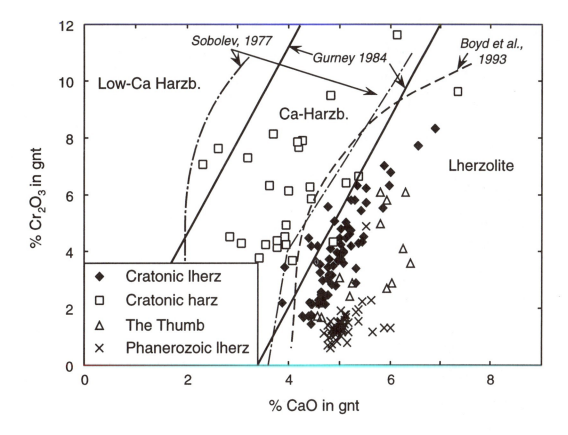

Figure 1. CaO-Cr$_2$O$_3$ plot for some garnet peridotite xenoliths (Table 2), illustrating the use of this plot to recognise rock type. Various suggestions for dividing lherzolitic from harzburgitic garnets are shown [Boyd et al., 1993; Gurney, 1984; Sobolev, 1974, 1977]. Also shown are a suggested division between low-Ca harzburgites and calcic harzburgites, based on a line enclosing 50% of African diamond-inclusion garnets (Gurney, 1984) and a similar line marking the low-Ca limit of garnets from extremely Na-rich lherzolites [Sobolev et al., 1973; Sobolev, 1977].

unusually Na-rich, and hence Ca-poor. We further subdivide the field of harzburgitic garnets in Figure 1 into calcic harzburgite and low-Ca harzburgite. This line is not completely arbitrary; it encloses 50% of diamond-inclusion garnets from South Africa [Gurney, 1984; Griffin et al., 1992], and thus highlights an important class of extremely depleted garnets, genetically associated with diamonds. A plot of typical garnet suites from Archon, Proton and Tecton settings (Figure 2) illustrates the classification of garnets using Figure 1.

The xenolith database assembled here (Table 2) is not intended to be comprehensive or definitive, but to allow correlations between rock composition and garnet composition. Samples have been chosen from a variety of tectonic settings and a range of ultramafic rock types. In each case, the data include whole-rock chemical analyses and mineral analyses. Where available, visually estimated modal-composition data have been used; otherwise, modes have been calculated from rock and mineral analyses using a least-squares mixing program.

3. XENOLITH DATA

Some compilations of compositional data on major-element compositions of mantle xenoliths have indicated significant differences between garnet peridotites and spinel peridotites, which could be interpreted as reflecting compositional stratification within the lithosphere [e.g., Maaloe and Aoki, 1977]. These studies did not recognise the potential importance of age differences between the garnet peridotites (drawn largely from cratonic settings) and the spinel peridotites (drawn almost entirely from Phanerozoic settings).

There are, however, fundamental differences in the composition of Archean lithospheric mantle, as represented by xenoliths in kimberlites from the Kaapvaal and Siberian Cratons, and Phanerozoic mantle, as represented by both ocean-floor peridotites and xenoliths in young (mostly intraplate) basalts erupted through Phanerozoic crust (Figure 3). These differences have been carefully documented in a series of papers by Boyd and co-workers

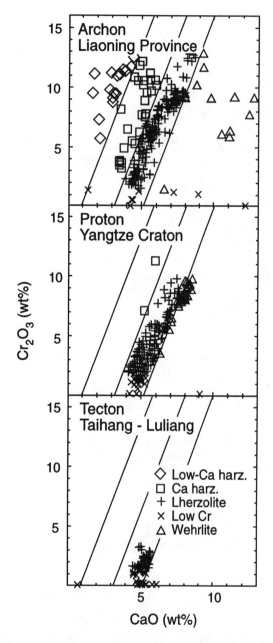

Figure 2. CaO-Cr_2O_3 plots of typical garnet suites from Archon, Proton and Tecton settings [Zhang et al., 1996]; dividing lines of Gurney [1984] from Figure 1 outline fields of garnets from wehrlites, lherzolites, etc. as shown by the different symbols.

lower bulk Mg/Si at any mg# (Figure 3). These features are not limited to garnet peridotites, but are shared by spinel peridotites from Archean settings [Carswell et al., 1984; F.R. Boyd, personal communication, 1996].

The xenoliths from Phanerozoic terrains (in both oceanic and continental settings) follow a compositional trend defined by ocean-floor peridotites: progressive depletion in a basaltic component, as measured by decreasing Ca and Al contents, leads to higher olivine contents and more magnesian olivine (Figure 3), but lower opx/olivine ratios than seen in Archean xenoliths. The levels of depletion in these suites vary considerably, but many are much less depleted than typical cratonic xenoliths, and on average they are only mildly depleted in Ca and Al compared with estimated primitive upper-mantle compositions (Figure 4). The Phanerozoic xenolith suites are interpreted by Boyd [1996] as the products of low-pressure extraction of basaltic melts, by analogy with processes at mid-ocean ridges. One implication of these observations is that much of the lithospheric mantle beneath Phanerozoic terrains may be derived from accreted oceanic material, but this interpretation may not be unique. The nature of residues from melting in different portions of mantle plumes, and in subduction-related environments, needs further study.

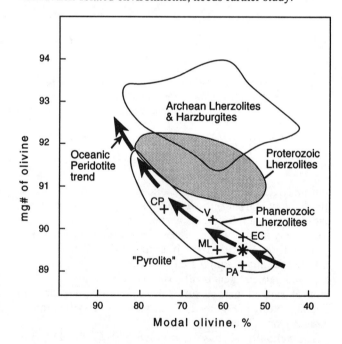

Figure 3. mg# (%Fo) of olivine, plotted against modal olivine content, for mantle-derived xenoliths from Archean, Proterozoic and Phanerozoic terrains, modified after Boyd [1996]. Archean xenoliths from Boyd [1996] and Boyd et al. [1993]; Proterozoic xenoliths from Hoal et al. [1995] and F.R. Boyd [personal communication]; field of Phanerozoic xenoliths from Boyd [1996]. Average values for selected Phanerozoic xenolith suites (Table 2): PA, Pali-Aike; EC, Eastern China; V, Vitim; CP, Colorado Plateau; ML, Malaita.

[Boyd and Mertzman, 1987; Boyd et al., 1993; Boyd, 1996]. Compared with xenoliths from Phanerozoic terrains, Archean garnet peridotite xenoliths are generally highly depleted in Ca and Al and have lower Ca/Al (Figure 4). They typically contain more magnesian olivine, but they also have distinctly lower olivine/opx ratios, reflecting a

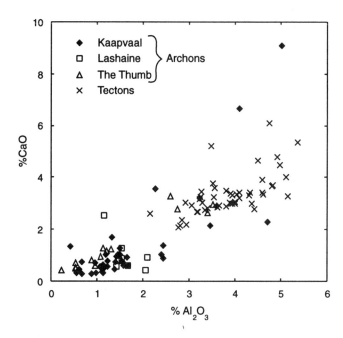

Figure 4. CaO vs Al$_2$O$_3$ for granular garnet peridotite xenoliths; data from Table 2. Note the low Ca/Al of xenoliths from Archons.

Most xenolith suites from Phanerozoic terrains are from the spinel facies, like the oceanic peridotites, indicating derivation from relatively shallow levels (<70 km) of the mantle. However, the observed differences between Phanerozoic and Archean xenoliths appear to extend to the garnet facies as well [Ionov et al., 1993]. Boyd [1996] showed that garnet peridotites from The Thumb, a minette on the Colorado Plateau, fall on the "oceanic" trend in Figure 3, despite a relatively high level of depletion in Ca and Al (Figure 4). Most garnet peridotite xenoliths from Phanerozoic terrains have high olivine/opx ratios, Al$_2$O$_3$ greater than 2.5% and CaO greater than 2%, like the spinel peridotite xenoliths from these terranes. Their Ca/Al ratios also are generally higher than those of the Archean xenoliths (Figure 4); this is reflected in their generally higher contents of (cpx+gnt) and higher cpx/gnt ratios (Figure 5). The Phanerozoic garnet peridotites used here (see Appendix 1) include samples from a variety of tectonic settings (Vitim, Paleozoic fold belt; Malaita, oceanic plateau; Pali-Aike, continental-margin arc; Mingxi, accreted continental margin). All except those from The Thumb show remarkably low degrees of depletion, even compared with Phanerozoic spinel lherzolite suites [cf Boyd, 1996]. However, this still may reflect the small number of localities and samples available at present.

The observed differences in Mg/Si and Ca/Al at any mg# suggest that Archean peridotites have not been formed by the same processes as the Phanerozoic peridotites. In addition, there are significant differences in the overall level of depletion between the Archean and Phanerozoic garnet peridotite suites, reflected in their abundances of garnet (gnt) and cpx (Figure 5). The Archean xenoliths generally have very low cpx contents and low (typically <0.5) cpx/gnt ratios; the Phanerozoic xenoliths have a broader range of garnet contents, higher contents of cpx and higher (typically ≥1) cpx/gnt ratios. Although the number of well-documented Phanerozoic garnet-peridotite suites is small, the high cpx/gnt ratios, and high modal cpx, are consistent features that appear to be independent of tectonic setting.

The clear differences between Archean and Phanerozoic mantle samples raise the question of when these differences first appeared in the geologic record. Unfortunately, there are few comparable data on xenoliths from Proterozoic terrains. Hoal et al.[1995] showed that garnet peridotite xenoliths from kimberlites in Namibia, which penetrate Proterozoic crust, are intermediate between Archean and Phanerozoic suites in terms of olivine abundance and composition (Figure 3). A xenolith suite from the Obnazhennaya kimberlite, which penetrates a Proterozoic terrane in northern Siberia [Rosen et al., 1994], contains a high proportion of pyroxenites, and the garnet lherzolite

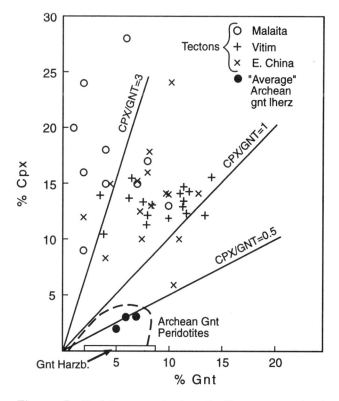

Figure 5. Modal garnet (gnt) and clinopyroxene (cpx) contents of xenoliths from Kaapvaal and Siberian kimberlites ("Archean"), Vitim, Malaita and eastern China (Table 2), illustrating the higher (cpx+gnt) contents and cpx/gnt of the Tecton peridotites. Three published estimates of average modes for the Archean garnet peridotites are shown as solid dots.

xenoliths have higher Mg/Si and higher (Al+Ca) than those from the kimberlites penetrating the Archean terranes further south [Spetsius, 1995]. Xenoliths from kimberlites in East Griqualand, bordering the Kaapvaal craton, have cpx contents up to 10%, and an average cpx/gnt ≈1 [Nixon, 1987; F.R. Boyd, personal communication]. These data suggest that Proterozoic mantle lithosphere is intermediate in composition between Archean and Phanerozoic mantle, with some localities tending toward Phanerozoic mantle compositions.

To investigate this question further, we can turn to a less precise, but far more abundant source of information: the heavy-mineral concentrates derived from kimberlites, lamproites and other volcanic rocks that have sampled the subcontinental lithosphere. These concentrates are interpreted as disaggregated mantle wall rock, and the minerals in them provide insight into the composition of the mantle over a much broader range of geographic and tectonic settings, and give a more continuous sampling through geologic time, than is possible with xenolith suites.

4. GARNET DATA

4.1 Distribution of Harzburgitic Garnets

The general restriction of harzburgitic garnets to Archon samples has been noted by many others [Sobolev, 1974; Boyd and Gurney, 1986; Schulze, 1995, and references therein] and has been used to map the presence or absence of "mantle roots" [Helmstaedt and Gurney, 1995; Schulze, 1995], on the assumption that such "roots" occur only beneath Archean cratons. However, there is general agreement that the harzburgitic garnets comprise a small proportion of the total garnet population, and in all suites of peridotitic garnets examined by us, lherzolitic garnets make up the majority of the sample.

The occurrence and relative abundance of harzburgitic garnets in concentrates from volcanic rocks is strongly controlled by tectonic setting. Garnets from low-Ca harzburgites (as defined in Figure 1) are restricted to Archons; they are present but very rare in some Archons reworked in Proterozoic time (Table 1), and none have been found in concentrates from Protons or Tectons. In most Archon samples, the low-Ca harzburgitic garnets show a relatively restricted range of T_{Ni}), indicating that they are confined to a relatively small stratigraphic range within the mantle section sampled by the kimberlite. For example, beneath the Daldyn kimberlite field in Siberia, these garnets are strongly concentrated in the depth range 140-180 km, where they make up as much as 35% of the garnet population [Griffin et al., 1996a]; the total lithosphere thickness is estimated at ca 230 km. Beneath the Guyana craton in Venezuela, such garnets make up more than 50% of the sample at depths of 160-180 km [Nixon et al., 1994].

Garnets from calcic harzburgites (as defined in Figure 1) are quite abundant in Archon samples (Table 1), and typically show a stratigraphic distribution similar to, but somewhat broader than, that of the garnets from low-Ca harzburgites [Boyd et al., 1993; Griffin et al., 1996a]. These garnets also occur in Proton samples, but are much less abundant and generally lie in the Ca-rich part of the harzburgite field, close to the lherzolite/harzburgite transition (Figure 2). As noted above, at least some of these garnets may be derived from Na-rich lherzolites, rather than harzburgites [Sobolev, 1974]. The stratigraphic distribution of the calcic-harzburgite garnets in Proton mantle sections also is different; they tend to be spread over a greater stratigraphic interval than in Archon sections, and to be more abundant toward the shallower levels. These garnets generally are not found in samples from Tectons, with the exception of rare grains in samples from the Colorado Plateau.

Schulze [1995] estimated that "G10" garnets (corresponding to all harzburgite garnets in Figure 1) make up ca 5% of the garnet population in Group 1 kimberlites from the Kaapvaal Craton, a lower proportion than shown in Table 1. Similarly, Sobolev and Nixon [1987] show estimates of the abundance of harzburgitic garnets for several fields in Siberia that are either higher or lower than those reported here. These different estimates may reflect differences in definition, the vagaries of selection of material, and the individual pipes sampled. However, the basic picture of a high abundance of harzburgitic garnets in Archon samples, and a very low abundance in samples from Protons, and especially Tectons, appears to be a worldwide phenomenon.

4.2 Compositional Variations in Lherzolitic Garnets

There are significant differences in the mean major- and trace-element composition of the average lherzolitic garnets from Archon, Proton and Tecton suites. There is a steady decrease in the mean Cr_2O_3 content of the garnet suites through time, and this is accompanied by an increase in mean Y content (Figure 6). There is some overlap between Archon and Proton suites; the garnet suite from Malo-Botuobiya field in Siberia has lower Cr_2O_3 than other Archon suites, while garnets from lamproites in Hunan, on the Proterozoic Yangtze Craton [Zhang et al., 1996], have unusually high mean Cr_2O_3 contents and low Y contents. There is no overlap between the Proton and Tecton suites; the Colorado Plateau lies in a transitional position, perhaps reflecting the Phanerozoic reworking of a Proterozoic mantle (Appendix 1). Although the standard deviations on the mean values commonly are large (Table 1), the ranges shown in Figure 6 are outside these uncertainties; a similar pattern is shown by the median values (Table 1).

A large range of variation also is shown by the mean Zr/Y and Y/Ga ratios of the garnet suites (Figure 7). Mean

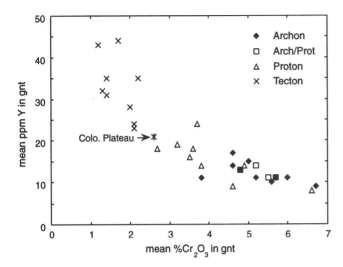

Figure 6. Mean Cr₂O₃ vs Y for lherzolitic garnets from different tectonic settings (Table 1).

Zr/Y decreases from Archon to Proton, with some overlap. Several (generally older) Proton suites have low Y/Ga like that of the Archon suites, while a (generally younger) group has both lower Zr/Y and higher Y/Ga. Tecton garnets have higher Y/Ga, and generally lower Zr/Y, than the Proton garnets.

The mean TiO_2 of lherzolitic garnets shows a broad negative correlation with Y/Ga, decreasing from Archon to Proton to Tecton (Figure 8a). This pattern is in part obscured by a group of high-Ti Tecton suites; all of these are from areas with high geotherms, in which all of the garnets have relatively high T_{Ni}) (Table 1). We interpret the high Ti contents of these garnets as reflecting high-T partitioning of Ti between garnet and coexisting cpx (see below); cooling of these suites to the lower mean T recorded by the other Tecton suites would reduce the Ti contents of the garnets to similar levels. There also is a general decrease in Ti/Y, correlated with decrease in mean Cr_2O_3, from Archon to Proton to Tecton (Figure 8b).

4.3 Modeling of Trace-Element Variations in Lherzolite Garnets

To interpret the compositional data on lherzolitic garnets in terms of rock composition, we will take two approaches: modeling of the garnet compositions as functions of cpx/gnt ratios, and comparison with known xenolith suites.

The Zr-Y-Ga relationship is especially interesting because these elements are strongly concentrated in garnet and clinopyroxene, but behave differently with regard to partitioning. Figure 9 shows the distribution of these elements between gnt and cpx in a large suite of lherzolite xenoliths, as a function of equilibration T. These relationships are similar to those seen in eclogite and pyroxenite xenoliths [O'Reilly and Griffin, 1995]. Zr is concentrated in cpx relative to gnt at low T, but is increasingly partitioned into gnt with increasing T. Ga is distributed roughly equally between gnt and cpx, and the partitioning is not strongly T-sensitive. Y is strongly concentrated in garnet, regardless of T. These relationships suggest that changes in the cpx/gnt ratio of the rock (equivalent to changing the Ca/Al ratio) at constant whole-rock trace-element concentration and T, would change the Zr/Y and Y/Ga ratios of the garnet, as greater or lesser proportions of the total Zr and Ga are sequestered in the cpx.

This effect is illustrated in Figure 10. Increasing the cpx/gnt ratio at constant T (by raising modal cpx content at constant garnet content) decreases Zr/Y in the garnet and increases Y/Ga, while a decrease in cpx/gnt has the opposite effect; the change in Zr/Y is most pronounced at low T. This effect can produce most of the observed variation in these ratios in the garnet suites studied here. As shown in Figure 5, typical Phanerozoic garnet peridotites have much higher cpx contents, and cpx/gnt ratios, than typical Archean xenoliths, as would be predicted by this modeling. This model also can explain the covariation of mean Y and Zr among the Archon suites, or among the Tecton suites (Figure 11). However, it does not provide a mechanism for producing the rise in the mean Y content of the garnets from Archon to Proton to Tecton (Figure 6), because even though Y is preferentially concentrated in the garnet, some must go into the cpx, and an increase in cpx/gnt ratio will

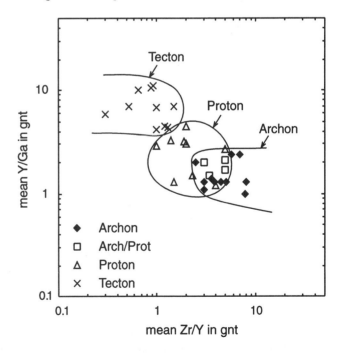

Figure 7. Mean Zr/Y vs Y/Ga for lherzolitic garnets from different tectonic settings (Table 1.)

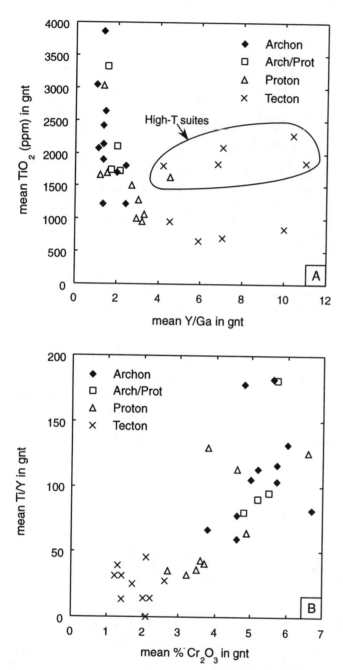

with the generally less depleted nature of known Phanerozoic garnet lherzolite xenoliths, as shown by their higher Ca and Al contents (Figure 4).

4.4 Comparison with Xenolith Suites

Data from garnet peridotite xenoliths show strong correlations between garnet composition and various indices of depletion or fertility (Figure 12). On a worldwide scale, the Cr_2O_3 content of garnet shows a strong inverse correlation with the Al_2O_3 content of its host rock (Figure 12a), as noted for the garnet-harzburgite suite by Boyd et al. [1993]. Xenoliths with high whole-rock Al contents also, with few exceptions, contain high CaO (Figure 4), so that the Cr contents of garnets correlate inversely with modal (cpx+gnt) abundance (Figure 12b) and (less strongly) with modal cpx/gnt (Figure 12c). As might be expected, the Cr contents of garnets also correlate inversely with another index of depletion, the mg# of the rock (Figure 12d), and with the Fo content of coexisting olivine (Table 1).

Both the modeling of the trace-element data and comparison with known xenoliths therefore indicate that the trends in mean Cr and Y content, and Zr/Y and Y/Ga ratios, of lherzolite garnet suites from Archon to Proton to Tecton reflect a decrease through time in the average degree of depletion of the subcontinental mantle, as expressed in higher (cpx+gnt), higher cpx/gnt, and lower mg#.

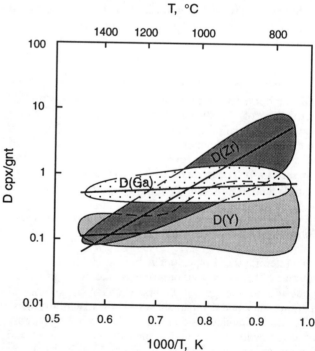

Figure 8. Mean Ti contents of lherzolitic garnets from different tectonic settings (Table 1). A: TiO_2 vs Y/Ga; garnet suites from areas with high geotherms (encircled) show anomalously high Ti contents due to equilibrium with cpx. B: Ti/Y vs Cr_2O_3.

Figure 9. Distribution of Zr, Y and Ga between gnt and cpx in garnet peridotites, as a function of temperature. T estimated by olivine-garnet thermometry [O'Neill and Wood, 1979] or Nickel Thermometry [Ryan et al., 1996].

inevitably reduce the Y content of the remaining garnet. Thus the higher Y contents and lower Zr/Y of the Tecton garnets appear to require both an increase in the cpx/gnt ratio, and an increase in the Y content of the whole rock, relative to Archon or Proton lherzolites. This is consistent

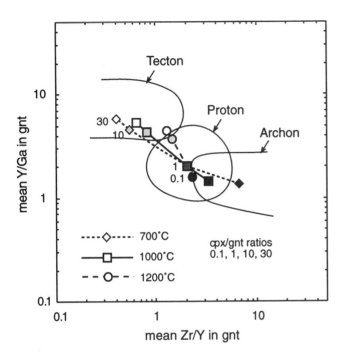

Figure 10. Changes in Zr/Y and Y/Ga of garnet, resulting from variations in cpx/gnt at constant T, Zr, Y and Ga. An average Proton garnet composition, a cpx calculated from partition coefficients shown in Figure 9, and an assumed cpx/gnt = 1 were taken as the starting point.

5. DISCUSSION

5.1 Vertical or Lateral Variation in Composition?

The lithosphere beneath Archean terrains typically is from 180-220 km thick, while that beneath Phanerozoic terrains is typically not more than 100 km thick; Proton lithosphere ranges from 140-180 km thick [Griffin and Ryan, 1995]. This difference in thickness raises the possibility that the differences in depletion of mantle lithosphere beneath terrains of different ages reflect sampling of progressively shallower levels of the mantle, rather than temporal differences in the composition of the whole mantle lithosphere. However, the ranges of T_{Ni}) represented by Archon and Proton garnet suites (Table 1) show that in these situations, the garnet concentrates have sampled the lithospheric column from its base to within at least a few tens of kilometres of the crust-mantle boundary. Furthermore, the chemical characteristics of Archean spinel peridotites, which might represent the uppermost part of the lithospheric mantle, are similar to those of the deeper garnet peridotites [Carswell et al., 1984; F.R. Boyd, personal communication]. The differences between Archon and Proton mantle therefore cannot be ascribed to differences in the depth range sampled by their respective garnet suites.

Phanerozoic garnet peridotites typically are drawn from the lower parts of their respective lithosphere sections, while most of these sections are in spinel facies. However, as discussed above, the fertile nature of these garnet peridotites is duplicated by the bulk compositions of the more common spinel peridotites, so that a generally low degree of depletion appears to be typical of the entire lithospheric section beneath Phanerozoic terrains, at least in the areas sampled by Cenozoic volcanism.

5.2 Mantle Evolution Through Time

The most significant feature of Archean mantle, which clearly distinguishes it from the mantle beneath younger terrains, is the presence of very depleted harzburgites with strongly subcalcic garnets (reflecting low Ca/Al) and high opx/oliv ratios (reflecting low Mg/Si). These harzburgites are interlayered with depleted lherzolites that have similarly low Mg/Si [Boyd, 1996]. Cox et al. [1987] used detailed petrographic data to argue that these rocks originated as high-T olivine+opx rocks, formed either as cumulates from ultramafic magmas or as residues from a process of extreme depletion; the opx of these high-T rocks then exsolved gnt ± cpx on cooling to ambient mantle temperatures. Most of the Archean lherzolites contain less than 3% cpx (Figure 3), and the compositional differences between harzburgites and lherzolites thus are small. The plausibility of this model

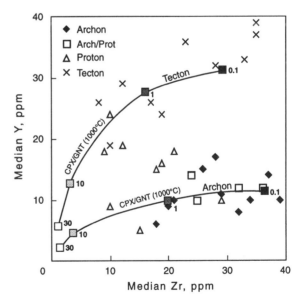

Figure 11. Changes in Y and Zr contents of garnet, resulting from varying cpx/gnt at constant T. Starting points were average Archon and Tecton garnets, cpx calculated from partition coefficients in Figure 9, and cpx/gnt = 1. The curves can describe the general trend of variation within each suite, but variation in cpx/gnt cannot account for the different average Y contents of Archon and Tecton garnets.

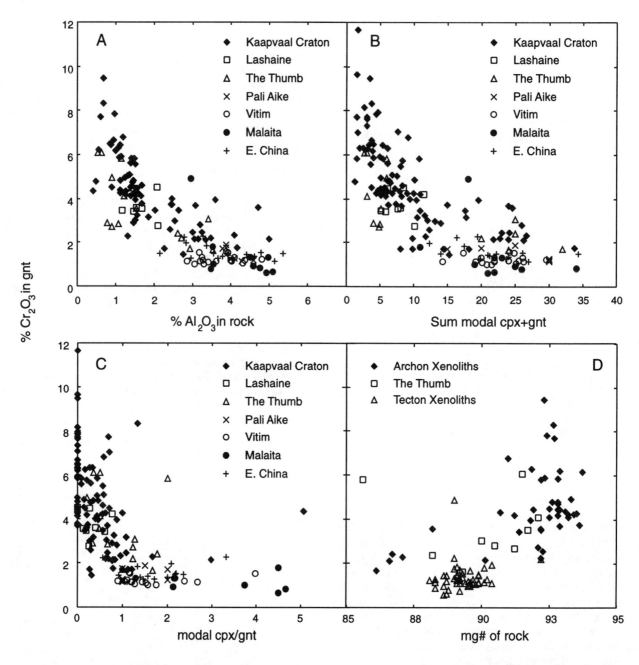

Figure 12. Relationships of garnet Cr_2O_3 contents to measures of fertility/depletion, in mantle-derived garnet peridotite xenoliths (Table 2), data include both granular an sheared xenoliths, except for Figure 12d (granular xenoliths only). A: Cr_2O_3 in gnt vs Al_2O_3 in rock; B: Cr_2O_3 in gnt vs modal (cpx+gnt); C: Cr_2O_3 in gnt vs modal cpx/gnt; D: Cr_2O_3 in gnt vs mg# (atomic 100 Mg/(Mg+Fe)) in rock.

has been experimentally verified by Canil [1991]. It also is consistent with the extremely depleted nature of many garnets in Archean xenoliths; the low Y, Zr and HREE contents of these garnets can be explained by exsolution of the garnets from orthopyroxene.

However, the nature of the depletion process that produced the Archean mantle, with its unusually low Mg/Si at any mg#, is not clear. The harzburgites (and presumably lherzolites) commonly are interpreted as the residue from extraction of komatiites, which are especially abundant in some Archean terrains. However, this model is not satisfactory because it does not produce residues with low enough Mg/Si, nor provide a residual bulk composition with low Ca/Al, that will crystallize strongly subcalcic garnets [Canil and Wei, 1992]. Schulze [1986] has suggested subduction of altered ocean-floor peridotites to provide subcalcic bulk compositions, but this process is not clearly restricted to the Archean, and hence does not explain the unique nature of Archean mantle. Whether the Archean peridotites are regarded as residues or cumulates, their derivation from a primitive upper mantle composition appears to involve an as yet undefined type of high-degree melting, probably at pressures corresponding at least to the present thickness of the lithosphere (60-80 kb). Alternatively, the source rock may have had a composition more Si-rich than accepted models for primitive upper-mantle compositions.

The Archean/Proterozoic boundary clearly represents a major discontinuity in the processes that produced subcontinental mantle; after approximately 2.5 Ga, these processes no longer produced strongly depleted subcalcic harzburgites, and even calcic harzburgites became much less common. The garnet data suggest that strongly depleted lherzolites were produced after 2.5 Ga, at least in some areas. However, there are few data to determine whether these Proterozoic lherzolites have the low Ca/Al and Mg/Si at high mg#, characteristic of Archean mantle. The xenoliths described by Hoal et al. [1995], which give Re-Os depletion ages less than 2.2 Ga, are intermediate between Archean and Phanerozoic xenoliths in terms of Mg/Si vs mg#, as are some from the kimberlites of east Griqualand [Nixon, 1987; F.R. Boyd, personal communication] (Figure 3). Similarly, xenoliths from the Birekte terrane of northern Siberia (probable crustal age ca 1.9 Ga [Rosen et al., 1994]) have higher Mg/Si (at the same mg#) than those from kimberlites within the Archean part of the same craton [Spetsius, 1995].

The data from the lherzolitic garnets suggest that, from at least mid-Proterozoic time to the present, there has been an evolution of subcontinental mantle toward less-depleted average compositions. This has resulted in a (quasi-) continuous rise in the average Ca and Al contents and Ca/Al (expressed as modal cpx/gnt and (cpx+gnt)) and a decrease in the average mg# of the lithospheric mantle. The trend deduced from the garnet compositions therefore is consistent with the xenolith data [Boyd and Mertzman, 1987; Boyd, 1996; Ionov et al., 1993], but the garnet data provide a more complete picture of a sharp break at the Archean/Proterozoic boundary (reflected in the disappearance of subcalcic harzburgites), followed by an apparently continuous evolution in process up to the present day.

The higher average Ti contents (Figure 8), and to some extent Zr contents (Figure 11), of the Archean garnet suites are not consistent with the otherwise strongly depleted nature of these garnets or their host rocks. We interpret this decoupling as the result of metasomatism at relatively low T (900-1100°C), involving the introduction of Zr and Ti with little accompaniment by Y or Ga. This style of metasomatism, which contrasts with that observed in high-T sheared lherzolite xenoliths, has been observed in xenoliths from the Wesselton Kimberlite in the Kimberley area of the Kaapvaal Craton [Shee et al., 1993; Griffin et al., 1996c; Griffin et al., 41997], and from the Siberian Craton [Shimizu et al., 1993]. It is accompanied by the introduction of Ca, producing garnets zoned from harzburgitic cores to lherzolitic rims, as documented by Shee et al. [1993], Schulze [1995], and Griffin et al. [1996c, 1997]. This metasomatism may account for the decrease in the abundance of low-Ca garnets from the older to the younger kimberlites in the Kaapvaal Craton (Table 1).

The progression of mean Zr/Y with tectonothermal age suggests that this style of metasomatism is not a major factor in modifying the composition of Proton or Tecton garnet suites. This could have several explanations: (1) the process is relatively uncommon in post-Archean time; (2) the Archean suites have been exposed to this sort of activity for longer time; (3) the Archean rocks are more susceptible to metasomatic alteration because their low cpx and gnt contents provide little buffering capacity. Modeling of zoned garnets [Griffin et al., 1996c, 1997] suggests that the process has operated in Phanerozoic time, so that option (1) seems unlikely. While the data are not sufficient to settle this problem at present, we tend to prefer the third option.

The proposed evolution of the subcontinental mantle is summarised in Figure 13, where the effects of low-T metasomatism in changing Zr/Y (and Ti/Y) are distinguished from the evolution in major-element and trace-element composition due to a decreasing degree of depletion in a basaltic component.

5.3 Crust-Mantle Coupling

The composition of the lithospheric mantle, as deduced from garnet concentrates, varies in a regular manner, related to the age of the overlying crust, even though the volcanic rocks that have provided our samples come from a relatively narrow time range (mostly within the Phanerozoic; Table 1). This covariation of crustal age and mantle composition has several important implications. The first major conclusion, as noted above, is that the processes that produce subcontinental mantle have changed throughout the history of the Earth.

The second major implication is that most volumes of crust have formed together with their underlying mantle lithosphere, and that crust and mantle have in most cases remained coupled together since that time. Mantle keels extending to depths of 200-300 km beneath Archean cratons apparently have been able to survive large-scale movement of cratons and cratonic fragments over large distances across

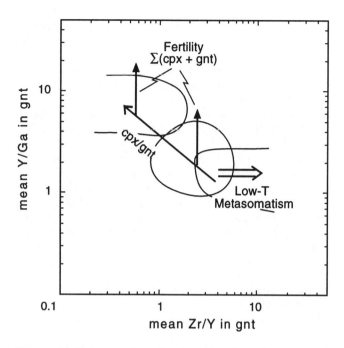

Figure 13. Interpretation of variations in average lherzolitic garnet compositions through time, as involving a secular increase in both cpx/gnt and (cpx+gnt). Low-T metasomatism of depleted Archean mantle probably has increased the average Zr/Y of some sections, without obvious change in Y/Ga.

the face of the Earth, without being detached from their crust. The data also suggest that boundaries between Archon and Proton blocks are steep, and translithospheric. It has been suggested that diamondiferous kimberlites or lamproites in craton-margin Proton settings might be rooted in underlying Archon mantle, and penetrate horizontally displaced Proton crust, but in our data set, no intrusive within a Proton setting contains a clearly Archon garnet suite.

The differences between Archean and Proterozoic crust have been emphasised by many authors (see Taylor and McLennan [1985] and Martin [1994] for useful summaries). In general, Archean crust is characterised by greenstone belts and the low-K felsic rocks of the trondjhemite-tonaiite-greenstone (TTG) suite; the Archean-Proterozoic boundary corresponds to a marked change as K-rich granites became a major component of the crust. In at least some regions, such as southern Africa, there is also a clear difference in the nature of the lower crust; xenoliths of mafic granulite and eclogite are abundant in kimberlites in Proton settings, but absent in those on the Archon [Griffin et al., 1979]. In areas such as northern Siberia, where granulite xenoliths are common in some kimberlites, the xenoliths probably represent upper-crustal rocks (analogous to those exposed in the Anabar Shield) at the time of kimberlite emplacement [Rosen et al., 1994]. In this respect, Proterozoic crustal processes may be more similar to those in the Phanerozoic, where mafic granulites are a major constituent of the lower crust in many regions [Griffin and O'Reilly, 1987; Rudnick and Fountain, 1995].

We suggest that the changes in crustal generation processes from Archean to present are directly related to the changes in mantle processes, implied by the garnet and xenolith data discussed here. This suggestion implies that, in addition to the recognised dichotomy in crustal processes at the Archean/Proterozoic boundary, there has been a progressive evolution in these processes since 2.5 Ga. If this is so, then actualistic models that interpret ancient continental-crust generation in terms of modern arc-accretion processes [Taylor and McLennan, 1985] or plume-related models [Campbell and Griffiths, 1992; Stein and Goldstein, 1996; Abbott, 1996] may be inadequate even for Proterozoic time, and especially for Archean time.

While the present data suggest that continental crust and its underlying lithospheric mantle have in general remained coupled to one another, decoupling (or "lithosphere erosion") can be documented in several regions, where xenolith-bearing volcanic rocks of different ages are present. One such area is the eastern part of the Sino-Korean Craton [Griffin et al., 1996b], where Paleozoic kimberlites in Liaoning and Shandong provinces have sampled typical Archean sections (Table 1). Mesozoic-Tertiary kimberlitic rocks that penetrate the same craton carry garnets of typically Tecton character (Taihang and Teiling localities, Table 1), and Tertiary basalts carry Tecton garnets and xenoliths of fertile spinel and garnet peridotite [Xu et al., 1996] (Nushan locality, Table 1) that lie on the "oceanic" trend of Boyd [1996] (Figure 1). The removal of the Archean keel, and its replacement by a thin, hot lithosphere, is believed to have occurred in Jurassic-Cretaceous time, accompanied by rifting and uplift [Griffin et al., 1996b].

A similar case has been documented in the Colorado-Wyoming region, where the Paleozoic State Line kimberlites sampled an Archean/Proterozoic type of section with a low geotherm (Table 1), while Tertiary kimberlites and other volcanic rocks carry xenoliths that define an elevated geotherm and a thin lithosphere [Eggler et al., 1988]. A few low-T garnets from one of these younger intrusions, the Williams Kimberlite of the Missouri Breaks area [Hearn and McGee, 1984], are typical of garnets from fertile mantle (mean Zr/Y = 1, Y/Ga = 9), suggesting that they may represent the younger Tecton mantle that has replaced the Archean keel. Eggler et al. [1988] ascribe the destruction of the keel to extension in a back-arc environment, prior to the Laramide Orogeny.

A third example is provided by the Colorado Plateau, which has a Proterozoic basement and forms a resistant core within the Laramide Orogen, but has been uplifted by more than 2 km during the Tertiary [Parsons and McCarthy, 1995]. The mantle garnets sampled by the Tertiary minettes and "kimberlitic" diatremes are typical of Tecton garnets in most respects, but transitional toward Proton suites (Table 1; Figures 6, 12), and garnet peridotites from The Thumb lie on the Phanerozoic "oceanic" trend of Boyd [1996] (Figure 3). The lithospheric mantle therefore appears to be younger than the tectonothermal age (mid-Proterozoic) of the basement crust (Appendix 1). This is consistent with Tertiary Sm-Nd mineral ages on eclogite xenoliths [Wendlandt et al., 1996]. However, shallow low-T garnets (T_{Ni}) <600°C) have mean Zr/Y = 3 and Y/Ga = 2, and thus fall into the Proton field on Figure 7, suggesting that older Proterozoic mantle may be preserved at depths of 40-60 km.

Garnet data from kimberlites in the Kaapvaal Craton show a thinning of the lithosphere by ca 30 km, accompanied by a rise in the modeled conductive geotherm from ca 35 mW m^{-2} to 40 mW m^{-2} at ca 90 Ma [Griffin et al., 1995]. The younger lithosphere also contains a lower proportion of low-Ca harzburgitic garnets and shows a higher proportion of low-T metasomatic effects. This lithosphere-thinning event corresponds to a major uplift episode in the Kaapvaal Craton [Brown et al., 1996], which is an expected consequence of lithosphere thinning and consequent heating.

Decoupling and removal of an Archean keel, and its replacement by younger mantle, will have severe tectonic consequences, both because of the differences in intrinsic density between the two volumes (see below), and because it will be accompanied by a rise in isotherms. The results are likely to include uplift, rifting, heating, magmatism, and extensive fluid movement within the crust, with implications for the formation of ore deposits.

5.4 Lithosphere Evolution and the Thermal History of the Earth

The evolution of lithospheric mantle composition through time must reflect fundamental changes in the way the Earth functions. In one sense these changes can be viewed as continuous: on average, the lithospheric mantle produced at each stage of Earth history has had less "crust" extracted from it than that formed in the previous stage. Such a simple decrease in the degree of depletion might be related to the secular cooling of the Earth, resulting in less vigorous convection and lower degrees of partial melting at progressively shallower depths. Herzberg [1995] has demonstrated secular variation in the geochemistry of komatiites and other anhydrous high-degree melts such as picrites, and has argued that these variations reflect changes in the thermal properties of mantle plumes as the Earth cooled. It is not obvious why this process would produce a distinct change in the lithospheric mantle composition near the Archean-Proterozoic boundary. However, while most plumes appear to represent a temperature excess of ca 200°C above the ambient gradient, there are some arguments for the occurrence of much hotter plumes, with a temperature

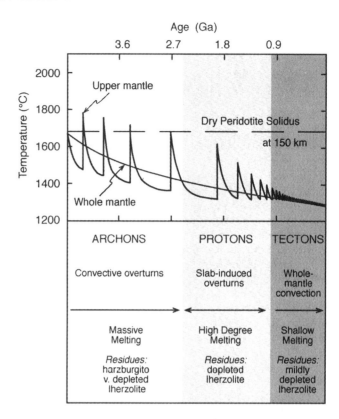

Figure 14. Model of the thermal evolution of a layered Earth, after Davies [1995]. Periodic overturns of the whole mantle early in Earth's history may provide a mechanism for producing opx-rich (low Mg/Si) Archean lithospheric mantle, by raising the temperature of the upper mantle well above the dry peridotite solidus at 150-200 km depth. These overturns end near the Archean-Proterozoic boundary; an increase in the frequency of overturns is accompanied by a decline in temperature difference, and hence the degree of depletion; this merges gradually into the modern plate-tectonic regime characterised by relatively small average degrees of depletion.

excess as high as 400°C [Herzberg, 1993, 1995; Herzberg and Zhang, 1996]. Herzberg [1993, 1995] has argued that Archean peridotites can represent harzburgitic cumulates from plume material melted to more than 50%, at depths of ca 600 km, and postulates that "the first continental lithosphere ... may have formed from a gigantic plume within which melting was deeper and more extensive than at any other time in Earth history." This interesting suggestion raises the question of the cause of such high-T plumes.

Davies [1995] has proposed a model for the thermal and dynamic evolution of the Earth, that offers a qualitative explanation for the evolution of subcontinental mantle documented here. The model (Figure 14) is based on the secular cooling of a two-layered mantle, with the interlayer boundary corresponding to major phase changes. In

Archean time, the upper mantle cools rapidly through heat loss at the Earth's surface, while the lower mantle warms due to heating from the core, and inefficient heat transfer across the boundary between the two layers. Over time, this divergence of temperatures leads to periodic massive overturn, bringing very high-T lower mantle rapidly to upper-mantle depths; this overturn could induce massive melting at relatively great depth as the rising material oversteps the dry peridotite solidus. We suggest that these processes might be responsible for the opx-rich, highly depleted residues (or cumulates] that characterize Archean mantle. These overturns become less frequent as the Earth cools, and the maximum temperature reached by the upper mantle becomes lower with each cycle.

Davies [1995] model predicts a Proterozoic regime with several large overturns; these are induced by the beginning of slab penetration into the lower mantle, and are less intense than those in Archean time. The maximum temperatures reached by the upper mantle in these overturns would still be great enough to produce extensive melting, but at shallower depths than in the Archean. We suggest that this process would produce depleted lherzolites, but not the opx-rich residues or cumulates produced earlier. This regime would gradually evolve to the modern regime of whole-mantle convection, producing relatively low average degrees of depletion at sites of lithosphere generation. Davies' [1995] model provides for a "catastrophic" regime of mantle (and crust) formation within the Archean, the disappearance of this regime near the Archean/Proterozoic boundary, and a gradual evolution in process through the Proterozoic and into the Phanerozoic. It thus is consistent with both the marked contrast between Archean and younger subcontinental mantle, and the observed progression toward lower degrees of depletion in lherzolitic mantle from the Proterozoic to the present.

This model is similar in several respects to the MOMO (mantle overturn-major orogenies) model of Stein and Hofmann [1994], which envisions alternating periods of Wilson-cycle activity and major overturn (in the form of deep slab penetration and rising plumes. If this model is combined with the effects of secular cooling of the mantle, so that each successive overturn resulted in plumes that produced less melt at shallower levels, it can explain the post-Archean evolution of lithospheric composition that we have described here.

Finally, there remains the possibility that Archean mantle lithosphere is different from younger mantle because its source material was different. If the Earth's lower mantle originally was more Si-rich than the upper mantle, melting of material ascending from the lower mantle (in either "plumes" or "overturns") could produce an opx-rich residue. With time, mixing between upper and lower mantle (by either overturns or more conventional plate tectonics) may have eliminated this reservoir, and changed the character of the source material from which mantle lithosphere is formed. Superimposed on secular cooling of the Earth, this process could explain both the lack of post-Archean subcalcic harzburgites, and the continuous evolution to lower degrees of depletion in lithospheric mantle.

5.5 Geophysical Consequences

We have used the average Cr_2O_3 content of the garnets in each age group (Figure 6), and the trends shown in Figures 4 and 12, to estimate average modal composition and CaO content for Archean, Proterozoic and Phanerozoic mantle lherzolites (Table 3). Average compositions for opx and cpx have been taken from our database of Archean and Phanerozoic xenoliths, and extrapolated to reasonable values for the minerals in Proterozoic lherzolites, given the average garnet composition. Rock densities have then been calculated using data on end-member compositions [Smyth and McCormick, 1995], and extrapolated to a range of temperatures using the thermal expansion coefficients of Fei [1995]. From the densities and compositions of the average rocks, we have calculated seismic velocities (V_p and V_s) using the algorithms of Simmons [1964] and Anderson and Sammis [1970]. Results are given in Table 3.

Boyd and McAllister [1976] used XRD data on individual minerals, combined with careful modal estimates, to calculate room-T densities for a strongly depleted Archean garnet lherzolite (sample PHN1569; density 3.30±02 t m^{-3}) and a high-T sheared peridotite (sample PHN1611; density = 3.39±02 t m^{-3}). The density estimate for PHN1569 agrees well with our calculated average for Archean lherzolites and harzburgites. The estimate for PHN1611 is higher than our average Archean high-T lherzolite. However, the garnet and cpx content of PHN1611 has been greatly increased by high-T melt-related metasomatism [Smith et al., 1993], so that its composition is extreme even among this class of xenoliths. The bulk composition of PHN1611 is very similar to some "pyrolite" compositions, and its density is identical within error to our estimated density for pyrolite, and that of Jordan [1979]. Jordan's [1979] higher estimated density (3.35 t m^{-3}) for "average continental garnet lherzolite" included many high-T sheared lherzolites like PHN1611; as noted above, we do not consider these to be representative of any large-scale or long-term mantle reservoir, and their inclusion in an "average mantle" is not warranted.

On average, Archean lherzolite is ca 0.04 t m^{-3} less dense than Phanerozoic lherzolite at any T (Figure 15). The density of Archean lherzolite is insensitive to variations in the small amounts of gnt and cpx present, and the density of average Archean harzburgite is not significantly different from that of the average lherzolite. Thus small variations in cpx/opx/gnt ratios with depth, due to changes in the composition of solid solutions, will not significantly affect V_p or V_s in the lithosphere. Despite its higher density, Phanerozoic mantle will have a V_p ca 5% lower than that of Archean mantle at the same T; this is because of the large

Table 3. Estimates of average modal composition, density and seismic velocity

	% oliv	% opx	%cpx	%gnt	%CaO	Density 20°C	Thermal Expansion Coefficients			Densities			Vp, km/s 1000 °C	Vs, km/s 1000 °C
							400 °C	800 °C	1225 °C	400 °C	800°C	1225 °C		
Archean lherz	62 Fo93	31 En92	2 cr-diopside*	5 Py70Alm15Uv15	1.06	3.310	0.012	0.024	0.037	3.269	3.229	3.186	8.24	4.38
Archean Harz	65 Fo94	30 En92	0 cr-diopside*	5 Py70Alm10Uv20	1.00	3.309	0.012	0.024	0.038	3.268	3.228	3.184	8.20	4.37
Proterozoic	70 Fo92	18 En91	7 cr-diopside*	5 Py70Alm15Gr15	1.75	3.330	0.012	0.024	0.038	3.289	3.248	3.205		
Phanerozoic	65 Fo90	15 En90	13 cr-diopside*	7 Py70Alm15Gr15	3.42	3.354	0.012	0.024	0.038	3.313	3.272	3.228	7.79	4.24
Archean Hi-T	75 Fo90	16 En90	3 cr-diopside*	6 Py70Alm15Gr15		3.354	0.012	0.024	0.038	3.313	3.272	3.228		
Pyrolite	59 Fo90	13 En90	14 cr-diopside*	14 Py70Alm15Gr15	3.80	3.378	0.012	0.024	0.038	3.337	3.296	3.251		

* Di75 Hd5Jd10Cc2En8

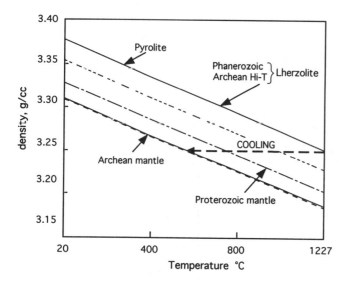

Figure 15. Calculated densities of average mantle compositions vs T; data from Table 3. Horizontal arrow shows increase in density as T decreases upward in a section.

effect of CaO on the calculated V_p. The difference in V_s is smaller (ca 3%) but still significant. Sobolev et al. [1995], using a smaller xenolith database (largely of spinel lherzolites) and a different procedure, estimated a smaller difference in V_s (1.2%) between depleted and fertile lithospheric mantle.

These data have consequences for the interpretation of seismic tomography. The range of V_s observed at 200 km depth in the subcontinental upper mantle by Su et al. [1994] is ca 6%, and this has commonly been interpreted as due to differences in T between Archean continental roots and the mantle beneath younger areas. The data presented here suggest that up to half of this variation may be ascribed to the differences in composition between Archean and Phanerozoic mantle; the remainder may represent real differences in the average geotherm between cratonic regions and mobile belts [cf Jordan, 1988].

5.6 Geodynamic Consequences

Phanerozoic lithosphere is only marginally buoyant relative to asthenosphere (taken as a generalised "pyrolite" composition) at the same temperature (Table 3). Cooling the Phanerozoic lithosphere by 200°C, corresponding to a thickness of only 35 km above the lithosphere-asthenosphere boundary on a typical elevated Phanerozoic geotherm [O'Reilly and Griffin, 1985], will produce neutral buoyancy (Figure 15). This means that a lithospheric mantle more than a few tens of km thick will begin to be negatively buoyant relative to the asthenosphere, and a thick lithosphere composed of Phanerozoic mantle therefore will be gravitationally unstable. Houseman and Molnar [1996] have presented mathematical models for the detachment and foundering of cooling continental lithosphere, using density parameters that are appropriate to Phanerozoic lithosphere as defined here.

The Ontong-Java oceanic plateau is interpreted as the result of plume-related volcanism, and its thick crust and relative elevation are inferred to reflect a high degree of depletion in the underlying mantle [Abbott, 1996, and references therein]. However, the garnet concentrates and xenoliths from Malaita [Neal, 1988] (Tables 1,2), which are believed to represent a sample of the lithospheric mantle beneath the Ontong-Java Plateau, are typically Tecton in character (low Cr and Zr/Y, high Y/Ga) and imply a relatively fertile mantle with a high cpx and garnet content (even if samples with more than 15% cpx (Figure 5) are excluded). These data suggest that the mantle beneath this oceanic plateau is no more depleted than that found under areas such as eastern China, eastern Australia and the Baikal region (Figure 5). There thus appears to be little evidence that plume-related volcanism involves a higher average degree of depletion of the residual mantle than other types of intraplate volcanism. If this is true, then oceanic plateaux may not be resistant to subduction, compared with normal oceanic lithosphere [cf Abbott and Mooney, 1995].

Archean lithosphere, because of its low density, must be at least 600°C cooler to attain neutral buoyancy relative to asthenosphere (Figure 15). Using the same lithosphere/asthenosphere boundary temperature of 1250°C, and a typical cratonic conductive geotherm corresponding to a surface heat flow of 40 mW/m^{-2} [Pollack and Chapman, 1977], we find that the lower 120 km of the mantle section will be buoyant relative to the asthenosphere. Our data indicate a wide range of lithosphere thicknesses beneath Archons, ranging from ca 160-230 km [Griffin et al., 1996a, and unpublished data]. Thus only the upper 30-80 km of a typical Archon mantle section will be negatively buoyant; this situation helps to ensure the long-term stability of Archean lithosphere, and the preservation of Archon keels.

Proterozoic lithospheric mantle is intermediate in composition and density between Archean and Phanerozoic mantle, requiring ca 400°C temperature difference in order to have neutral buoyancy relative to asthenosphere (Table 3). Assuming a temperature of 1250°C at the lithosphere/asthenosphere boundary, and a typical Proton conductive geotherm (45 mW/m^{-2} model of Pollack and Chapman [1977]), this point of neutral buoyancy would be reached ca 60 km above the lithosphere/asthenosphere boundary. Analysis of many Proton areas suggests typical lithosphere thicknesses on the order of 140-160 km [Zhang et al., 1996; Griffin, O'Reilly and Ryan, unpublished data), and the lack of thicker Proterozoic lithosphere sections may reflect their greater density, relative to Archean sections.

The problem is thus, not how or why Archon keels survive through time, but the mechanism by which they can be destroyed or removed in areas such as eastern China and Colorado-Wyoming, as described above. Heating ("thermal erosion") is not an adequate mechanism, because heating of the Archean lithosphere will only make it more buoyant relative to asthenosphere. Because of its refractory composition, it will also resist partial melting. Metasomatism, as observed in the high-T sheared peridotite xenoliths in kimberlites, can greatly increase the density of the lithosphere, to the point where it is indistinguishable from asthenosphere and thus at least negatively buoyant. In this case, it will certainly be easier to remove, though it is not clear under what circumstances foundering would occur.

One answer may lie in the mechanism recently proposed by Yuan [1996], based on detailed seismic tomography beneath the eastern part of the Sino-Korean craton. This study suggests that the Archean root has been pulled apart during extension, and asthenospheric material has risen along linear zones, spreading out at levels of 75-100 km. Below this level, blocks of Archean mantle alternate with vertical zones of asthenospheric mantle; these are still recognisable by their higher seismic velocity, which is due to compositional differences, despite the entire zone being essentially at high T. With time, mechanical mixing, intrusion of magmas, and metasomatism by melt infiltration, may change the bulk composition of these relict Archean blocks, making them physically indistinguishable from the surrounding asthenosphere.

6. CONCLUSIONS

Major- and trace-element data on concentrates of mantle-derived garnet from volcanic rocks show strong correlations between mean garnet composition and the tectonothermal age of the crust penetrated by the eruptions. These data complement those obtained from Archean and Phanerozoic xenolith suites, and indicate that the composition of newly created subcontinental lithospheric mantle has changed significantly through the history of the Earth. Subcontinental lithosphere produced prior to ca 2.5 Ga consists of highly depleted lherzolites, harzburgites, and subcalcic harzburgites, all with relatively low Mg/Si ratios at high mg#. In Proterozoic time, subcalcic harzburgites were no longer produced, the lherzolitic component became dominant, the lherzolites and calcic harzburgites were generally less depleted than those formed in Archean time, and the mean Mg/Si ratio probably increased relative to Archean mantle. Since ca 1 Ga, the proportion of cpx-free harzburgite in subcontinental mantle has declined to essentially zero, and the overall degree of depletion of lithospheric mantle has declined still further.

These data and observations suggest that after a marked change at the Archean-Proterozoic boundary, there has been a continual change in mantle melting and crustal generation processes, to produce progressively less depleted subcontinental mantle through time. These changes may reflect an evolution in tectonic style, from periodic whole-mantle overturns in the Archean, to less violent, more nearly continuous whole-mantle convection today, as proposed by Davies [1995]. They might also reflect a secular change in the compositions of both upper and lower mantle, due to mixing of two layers with initially different compositions. The progressive evolution of lithosphere-forming processes since the time of the Archean-Proterozoic boundary implies that actualistic models of lithosphere formation based on modern processes may be inadequate, even for Proterozoic time, and are almost certainly not applicable to Archean time.

The correlation between mantle type and crustal age indicates that crustal volumes and the underlying lithospheric mantle have been formed together, and generally have stayed coupled together thereafter. The differences between Archean and Proterozoic lithospheric mantle probably are linked to the known dichotomy in crustal processes at the Archean-Proterozoic boundary, and understanding of the differences in mantle processes across this boundary can help to constrain models for crustal formation processes.

The stability and thickness of Archean mantle is directly related to its low density, which in turn reflects both its high degree of depletion in basaltic components, and its low Ca/Al and Mg/Si. These chemical characteristics produce high seismic velocities, so that compositional factors may account for at least half of the velocity contrasts between Archean and younger areas, seen in seismic tomography. The generally higher density, and lower degree of depletion, of Proterozoic mantle sections may account for the generally thinner lithosphere and commonly higher mantle heat flow of Proterozoic terrains. Phanerozoic mantle, with the high density resulting from its low degree of depletion, will not sustain a thick lithosphere section.

Acknowledgments. Numerous critical samples and data have been provided by Joe Boyd, John Gurney, Peter Nixon and Doug Smith, as well as Stockdale Prospecting Ltd., RTZ-CRA Exploration Pty Ltd., Kennecott Canada Inc. and Ashton Mining. This research has been supported financially by the Australian Research Council, Macquarie University, and the companies mentioned above. We also have benefited from many discussions with Joe Boyd, Buddy Doyle, Bram Janse, Simon Shee, and Bruce Wyatt. We are especially grateful to Joe Boyd, Bruce Wyatt and Simon Shee for thoughtful reviews of earlier versions of the manuscript. The final version was improved through thoughtful reviews by Michael Walter and Alexei Goncharov. This is publication number 90 from the National Key Centre for Geochemical Evolution and Metallogeny of Continents (GEMOC).

APPENDIX 1. NOTES ON LOCALITIES

Russia

The terrane structure of the Siberian Platform is summarized by Rosen et al. [1994]. The Malo-Botuobiya kimberlite field lies in the Magan terrane, while the Daldyn,

Alakit and Upper Muna fields lie in the Daldyn terrane; both of these are interpreted as Archean blocks. The Merchimden and Upper Molodo fields lie in the Proterozoic (1.9-2.3 Ga) Habschan terrane, and the Kuoika field in the Proterozoic Birekte terrane. All of these terranes were welded together to form the craton at ca 1.9 Ga. The Vitim basaltic field lies 100-200 km west of the Baikal Rift zone, in a Paleozoic fold belt with extensive Mesozoic granite magmatism.

China

The Ordovician kimberlites of Liaoning and Shandong provinces intrude the Archean Liaolu nucleus of the Sino-Korean Craton [Zhang et al., 1994, 1996]. The Paleozoic to Mesozoic lamproites of Guizhou, Hubei and Hunan Provinces intrude the Proterozoic Yangtze craton. The distribution of crustal ages within this craton is poorly known. The Hunan lamproites lie near the suture between the Yangtze and Cathaysia blocks, which are believed to have been welded together at ca 1 Ga. The Hubei lamproites lie near the northern border of the craton, in a region that may have been affected by the Triassic collision between the Yangtze and Sino-Korean Cratons. The Cretaceous-Tertiary kimberlites grouped here as "Taihang" intrude a Proterozoic (1.8 Ga?) mobile belt that crosses the Sino-Korean Craton, dividing the Liaolu nucleus from the Ordos nucleus to the west. This belt also is the site of Mesozoic-recent rifting, and hence is classed as a Tecton. The Teiling kimberlites are believed to be of Cretaceous-Tertiary age [Griffin et al., 1996b], and intrude the eastern part of the Sino-Korean Craton, where it is affected by Mesozoic-recent extension, rifting and volcanism. The Nushan locality is a Pleistocene basaltic volcano near Nanjing, in the same zone of extension. The Mingxi locality is a basaltic diatreme located in the coastal mountains of southeast China, an area affected by Cenozoic subduction and extension [Xu et al., 1996].

Australia

The "Tasman Line" divides cratonic central and western Australia from the Paleozoic accretionary terranes and fold belts of the east. The kimberlites grouped here as "South Australia" intrude Proterozoic crust on the eastern margin of the craton. The Miocene Ellendale lamproites intrude the edge of the Proterozoic King Leopold Fold Belt in the northern part of Western Australia. The White Cliffs and Renmark localities are basaltic diatremes in the Paleozoic mobile belt, just east of the Tasman Line, while the Jugiong basaltic diatremes lie further east, well within the fold belts. Re-Os depletion ages on spinel peridotite xenoliths from eastern Australia range (with one exception) from Mesozoic to uppermost Proterozoic [McBride et al., 1996].

North America

The Slave Province of Canada is one of the oldest Archon fragments, and the scene of intense diamond exploration activity [Pell, 1995]. The Saskatchewan kimberlites lie within the Proterozoic Trans-Hudsonian Orogen, but may intrude an Archean fragment that is overthrust on all sides [Leahy and Taylor, 1995]. The Upper Peninsula of Michigan includes Archean crust, intruded by abundant Proterozoic (1.8 Ga) granites. It may also have been affected by the Later Proterozoic Keweenawan Rift system. The State Line district is intruded into a Proterozoic mobile belt on the reworked margin of the Wyoming Craton. The Arkansas lamproites are believed to be intruded into the Proterozoic mobile belts that surround the North American composite cratonic nucleus. The Colorado Plateau (represented here by The Thumb, a minette plug) has largely Proterozoic basement [Wendlandt et al., 1996], but has been uplifted (with some accompanying basaltic volcanism) ca 2 km since Cretaceous time. Parsons and McCarthy [1995] conclude that this uplift requires an anomalously light upper mantle, corresponding to the elevated geotherm demonstrated by xenolith studies [Smith et al., 1991, and references therein]. The Mt. Horeb, Virginia kimberlite intrudes the Paleozoic Appalachian Fold Belt.

South Africa

The Kaapvaal craton is a classic Archon. The kimberlites used in this study have been divided into two groups, with ages greater than 90 Ma and less than 90 Ma. This age corresponds to a significant change in the thickness and thermal structure of the lithosphere [Griffin et al., 1992] accompanied by a surface uplift of more than 3 km [Brown et al., 1996].

Other Areas

The Proterozoic (1.7-1.8 Ga) kimberlites of the Guaniamo area in Venezuela intrude Proterozoic sediments overlying the Guyana Craton [Nixon et al., 1994]. The Finnish kimberlites intrude the edge of the Baltic Shield, reworked during middle Proterozoic; juvenile Proterozoic crust is found ca 50 km further from the craton margin [A.J.A. Janse, personal communication, 1996]. The Malaita alnoites intrude part of the Ontong-Java oceanic plateau, obducted in the Solomon Islands; they are believed to have sampled the plume-related mantle beneath the plateau [Neal, 1988]. The Majhgawan kimberlite/lamproite intrudes Upper Proterozoic sediments overlying granites of 2.5 Ga age [Chatterjee and Rao, 1995] near the edge of the Arivalli Craton in northern India. The Pali-Aike locality represents basaltic volcanism in the southern tip of the Andes, providing a sample of mantle in a collisional environment [Stern et al., 1989].

REFERENCES

Abbott, D. H., Plumes and hotspots as sources of greenstone belts, *Lithos, 37*, 113-127, 1996.

Abbott, D. H., and W. D. Mooney, Crustal structure and evolution: support for the ocean plateau model of continental growth, *Natl. Rept. IUGG*, 231-24, 1995.

Anderson, D. L., and C. Sammis, Partial melting in the upper mantle, *Phys. Earth Planet. Interiors, 3*, 41-50, 1970.

Boyd, F. R., Origin of peridotite xenoliths: major element considerations, in *High pressure and high temperature research on lithosphere and mantle materials*, edited by G. Ranalli, F. Ricci Lucchi, C. A. Ricci, and T. Trommsdorff, University of Siena, in press, 1996.

Boyd, F. R., and J. J. Gurney, Diamonds and the African lithosphere, *Science, 232*, 472-477, 1986.

Boyd, F. R., and R. H. McAllister, Densities of fertile and sterile garnet peridotites, *Geophys. Res. Lett., 3*, 509-512, 1976.

Boyd, F. R., and S. A. Mertzman, Composition and structure of the Kaapvaal lithosphere, southern Africa, in *Magmatic Processes: Physicochemical Principles.*, edited by B.O. Mysen, *Geochem. Soc. Spec. Publ.., 1*, 13-24, 1987.

Boyd, F. R., D. G. Pearson, P. H. Nixon, and S. A. Mertzman, Low-calcium garnet harzburgites from southern Africa: their relations to craton structure and diamond crystallization, *Contr. Mineral. Petrol.. 113*, 352-366, 1993.

Brown, R., D. Belton, K. Gallagher, and R. Harmon, Extraordinary denudation rates for cratons inferred from apatite fission track data: Implications for the long-term stability of cratonic lithosphere, *EOS, Trans. Amer. Geophys. Un., 77*, w152, 1996.

Campbell, I. H., and R. W. Griffiths, The changing nature of mantle hotspots through time: implications for the chemical evolution of the mantle, *J. Geol., 92*, 497-523, 1992.

Canil, D., Experimental evidence for the exsolution of cratonic peridotite from high-temperature harzburgite, *Earth Planet. Sci. Lett., 106*, 64-72, 1991.

Canil, D., and K. Wei, Constraints on the origin of mantle-derived low Ca garnets, *Contr. Mineral. Petrol., 109*, 421-430, 1992.

Carswell, D. A., W. L. Griffin, and P. Kresten, Peridotite nodules from the Ngopetsu and Lipelaneng Kimberlites, in *Kimberlites II, the Mantle and Crust-Mantle Relationships.* Developments in Petrology, 11B, edited by J. Kornprobst, pp 299-343, Elsevier, Amsterdam, 1984.

Chatterjee, A. K., and K. S. Rao, Majhgawan diamondiferous pipe, Madhya Pradesh, India - a review, *J. Geol. Soc. India, 45*, 175-189, 1995.

Cox, K. G., J. J. Gurney, and B. Harte, Xenoliths from the Matsoku pipe, in *Lesotho Kimberlites*, edited by P. H. Nixon, Lesotho National Development Corporation, Maseru, pp. 76-100, 1973.

Cox, K. G., M. R. Smith, and S. Beswetherick, Textural studies of garnet lherzolites: evidence of exsolution origin from high-temperature harzburgites, in *Mantle Xenoliths*, edited by P.H. Nixon, pp 537-550, Wiley and Sons, Chichester, 1987.

Danchin, R.V., Mineral and bulk chemistry of garnet lherzolite and garnet harzburgite from the Premier mine, South Africa, in *The mantle sample: inclusions in kimberlites and other volcanics*, edited by F. R. Boyd and H. A. O. Meyer, pp. 104-126, American Geophysical Union, Washington, D.C., 1979.

Davies, G. F., Punctuated tectonic evolution of the Earth, *Earth Planet. Sci. Lett., 136*, 363-379, 1995.

Eggler, D. H., J. K. Meen, R. Welt, F. O. Dudas, K. P. Furlong, M. E. McCallum, and R. W. Carlson, Tectonomagmatism of the Wyoming Province, *Colorado School of Mines Quarterly, 83*, 25-40, 1988.

Ehrenberg, S. N., Petrogenesis of garnet lherzolite and megacrystalline nodules from The Thumb, Navajo volcanic field, *J. Petrology, 23*, 507-547, 1982

Fan, Q. C., and P. R. Hooper, The mineral chemistry of ultramafic xenoliths of eastern China: implication for upper mantle composition and the paleogeotherms, *J. Petrology, 30*, 1117-1158, 1989.

Fei, Y., Thermal expansion, in *Mineral Physics and Crystallography: a Handbook of Physical Constants.* edited by T.J. Ahrens, American Geophysical Union, Washington, D.C., pp. 29-44, 1995.

Griffin, W. L., and S. Y. O'Reilly, The composition of the lower crust and the nature of the continental Moho: xenolith evidence, in *Mantle Xenoliths*, edited by P. H. Nixon, pp. 413-430, Wiley and Sons, Chichester, U.K., 1987.

Griffin, W. L., and C. G Ryan, Trace elements in indicator minerals: area selection and target evaluation in diamond exploration, *J. Geochem. Explor., 53*, 311-337, 1995.

Griffin, W. L., D. A. Carswell, and P. H. Nixon, Lower crustal granulites and eclogites from Lesotho, southern Africa, in *The Mantle Sample: Inclusions in Kimberlites and Other Volcanics*, edited by F. R. Boyd and H. O. A. Meyer, pp. 59-86, American Geophysical Union, Washington, D.C., 1979.

Griffin, W. L., D. Smith, F. R. Boyd, D. R. Cousens, C. G. Ryan, S. H. Sie, and G. F. Suter, Trace element zoning in garnets from sheared mantle xenoliths, *Geochim. Cosmochim. Acta,. 53*, 561-567, 1989a.

Griffin, W. L., C. G. Ryan, D. C. Cousens, S. H. Sie, and G. F. Suter, Ni in Cr-pyrope garnets: a new geothermometer, *Contr. Mineral. Petrol., 103*, 199-202, 1989b.

Griffin, W. L., J. J. Gurney, and C. G Ryan, Variations in trapping temperatures and trace elements in peridotite-suite inclusions from African diamonds: evidence for two inclusion suites, and implications for lithosphere stratigraphy, *Contr. Mineral. Petrol., 110*, 1-15, 1992.

Griffin, W. L., F. V. Kaminsky, C. G. Ryan, S. Y. O'Reilly, T. T. Win, and I. P. Ilupin, Thermal state and composition of the lithospheric mantle beneath the Daldyn kimberlite field, Yakutia, *Tectonophysics, 262*, 19-33, 1996a.

Griffin, W. L., A. Zhang, S. Y. O'Reilly, and C.G. Ryan, Phanerozoic evolution of the lithosphere beneath the Sino-Korean Craton, in *Mantle Dynamics and Plate Interactions in East Asia*, edited by M. Flower, S. L. Chung, C. H. Lo, and T. Y. Lee, Amer. Geophys. Un., Spec. Publ.., in press, 1996b.

Griffin, W. L., D. Smith, C. G. Ryan, S. Y. O'Reilly, and T. T. Win, Trace element zoning in mantle minerals: metasomatism and thermal events in the upper mantle, *Canad. Mineral., 34*, 1179-1193, 1996c.

Griffin, W. L., S. R. Shee, C. G. Ryan, T. T. Win, and

B. A. Wyatt, Harzburgite to lherzolite and back again: metasomatic processes in ultramafic xenoliths from the Wesselton Kimberlite, submitted to Contr. Mineral. Petrol., 1997.

Gurney, J. J., A correlation between garnets and diamonds, in *Kimberlite Occurrence and Origins: a Basis for Conceptual Models in Exploration*, edited by J. E. Glover and P. G. Harris, pp. 143-166, Geology Department and University Extension, University of Western Australia, Publ. No. 8, 1984.

Hearn, B. C. Jr, and E. S. McGee, Garnet peridotites from Williams Kimberlites, north-central Montana, U.S.A., in *Kimberlites II: The Mantle and Crust-Mantle Relationships*, edited by J. Kornprobst, pp. 57-70, Elsevier, Amsterdam. 1984.

Helmstaedt, H. H., and J. J. Gurney, Geotectonic controls of primary diamond deposits: implications for area selection, *J. Geochem. Explor., 53*, 125-144, 1995.

Herzberg, C., Lithosphere peridotites of the Kaapvaal craton, *Earth Planet. Sci. Lett., 120*, 13-29, 1993.

Herzberg, C., Generation of plume magmas through time: an experimental perspective, *Chemical Geology, 126*, 1-16, 1995.

Herzberg, C., and J. Zhang, Melting experiments on anhydrous peridotite KLB - 1: Compositions of magmas in the upper mantle and transition zone, *J. Geophys. Res., 101*, 8271-8295, 1996.

Hoal, B. G., K. E. O. Hoal, F. R. Boyd, and D. G. Pearson, Age constraints on crustal and mantle lithosphere beneath the Gibeon kimberlite field, Namibia, *S. Afr. Tydskr. Geol., 98*, 112-118, 1995.

Houseman, G. A., and P. Molnar, The mechanical stability of continental lithosphere, *EOS, Trans. Amer. Geophys. Un., 77*, w15, 1996.

Ionov, D. A., I. V. Ashchepkov, H.-G. Stosch, G. Witt-Eickschen, and H. A. Seck, Garnet peridotite xenoliths from the Vitim volcanic field, Baikal region: the nature of the garnet-spinel transition zone in the continental mantle, *J. Petrology, 34*, 1141-1175, 1993.

Janse, A. J. A., Is Clifford's Rule still valid? Affirmative examples from around the world, in *Diamonds: Characterization, Genesis and Exploration*, edited by H. O. A. Meyer, and O. Leonardos, pp. 215-235, CPRM Spec. Publ. 1A/93, Dept. Nacional da Prod. Mineral., Brazilia, 1994.

Jordan, T. H., Mineralogies, densities and seismic velocities of garnet lherzolites and their geophysical implications, in *The Mantle Sample: Inclusions in Kimberlites and Other Volcanics*, edited by F. R. Boyd and H. O. A. Meyer, pp. 1-14, American Geophysical Union, Washington, D.C. 1979.

Jordan, T. H., Structure and formation of the continental lithosphere, *J. Petrol., Spec. Lithosphere Issue*, 11-37, 1988.

Leahy, K., and W. R. Taylor, The influence of the deep structure of the Glennie Domain on the diamonds in Saskatchewan Kimberlites (abstract), 6th International Kimberlite Conference, Novosibirsk, 314-316, 1995.

Liu, R. X., and Q. C. Fan, Major and trace element geochemistry of ultramafic xenoliths in east China in *Characteristics and Geodynamics of the Arc Mantle in China*, pp. 45-61, Seismology Press, Beijing, 1990.

Maaloe, S., and K. Aoki, The major element composition of the upper mantle estimated from the composition of lherzolites, *Contrib. Mineral. Petrol. , 63*, 161-173, 1977.

McBride, J. S., D. D. Lambert, A. Greig, and I. A. Nicholls, Multistage evolution of Australian subcontinental mantle: Re-Os constraints from Victorian mantle xenoliths, *Geology, 24*, 631-634, 1996.

Martin, H., The Archean grey gneisses and genesis of continental crust, in *Archean Crustal Evolution,* edited by K. Condie, pp. 205-259, Elsevier, Amsterdam, 1994.

Morgan, P., Diamond exploration from the bottom up: regional geophysical signatures of lithosphere conditions favorable for diamond exploration, *J. Geochem. Explor., 53*, 145-165, 1995.

Neal, C. R., The origin and composition of metasomatic fluids and amphiboles beneath Malaita, Solomon Islands, *J. Petrology, 29,* 149-179, 1988.

Nixon, P. H., Kimberlitic xenoliths and their cratonic setting, in *Mantle Xenoliths*, edited by P.H. Nixon, pp. 215-239, Wiley and Sons, Chichester, 1987.

Nixon, P. H., and F. R. Boyd, Garnet-bearing lherzolites and discrete nodule suites from the Malaita Alnoite, Solomon Islands, S.W. Pacific, and their bearing on oceanic composition and geotherm, in *The Mantle Sample: Inclusions in Kimberlites and Other Volcanics*, edited by F. R. Boyd and H. A. O. Meyer, pp. 400-422, American Geophysical Union, Washington, D.C., 1979.

Nixon, P. H., W. L. Griffin, G. R. Davies, and E. Condliffe, Cr-garnet indicators in Venezuela kimberlites and their bearing on the evolution of the Guyana Craton, in *Kimberlites, Related Rocks and Mantle Xenoliths*, edited by H. O. A. Meyer and O. H. Leonardos, pp 378-387, CPRM Spec. Publ. 1A/93, Dept. Nacional da Prod. Mineral., Brazilia, 1994.

Norman, M. D., N. J. Pearson, A. Sharma, and W. L. Griffin, Quantitative analysis of trace elements in geological materials by laser ablation ICPMS: instrumental operating conditions and calibration values of NIST glasses, *Geostandards Newsl., 20*, 247-261, 1996.

O'Neill, H. St.C., and B. J. Wood, An experimental study of Fe-Mg partitioning between garnet and olivine, and its calibration as a geothermometer, *Contrib. Mineral. Petrol., 70*, 59-70, 1979.

O'Reilly, S. Y., and W. L. Griffin, A xenolith-derived geotherm for southeastern Australia and its geophysical implications, *Tectonophysics, 111*, 41-64, 1985.

O'Reilly, S. Y., and W. L. Griffin, Trace element partitioning between garnet and clinopyroxene in mantle-derived pyroxenites and eclogites: P-T-X controls, *Chem. Geol., 121*, 105-130, 1995.

Parsons, T., and J. McCarthy, The active southwest margin of the Colorado Plateau: uplift of mantle origin, *Geol. Soc. Amer. Bull., 107*, 139-147, 1995.

Pearson, D. G., R. W. Carlson, S. B. Shirey, F. R. Boyd, and P. H. Nixon, The stabilisation of Archaean lithospheric mantle: a Re-Os isotope study of peridotite xenoliths from the Kaapvaal and Siberian Cratons, *Earth Planet. Sci. Lett., 134*, 341-357, 1995.

Pell, J. A., Kimberlites in the Slave structural province, Northwest Territories, Canada: a preliminary review (abstract), 6th International Kimberlite Conference, Novosibirsk, 433-434, 1995.

Pollack, H. N., and D. S. Chapman, On the regional variation of heat flow, geotherms and lithosphere thickness, *Tectonophysics, 38,* 279-296, 1977.

Qi, Q., L. A. Taylor, and X. Zhou, Petrology and geochemistry of mantle peridotite xenoliths from SE China, *J. Petrology, 36,* 55-79, 1995.

Rhodes, J. M., and J. B. Dawson, Major and trace element chemistry of peridotite xenoliths from the Lashaine volcano, Tanzania, *Phys. Chem. Earth, 9,* 545-557, 1974.

Richardson, S. H., J. J. Gurney, A. J.Erlank, and J. W. Harris, Origin of diamonds in old enriched lithosphere, *Nature, 310,* 198-202, 1984.

Reid, A. M., C. H. Donaldson, R. W. Brown, W. I. Ridley, and J. B. Dawson, Mineral chemistry of peridotite xenoliths from the Lashaine volcano, Tanzania, *Phys. Chem. Earth, 9,* 525-543, 1974.

Rosen, O. M., K. C. Condie, L. M. Natapov, and A. D. Nozhkin, Archean and early Proterozoic evolution of the Siberian Craton: a preliminary assessment, in *Archean Crustal Evolution,* edited by K. Condie, pp. 411-459, Elsevier, Amsterdam, 1994.

Rudnick, R. L., and D. M. Fountain, Nature and composition of the continental crust: a lower crustal perspective, *Rev. Geophys., 33,* 267-309, 1995.

Ryan, C. G., D. R. Cousens, S. H. Sie, W. L. Griffin, and E. J. Clayton, Quantitative PIXE microanalysis in the geosciences, *Nucl. Instr. and Meth., B47,* 55-71, 1990.

Ryan, C. G., W. L. Griffin, and N. J. Pearson, Garnet geotherms: a technique for derivation of P-T data from Cr-pyrope garnets, *J. Geophys. Res., 101,* 5611-5625, 1996.

Schulze, D. J., Calcium anomalies in the mantle and a subducted metaserpentinite origin for diamonds, *Nature, 319,* 483-485, 1986.

Schulze, D. J., Low-Ca garnet harzburgites from Kimberley, South Africa: abundance and bearing on the structure and evolution of the lithosphere, *J. Geophys. Res., 100,* 12513-12526, 1995.

Shee, S. R., B. A. Wyatt, and W. L. Griffin, Major and trace element mineral chemistry of peridotite xenoliths from the Wesselton kimberlite, South Africa (abstract), *IAVCEI Abstracts,* p. 93, Canberra, 1993.

Shimizu, N., F. R. Boyd, and N. P Pokhilenko, Trace element zoning patterns of mantle garnets (abstract), *Geol. Soc. Amer. Abstr. Programs, 25,* A-36, 1993.

Simmons, G., Velocity of compressional waves in various minerals at pressures to 10 kilobars, *J. Geophys. Res., 69,* 1117-1121, 1964.

Smith, C. B., R. W. E. Green, M. Q. W. Jones, and K. S. Viljoen, Progress towards understanding the evolution of the Kaapvaal lithosphere; the mantle perspective (abstract), 6th International Kimberlite Conference, Novosibirsk, 343-346, 1995.

Smith, D., and F. R. Boyd, Compositional heterogeneities in a high-temperature lherzolite nodule and implications for mantle processes, in *Mantle Xenoliths,* edited by P. H. Nixon, pp. 551-562, Wiley and Sons, Chichester, U.K., 1987.

Smith, D., W. L. Griffin, C. G. Ryan, D. R. Cousens, S. H. Sie, and G. F. Suter, Trace-element zoning of garnets from The Thumb: a guide to mantle processes, *Contr. Mineral. Petrol., 107,* 60-79, 1991.

Smith, D., W. L. Griffin, and C. G Ryan, Compositional evolution of high-temperature sheared lherzolite, PHN1611, *Geochim. Cosmochim. Acta, 57,* 605-613, 1993.

Smyth, J. R., and T. C. McCormick, Crystallographic data for minerals, in *Mineral Physics and Crystallography: a Handbook of Physical Constants,* edited by T. J. Ahrens, pp. 1-17, American Geophysical Union, Washington, D.C., 1995.

Sobolev, N. V., *Deep Seated Inclusions in Kimberlites and the Problem of the Composition of the Upper Mantle* (in Russian), 1974.

Sobolev, N. V., *Deep Seated Inclusions in Kimberlites and the Problem of the Composition of the Upper Mantle,* Izdatel'stvo Mauka; English transl. by D.A. Brown, 279 pp., American Geophysical Union, Washington, D.C., 1977.

Sobolev, N. V., and P. H. Nixon, Xenoliths from the USSR and Mongolia: a selective and brief review, in *Mantle Xenoliths,* edited by P.H. Nixon, pp. 159-166, Wiley and Sons, Chichester, U.K., 1987.

Sobolev, N. V., Yu. G. Lavrent'ev, N. P. Pokhilenko, and L. V. Usova, Chrome-rich garnets from the kimberlites of Yakutia and their parageneses, *Contr. Mineral. Petrol., 40,* 39-52, 1973.

Sobolev, S. V., R. Widmer, and A. Yu. Babeyko, 3D temperature and composition in the upper mantle: constraint by global seismic tomography and mineral physics, (abstract), 6th International Kimberlite Conference, Novosibirsk, 561-563, 1995.

Spetsius, Z. V., Occurrence of diamond in the mantle: a case study from the Siberian Platform, *J. Geochem. Explor., 53,* 25-40., 1995.

Stein, M., and S. L. Goldstein, From plume head to continental lithosphere in the Arabian-Nubian Shield, *Nature, 382,* 773-778, 1996.

Stein, M., and A. W. Hofmann, Mantle plumes and episodic crustal growth, *Nature, 372,* 63-68, 1994.

Stern, C. J., S. Saul, M. A. Skewes, and K. Futa, Garnet peridotite xenoliths from the Pali-Aike alkali basalts of southernmost South America, *Geol. Soc. Australia Spec. Publ., 14,* 735-744, 1989.

Su, W., R. L. Woodward, and A. M. Dziewonski, Degree 12 model of shear velocity heterogeneity in the mantle, *J. Geophys. Res., 99,* 6945-6980, 1994.

Taylor, S. R., and S. M. McLennan, *The continental crust: its composition and evolution,* 312 pp., Blackwell Scientific, Oxford, U.K., 1985.

Wendlandt, E., D. J. DePaolo, and W. S. Baldridge, Thermal history of Colorado Plateau lithosphere from Sm-Nd mineral geochronology of xenoliths, *Geol. Soc. Amer. Bull., 108,* 757-767, 1996.

Xu, X., S. Y, O'Reilly, X. Zhou, and W. L. Griffin, A xenolith-derived geotherm and the crust-mantle boundary at Qilin, southeastern China, *Lithos, 38,* 41-62, 1996.

Yuan, X., Velocity structure of the Qinling lithosphere and the mushroom cloud model, *Science in China,* Series D39, 235-244, 1996.

Zhang, A., X. Dehuan, S. Siling, G. Lihe, Z. Jianzong, and W. Wuyi, The status and future of diamond exploration in China, in *Diamonds: Characterization, Genesis and Exploration,* edited by H. O. A. Meyer and O. H. Leonardos, pp. 268-284, *CPRM Spec. Publ. 1B/93,* 1994.

Zhang, A, W. L. Griffin, and D. Xu, Lithosphere mapping in eastern China with garnets and spinels from kimberlitic and lamproitic rocks (abstract), *Abstr. 30th Intern. Geol. Congress, Beijing, 2,* 398, 1996.

W. L. Griffin, S. Y. O'Reilly, O. Gaul, and D. A. Ionov, GEMOC National Key Centre, School of Earth Sciences, Macquarie University, NSW 2109, Australia.

W. L. Griffin and C. G. Ryan, CSIRO Exploration and Mining, P.O. Box 136, North Ryde, NSW 2113, Australia

Hypotheses Relevant to Crustal Growth

A. L. Hales

Research School of Earth Sciences, The Australian National University, Canberra, Australia

Estimates of the growth of the continental crust over the past 2000 Ma range from no growth to growth of the order of 60 - 70 percent. Most present day crust is younger than 2000 Ma, with the fractions formed new, and recycled, still debated. Elliston in 1968 proposed that granites were formed from wet sediments in ocean basins adjacent to continents, and that metamorphic rocks and consolidated sediments were less mature products of the same process. Chappell in 1967 hypothesized that there was some unmelted relict crustal material (later called restite) in the granites of the New England Fold Belt in Australia. Chappell and White in 1974 hypothesized that two source materials were involved in the generation of granites of the Lachlan Fold Belt (southeast Australia), and classified them as S (sedimentary) type and I (igneous) type. Zircon-core ages from SE Aust. and world-wide establish that there is a relict unmelted sedimentary component in many granites, consistent with Chappell's suggestion. Isotope studies of the Lachlan Fold Belt, and of the Peninsular Range batholith, California, make it clear that two source materials are involved in both suites of granites. Comparison of the pre-breakup configurations of the Atlantic and Gondwana continents, with Sclater et al.'s 1980 map of continental crust younger than 200 Ma, supports Elliston's suggestion that new crust is created from sediments in ocean basins adjacent to continents.

INTRODUCTION

Hutton [1788] recognised the role that sediments eroded from the continents and deposited in the oceans played in the growth of the continents. Elliston [1968, 1984, 1985] described his concept of the derivation of new continental crust from the wet sediments in ocean basins adjacent to the continents. In these papers Elliston [1968] suggested that colloidal processes in wet sediments offered an "alternative genesis" for many granites, porphyries and other "igneous looking" rocks, adding that "this does not imply or mean that there are no rocks of molten origin", as had been proposed by Hutton [1788] for the genesis of granite.

Chappell [1967] suggested that there was an unmelted relict crustal component in the rocks of the New England Fold Belt in New South Wales, Australia. Chappell and White [1974] suggested that two source materials were involved in the generation of the granites of the Lachlan Fold Belt in eastern Australia. There is abundant isotopic evidence for the Chappell [1967] and Chappell and White [1974] hypotheses. The main thrust of this paper will be on evidence that the new continental crust of the past 200 Ma was generated from the wet sediments in ocean basins adjacent to the continents. The isotopic and other evidence for the Chappell [1967] and Chappell and White [1974] hypotheses is strong. The hypotheses which will be discussed are summarised in the next section.

THE HYPOTHESES

The hypotheses to be considered are:
Hutton [1788]
Granites crystallize from molten magma.

Elliston [1968, 1984, 1985]
Granites are formed from the wet sediments in ocean basins adjacent to the continents. The sediments are hydrolysed. The granites are formed by crystallization from the colloidal state.

Chappell [1967]
There is an unmelted relict crustal component in granites (later called restite).

Chappell and White [1974]
Two source materials were involved in the formation of granites, one igneous (infracrustal), the other sedimentary. Granites were classified as I-type and S-type.

THE WET SEDIMENT-OCEAN BASIN HYPOTHESIS

In this section it is suggested that comparison of the conditions at the margins of the continents before and after the breakup of Pangea supports the hypothesis that new crust, granites, metamorphic rocks and consolidated sediment, are made from the wet sediments in ocean basins adjacent to the continents.

Figure 1 is based on the Bullard et al. [1965] pre-breakup configuration of the Atlantic continents and Figure 2 on the Smith and Hallam [1970] pre-breakup configuration of the southern continents, the Gondwanaland continents. In both figures the fitted margins, where the computer fits showed acceptably small overlaps and gaps at 500-fathom (914.4 m) depth level, are marked by a thick line, the pre-breakup ocean facing margins are marked by a zigzag line. The pre-breakup margins not defined by the computer fits, e.g. the eastern boundary of Europe and the northern margin of India, are shown dotted. It should be noted that the east coast of Africa is shown as a fitted margin following Smith and Hallam [1970] on both Figures 1 and 2.

Figures 3 and 4 are based on two segments of the Sclater et al. [1980] map of the ages of the continents. The margins are shown as fitted or ocean facing exactly as they were on Figures 1 and 2. The new crust, age less than 250 Ma, of Sclater et al. [1980] is shown dashed. Except for tiny areas in the Pyrenees and south of Gibraltar, there is no new crust on the fitted margins of the east coasts of North and South America, the west coasts of Europe and Africa, and also Antarctica. There is new crust along most of the ocean facing west coasts of North and South America, the Mediterranean coasts of Europe and Africa and the ocean facing segments of the coast of Antarctica. The northern boundary of India was not defined in the Smith and Hallam [1970] reconstruction, but it is likely that there was ocean between India and Asia, so that the new crust in the Himalayas was also formed from the sediments on the south coast of Asia and north coast of India. The old crust along the north coast of India appears to have extended north of the dotted line in Figures 2 and 4.

Summing up, the new crust along the fitted margins of the Bullard et al. [1965] and Smith and Hallam [1970] reconstructions shown in Figures 3 and 4 is much less than 1% of the area of new crust shown along the ocean-facing margins of Figures 3 and 4. This raises the question of what difference between the conditions at these margins is responsible for the development of new crust at ocean-facing margins but not at the fitted margins. The simplest explanation – and in fact the only reasonable one I can think of – is that the real continental margins along the ocean-facing coasts were further inland before the breakup of Pangea, and that ocean basins in which sediment had been accumulating for 100 Ma or more lay along the real margins. It is possible also that subduction zones had already developed along these margins.

My conclusion is that the major process responsible for the increase in the area of the continental crust during the past 250 Ma was the conversion of the wet sediment in the ocean basins adjacent to the real margins (ocean facing) of the continents before the breakup of Pangea into granite, metamorphic rocks and consolidated sediment as envisaged by Elliston [1968]. There are no doubt subsidiary changes such as sedimentary areas developing on continental shelves, in shallow seas and intrusions in rifted regions.

As will be discussed later, the development along the coast of southern California took place in a back-arc basin. Along the west coast of South America, accretion and subduction of sediment appear to be taking place at present, but some form of basin may have developed earlier. In the Mediterranean and Himalaya regions the process involved was collision of two continental plates. In the collision regions, the pattern of displacement of the sediments before consolidation and conversion to granite and metamorphic rock or consolidated sediment, would have been more complex than in southern California, and the pattern of the isotope data may not be as easy to interpret.

Developments at the Fitted Margins after the Breakup

It is widely accepted that the driving forces on the plates are the ridge-push force and the slab-pull force, and that the

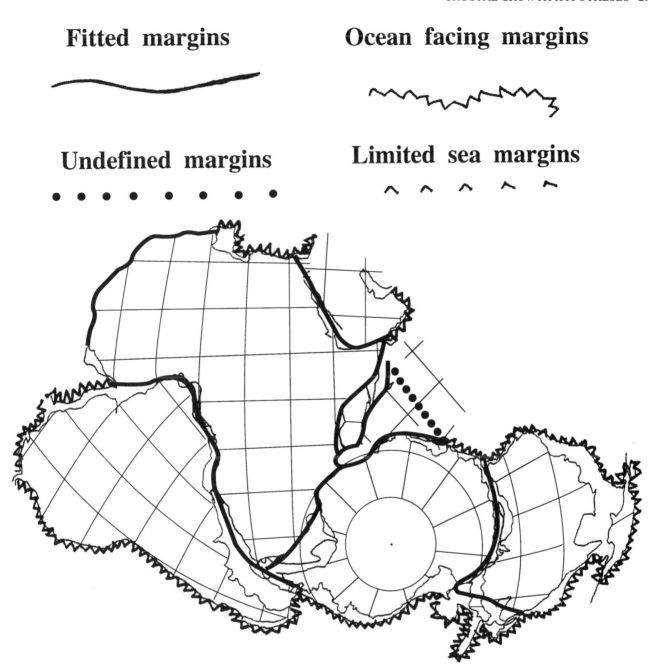

Figure 1. Based on the Bullard et al. [1965] pre-breakup configuration of the Atlantic continents. The fitted margins on either the Bullard et al. [1965] or the Smith and Hallam [1970] configuration of the Gondwanaland continents are marked by heavy lines, the ocean-facing margins by zigzag lines, and the margins not defined in the configurations by dotted lines.

slab-pull force is substantially the greater of the two forces [McKenzie, 1969; Fowler, 1990]. It is noteworthy that subduction zones have not yet developed along any of the margins shown as fitted margins in the pre-breakup configurations of Bullard et al. [1965] and Smith and Hallam [1970]. Thus it is difficult to ascribe any of the force responsible for the movement apart of the divergent margins of South America and Africa, and the divergent margins of North America and Europe, to slab-pull. It is probable that in time subduction will develop along these

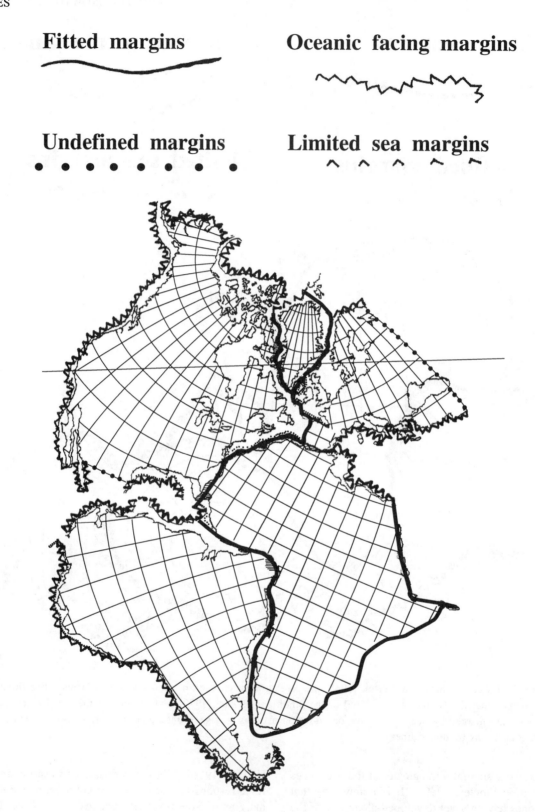

Figure 2. Based on the Smith and Hallam [1970] configuration of the Gondwanaland continents and marked in the same way as in Figure 1.

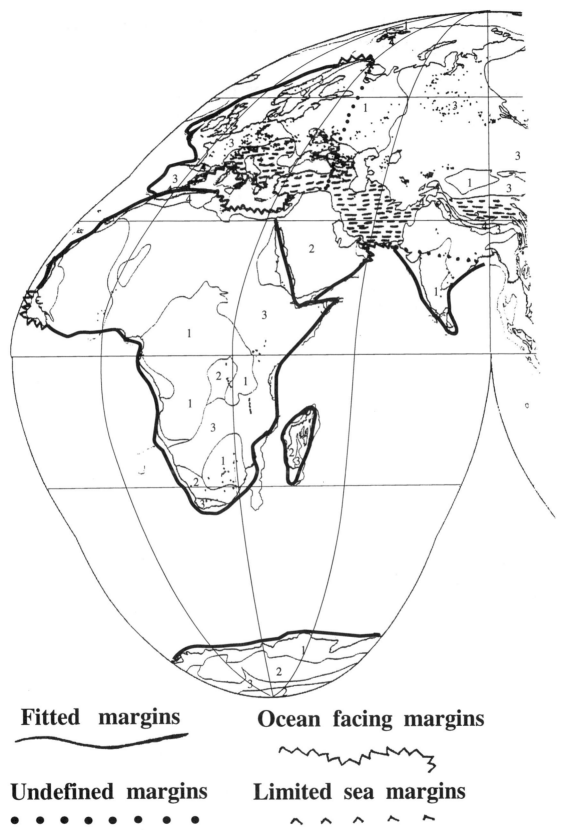

Figure 3. Based on a segment of the Sclater et al. [1980] map of the ages of the continents. The margins of the continents are marked as in Figures 1 and 2.

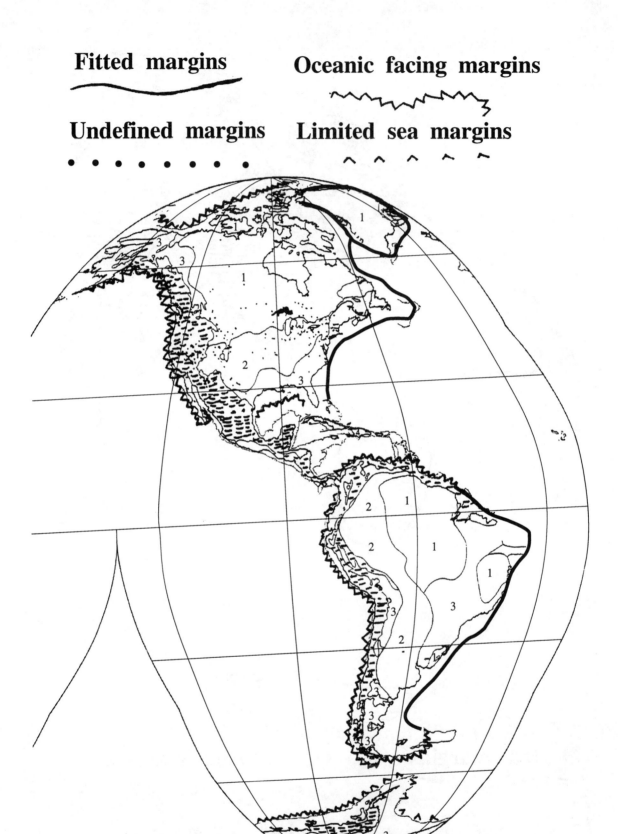

Figure 4. Based on another segment of the Sclater et al. [1980] map of the ages of the continents. Again the margins are marked as in Figures 1 and 2.

margins. However, in spite of the considerable discussion of subduction which has taken place in the past thirty years, almost nothing has been written about the initiation of subduction.

Consider the developments at the fitted margins after the breakup. Spreading ridges can be presumed to develop at the fitted margins during the breakup. The continent must move away from the spreading ridge, or the spreading ridge away from the continent. The initial configuration after the breakup is shown in Figure 5(a). Sediment will be deposited offshore at a rate depending on the topography of the interior and the development of rivers draining the continent. After some time, the oceanic crust and underlying lithosphere will be forced down to form an offshore basin, as shown in Figure 5(b). It is clear that subduction will not begin until the oceanic crust and underlying lithosphere has fractured. The most likely areas for the first fracture are at the points of maximum curvature, A, B or C, where the stress is greatest and tensional. However the exact point at which fracture begins in the maximum stress regions will depend to some extent on the zones of weakness, if any, which have developed during spreading. If the failure occurs at point C, i.e. at the ocean-continent margin, it is possible that a fore-arc basin will develop. However that has not happened at the eastern margins of North and South America, nor the western margins of Europe and Africa. If the fracture occurs in the maximum stress regions A or B, more than one fracture will be necessary before the slab (oceanic crust and some underlying lithosphere) can begin to descend below the sediment filled basin. In that case, several fragments of oceanic crust attached to fragments of the lithosphere will slide, or be pushed into the pile of wet sediment in the basin, and will become ophiolites in the consolidated crust which develops from the wet sediments at a later time. It is possible that melting, or partial melting, will occur in the lithosphere below the fracture zones, so that the initial arc magmatism, and the underthrusting of oceanic crust and lithosphere below the basin, will occur at about the same time.

Taylor and McLennan [1985] estimated the total annual load of sediment carried to the sea to be 3 km³. They also estimated the annual load carried to the sea for each of twelve major rivers. The major contributors arise in the Himalayas, and feed the back-arc basins along the east coast of Asia and the sediment pile in the Bay of Bengal [Hales, 1992]. Subduction and island arc magmatism are already well developed along the east coast of Asia suggesting that some ocean basin sediments had already been deposited there before the breakup of Pangea. The Bay of Bengal sediment pile could not have begun to develop until after the

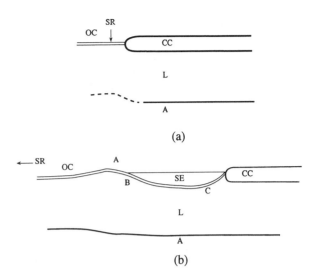

Figure 5. (a) A sketch showing a continental margin not long after breakup. The base of the lithosphere in the region of the spreading ridge is not defined and is dashed. SR indicates spreading ridge, OC oceanic crust, CC continental crust, L lithosphere and A asthenosphere. No vertical or horizontal exaggeration. (b) About 50 - 100 Myr after the breakup. SE indicates sediment.

breakup, but already the pile is thick, and Sykes [1970] has suggested a "nascent" subduction zone south of the sediment pile.

The Elliston concept of the genesis of new crust from the wet sediments in ocean basins adjacent to the continents (described above) is supported by the fact that, except for the possible developments in the Bay of Bengal, no subduction zones have so far developed on any of the margins which were fitted margins in the Bullard et al. [1965] and Smith and Hallam [1970] computer-fitted configurations. Some of those fitted margins have been exposed to the ocean for more than 100 Myr since the breakup of Pangea.

The Variation of Crustal Composition with Depth

Acceptance of the evidence for the hypothesis that new continental crust is formed in the main from the wet sediments in ocean basins adjacent to the continents, poses some questions about the changes in composition with depth which must occur in the process of conversion of the wet-sediment pile into consolidated continental crust consisting of consolidated sediment, metamorphic rock and granite.

It is probable that a considerable quantity of sediment eroded from the adjacent continent will have been deposited in the basin before the magmatic arc developed. The lowermost layers of the sedimentary pile will be of continental origin. The more mafic sediment from the magmatic arc and volcanic ash would be found only at higher levels in the sediment pile. However, the discussion of the seismic information world wide by Christensen and Mooney [1995] shows that the lower-crust seismic velocities are greater than the average velocities for the upper crust, and these authors interpreted the difference as meaning that the lower crust was more mafic. The discussion of models of the composition of the crust of various ages and environments by Rudnick and Fountain [1995] also shows that some changes in the distribution of material with depth are necessary.

THE RESTITE HYPOTHESIS: EVIDENCE FOR A RELICT UNMELTED COMPONENT IN GRANITE

White and Chappell [1977] pointed out that melts formed by ultrametamorphism are of granitoid composition. They argued that at the source partial melting results in a mixture of melt and solid residual material and that in systems containing quartz and feldspar the minimum melt is granitoid. White and Chappell went on to describe how these concepts could be used to estimate the proportion of relict source material in granites. Chappell et al. [1987] described how these concepts had been applied to the granites of the Lachlan Fold Belt. However it was the evidence from the U-Pb age data on zircons extracted from granites which established beyond all possible doubt that there was relict source material in many granites.

It has long been known that many zircons extracted from granites have cores of fractured or discolored zircon surrounded by rims of clear zircon. In the sixties and seventies, conventional solid-source mass spectrometry of multi-zircon samples showed that the U-Pb ages were frequently slightly older than the emplacement ages of the granite from which the zircon had been extracted. In a few cases, the ages found were so different from the emplacement ages that it was not possible to ascribe the difference to experimental error, for example Grauert and Arnold [1968] for the Swiss Alps, or Williams [1977] for the Lachlan Fold Belt. When high-sensitivity high-resolution ion-probe (SHRIMP) ages became available, it was found that many zircon cores from granites were significantly older than the rocks in which they had been found [Williams et al., 1988; Pidgeon and Compston, 1992; Paterson et al., 1992]. Williams et al. [1991] and Williams [1992] measured the ages of nearly 1000 zircons from the Lachlan Fold Belt granites. They found that the cores of zircons from the S-type granites of the Lachlan Fold Belt showed groups of ages 450-650 Ma, 800-1075 Ma, and a group with ages ranging up to 3350 Ma. The I-type granites zircon cores had the same groups in the same relative proportions though the number of zircons showing old cores was proportionally much less than in the S-type granite. The country Ordovician sediments also showed the groups of ages in the same relative proportions.

It follows that there is a relict sedimentary component in the I-type granites as well as the S-type thus confirming the isotopic evidence of mixing across the batholiths. It has been suggested that the sedimentary component of the granites could not have come from the country rock sediments, because the granites have greater abundances of sodium and calcium than the sediments. However, sea water contains significant amounts of sodium, and calcium-rich microfossils are readily dissolved in sea water at depths greater than 4 km.

Zircons with old cores have been found in other granites world wide and provide clues to the provenance of the sediments from which those granites were made.

THE MOLTEN MAGMA HYPOTHESIS

Many earth scientists, perhaps the majority are convinced that crystalline rocks, or "igneous looking rocks", can have been derived only from a magma, as suggested by Hutton [1788].

The evidence of the old cores of the zircons in many granites establishes that the melting was not complete in these granites. Thus only partial melting need be considered. The experiments of Wyllie and his associates on the melting of finely crushed granite in water show that partial melting occurs at temperatures which are conceivable at mid-crustal depths.

THE TWO SOURCE MATERIAL HYPOTHESIS

The initial Sr^{87}/Sr^{86} data of Early and Silver [1973], and the $\delta^{18}O$ data of Taylor and Silver [1978], for the Peninsular Ranges Batholith (PRB), southern California, the initial Sr^{87}/Sr^{86} data of Kistler and Peterman [1973, 1978] for the Sierra Nevada (SN), and the e_{Nd} data of De Paulo [1980], were discussed by De Paulo [1981], in the light of the contribution these data might make to better understanding of crustal growth and other processes in the crust and mantle. De Paulo [1981] concluded that the interpretation of the authors cited above, namely that two source materials, one recently arrived from the mantle and one derived from old continental sediment, were required to account for the isotope data. Furthermore, De Paulo [1981] showed that the variation of the isotope data across the batholiths was consistent with the mixing of mantle-derived magma with material derived from old crustal rock. He asserted that no other plausible model existed which was consistent with all the data.

De Paulo [1981] described three models of how the mixing process might occur. All involved the incorporation of old continental rock, either crystalline or sediment, in magma recently derived from the mantle. It is clear from the variation of the isotope data across the batholiths (see for example Figure 1 of Taylor and Silver [1978]) that the amount of mantle derived magma incorporated in the granites decreases across the batholiths, and that at the eastern edge of the PRB and SN, the proportion of sediment must be three or four times that of the melt. It is difficult to see how sediment could be assimilated in 0.2 or 0.3 of its own volume of melt and still result in melt which would crystallize as granite.

Hales [1992], following the hypothesis of Elliston [1968] that granites were formed from the wet sediments in ocean basins adjacent to the continents, proposed that the Peninsular Ranges Batholith and Lachlan Fold Belt granites were formed from the wet sediments in back-arc basins, the sedimentary pile being fed by erosion of the magmatic arc on one side, and by erosion of the sediment from old continental crust on the other. It should be added that volcanic ash from the arc volcanoes would also contribute to the sedimentary pile. In this process, the two source materials would be roughly mixed across the basin in the manner indicated by the isotope data. It is improbable that the mixing in collision regions, or even in accreted margins, would be as simple as in back-arc basins.

It should be noted that the criteria used by Chappell, White and their colleagues to distinguish between I- and S-type granites when mapping were mineralogical and geochemical, chiefly the presence of hornblende in the I-type granites but not in the S-type granites, and of cordierite in the S-type granites but not in the I-type granites.

The conversion of wet sediment to granite

Three solutions have been proposed for the formation of granite from wet sediments. These are:-
(1) the Elliston [1985] and earlier references, suggestion that the granites were formed by crystallization from the colloidal state;
(2) the suggestion that solution and recrystallization occurred as the water in the pile carrying colloidal material and minerals in solution moved upwards, either continuously or episodically, to regions of lower pressure and temperature;
(3) the partial melting process favoured by Chappell and most others.

Partial melting

Wyllie and his students carried out very many experiments on granites in the presence of excess water at pressures ranging from 1 kbar to 25 kbar and temperatures from 600°C to 1100°C. Piwinskii [1968] describes the procedure. The samples consisted of granite crushed fine with added water up to 15%. The Piwinskii experiments were carried out at pressures of 1, 2, and 3 kbar. The results of these experiments, and those of Boettcher and Wyllie [1968] to 25 kbar, are shown in Figure 1 of Stern and Wyllie [1981]. It is clear from this figure that partial melting would occur at temperatures between 600°C and 700°C for pressures greater than 1.5 kbar and for all greater pressures possible in the crust. Complete melting of wet sediment over the same pressure range requires temperatures greater than 700°C [Stern and Wyllie, 1981]. The conditions in the experiments cited above correspond to those in ocean basins, except that temperatures of 600°C are not likely over large areas at depths shallower than 10 km. Partial melting therefore is a possible process for the origin of granites. I am not yet convinced that partial melting of dry rock is probable at mid-crustal depths. The measurements of the ages of zircons in the Lachlan Fold Belt granites reported by Williams et al. [1991] and Williams [1992] showing that many of the zircons have cores much older than the emplacement age of the granites, and similar information from many granites world wide, show that total melting has not occurred for many granites.

Crystallization from the colloidal state and solution followed by recrystallization

The evidence offered by Elliston [1985] for crystallization from the colloidal state was a series of experiments reported by Raleigh [1965]. In these experiments, colloidal silica was heated to temperatures between 250°C and 600°C and pressures of 5 and 10 kbar, in the presence of water for times varying from 1 to 144 hours. In six runs at 10 kbar and one at 5 kbar, complete crystallization occurred for all temperatures above 300°C, and at temperatures above 250°C if the experiment ran for more than 5 hours. In still another experiment, small flint chips were included in the charge. When the charge was examined after the run it was found that a clear rim of quartz surrounded the flint chips. Zircons with old cores often have a rim of clear zircon of granite emplacement age round the old cores, which, on this hypothesis, is not necessarily melt-precipitated.

These experiments could be equally well explained in terms of solution of the quartz and recrystallization when the temperature was lowered and the pressure released at the end of the run. If crystallization directly from the colloidal state occurred, there would nevertheless have been some contribution to the end product from solution and recrystallization. Whatever process plays the major role in the conversion of wet sediment to granite, the temperatures at mid-crustal depths will be such that solution of between 0.5 and 1.0 percent of quartz will occur. If quartz goes into solution, so also may other minerals.

DISCUSSION

On the experimental evidence available at present it is not possible to reject any one of the three processes, partial melting, crystallization from the colloidal state, or solution and recrystallization, or some combination of the three.

Much of the uncertainty about the role of the three processes could be resolved by further experiments of the Raleigh type at temperatures less than 600°C. It is clear, however, that if granites originate from wet sediment in ocean basins then metamorphic rocks and consolidated sediment should be regarded as less mature products of the same process in which the granites are made.

The discussion of the initial isotope data has been based on the data from back-arc basins. In areas where the new crust was created from the wet sediments involved in the collision of two continental plates, e.g., in the Alps or the Himalayas, it would not be expected that the pattern of initial isotopes would be as simple. However, it is reasonable to expect that the processes involved in the conversion of wet sediment to granite would be the same as in the back-arc basins. It is also reasonable to consider whether similar processes occurred in older granites, perhaps as old as the Archean granites, and whether data from older granites could contribute to identification of former continental margins, and hence to further understanding of crustal evolution.

CONCLUSIONS

The simplest explanation of the fact that new crust less than 250 Ma old is found along most of the continental margins which were ocean facing before the breakup of Pangea, but that there is only an insignificant amount of new crust on the margins which were fitted together before the breakup, is that sedimentary basins had been developing along the ocean facing margins for more than 100 Myr before the breakup occurred.

The fact that zircons with cores older than the emplacement age of the granites are found in the I- and S-type granites and the country rock sediments of the Lachlan Fold Belt, eastern Australia, and in many granites world wide, shows that there is a relict unmelted component in the granites (restite).

The initial isotope data from the Peninsular Ranges and Sierra Nevada granites of California and from the Lachlan Fold Belt granites of eastern Australia, make it clear that two sources, one from material recently arrived from the mantle and one from old continental rocks, were involved in the petrogenesis of these granites.

It is proposed that granites were made from the wet sediments in ocean basins adjacent to the continents, and that one of the three processes, partial melting, crystallization from the colloidal state, solution followed by recrystallization or some combination of these processes, plays the major role in the conversion of the wet sediment to granite, metamorphic rock and consolidated sediment. Solution certainly occurs at mid crustal depths but it is not certain that it plays the major role.

Acknowledgments. I am grateful to Professor S.W. Carey, John Elliston and Bruce Chappell for discussion of their ideas about the origin of granites during the past twenty years. I also am grateful to colleagues at the Research School of Earth Sciences for discussion of the evidence from isotope geochemistry and SHRIMP ages of zircons, especially Bill Compston and Ian Williams, and to Lesley Wyborn, Chris Klootwijk and Jean Braun for reviews of the manuscript.

REFERENCES

Boettcher, A. L., and P. J. Wyllie, Melting of granite with excess water to 30 kilobars pressure, *J. Geol.*, 76, 235-244, 1968.

Bullard, E. C., J. E. Everett, and A. G. Smith, The fit of the continents round the Atlantic, *Phil. Trans. Roy. Soc. A.*, 258, 41- 51, 1965.

Chappell, B. W., PhD Thesis, Australian National University, 1967.

Chappell, B. W., and A. J. R. White, Two contrasting granite types, *Pac. Geol.*, 6, 173-174, 1974.

Chappell, B. W., A. J. R. White, and D. Wyborn, The importance of residual source material (restite) in granite petrogenesis, *J. Petrology*, 28, 1111-1138, 1987.

Christensen, N. I., and W. D. Mooney, Seismic velocity structure and composition of the continental crust: a global view, *J. Geophys. Res.*, 100, 9761-9788 1995.

De Paulo, D. J., Sources of continental crust: Neodymium isotope evidence from the Sierra Nevada and Peninsular Ranges, *Science, 209,* 684-687,.1980.

De Paulo, D. J., Neodymium and strontium isotopic study of the Mesozoic Calc-alkaline granitic batholiths of the Sierra Nevada and Peninsular Ranges, California, *J. Geophys. Res., 86,* 10470-10488, 1981.

Early, T. O., and L. T. Silver, Rb-Sr isotopic systematics in the Peninsular Ranges batholith of southern and Baja California, *EOS Trans. AGU, 54,* 494.1973.

Elliston, J. N., Retextured sediments, *Reports of the XXIII International Geological Congress, 8,* 85-104, 1968.

Elliston, J. N., Orbicules: an indication of the crystallization of hydrosilicates, I, *Earth Sci. Rev., 20,* 265-344, 1984.

Elliston, J. N., Rapakivi texture: an indication of the crystallization of hydrosilicates, II, *Earth Sci. Rev., 22,* 1-92, 1985.

Fowler, C. M. R., The Solid Earth, Cambridge University Press, 260-263, 1990.

Grauert, B., and A. Arnold, Deutung diskordanter Zirconalter der Silvrettadecke und des Gotthardmassivs (Schweizer Alpen), *Contrib. Mineral. Petrol., 20,* 34-56, 1968.

Hales, A. L., Speculations about crustal evolution, *J. Geodynamics, 16,* 55-64, 1992.

Hutton, J., Theory of the Earth, *Trans. Roy. Soc. Edinburgh, 1,* 209-304, 1788.

Kistler, R. W., and Z. E. Peterman, Variations in Sr, Rb, K, Na and initial Sr^{87}/Sr^{86} in Mesozoic granitic rocks and intruded wall rocks in Central California, *Geol. Soc. Am. Bull., 84,* 3489-3512, 1973.

Kistler, R. W., and Z. E. Peterman, A study of regional variation of initial strontium isotopic composition of Mesozoic granitic rocks in California, *U.S. Geol. Surv. Prof. Pap., 1071,* 1978.

McKenzie, D. P., Speculations on the consequences and causes of plate motions, *Geophys. J., 18,* 1-22, 1969.

Paterson, B. A., W. E. Stephens, G. Rogers, I. S. Williams, R. W. Hinton, and D. A. Herd, The nature of zircon inheritance in two granite plutons, *Trans. Roy. Soc. Edinburgh, Earth Sci., 83,* 459-471, 1992.

Pidgeon, R. T., and W. Compston, A SHRIMP ion microprobe study of four Scottish Caledonian granites, *Trans. Roy. Soc. Edinburgh, 83,* 473-483, 1992.

Piwinskii, A. J., Experimental studies of igneous rock series, *J. Geol., 76,* 548-570, 1968.

Raleigh, C. B., Crystallization and recrystallization of quartz in a simple piston-cylinder device, *J. Geol., 73,* 369-375, 1965.

Rudnick, R. L., and D. M. Fountain, Nature and composition of the continental crust: a lower crustal perspective, *Rev. Geophys, 33,* 267-309, 1995.

Sclater, J. G., C. Jaupart, and D. Galson, The heat flow through oceanic and continental crust and the heat loss of the Earth, *Rev. Geoph. Space Phys, 18,* 269-311, 1980.

Smith, A. G., and A. Hallam, The fit of the southern continents, *Nature, 225,* 139-144, 1970.

Stern, C. R., and P. J. Wyllie, Phase relationships of I-type granite with H_2O to 35 kilobars: the Dinkey Lakes biotite-granite from the Sierra Nevada batholith, *J. Geophys. Res., 86,* 10412-10422, 1981.

Sykes, L. R., Seismicity of the Indian Ocean and a possible nascent island arc between Ceylon and Australia, *J. Geophys. Res., 83,* 2233-2245, 1970.

Taylor, H. P., and L. T. Silver, Oxygen isotope relationships in plutonic igneous rocks of the Peninsular Ranges batholith, Southern and Baha California, *U.S. Geol. Surv. Open-File Rep., 78-701,* 423-436, 1978.

Taylor, S. R., and S. H. McLennan, *The continental crust: its composition and evolution*, Blackwell Scientific Publications, Oxford, 1985.

White, A. R., and B. W. Chappell, Ultra-metamorphism and granite genesis, *Tectonophysics, 43,* 7-22, 1977.

Williams, I. S., PhD Thesis, Australian National University, 1977.

Williams, I. S., Some observations on the use of zircon U-Pb geochronology in the study of granitic rocks, *Trans. Roy. Soc. Edinburgh, 83,* 447-458, 1992.

Williams, I. S., Y. D. Chen, B. W. Chappell, and W. Compston, Dating the sources of Bega Batholith granites by ion microprobe, *Geol. Soc. Aust, Abstr. 21,* 424, 1988.

Williams, I. S., B. W. Chappell, Y. D. Chen, and K. A. Crook, Inherited and detrital zircons - vital clues to the granite protoliths and early igneous history of southeastern Australia, *Bur. Miner. Resour. Aust. Rec., 1991/25,* 1-121, 1991.

A.L. Hales, Research School of Earth Sciences, Australian National University, Canberra, ACT, 0200, Australia.

Upper Mantle Structure beneath Australia from Portable Array Deployments

R.D. van der Hilst
Research School of Earth Sciences, The Australian National University, Canberra ACT 0200, Australia and Department of Earth, Atmospheric, and Planetary Sciences, Massachusetts Institute of Technology, Cambridge MA 02139, USA

B.L.N. Kennett
Research School of Earth Sciences, The Australian National University, Canberra ACT 0200, Australia

T. Shibutani
Research Centre for Earthquake Prediction, Disaster Prevention Research Institute, Kyoto University, Kyoto 611, Japan

The distribution of earthquakes at regional distances around Australia is particularly favourable for these natural events to be used as probes into the seismic structure of the lithosphere and upper mantle. The distribution of permanent seismic stations is too sparse for detailed continent-wide seismic imaging, and most information used in our studies has come from deployments of portable seismic recorders. These were initially used in northern Australia to define the radial variations in P and S velocity and attenuation, using a combination of short-period and broad-band observations. The whole Australian continent has now been covered in the SKIPPY experiment from 1993-1996, in which a sequence of deployments of up to 12 recorders at a time have been used to synthesize a continental-scale array of broad-band instruments. The major objectives of this project are to delineate the three-dimensional structure of the Australian lithosphere and underlying mantle by tomographic inversions for lateral variations in seismic wave speed and site-specific studies to map crustal thickness (receiver functions) and seismic anisotropy (shear-wave splitting). Surface-wave studies have begun to reveal the three-dimensional variations in shear-wave structure beneath the continent by exploiting the records from portable broad-band stations. Dense data coverage enables the imaging of wave-speed variations on length scales of 250 km, and larger beneath most of central and eastern Australia. Results of the wave-form inversion suggest that the eastern edge of the Proterozoic shields of central Australia is a complex three-dimensional surface, which does not have a simple relation to the conventional Tasman Line marking the separation of Precambrian and Phanerozoic outcrop. In the shallow lithosphere (80 km depth), the most pronounced lateral contrast in seismic wave speed occurs significantly further east than the conventional Tasman Line. However, in northern Australia the eastern boundary of exposed Precambrian basement coincides with a transition from moderately high wave speeds to fast wave propagation that is particularly prominent at a depth of about 140 km. At larger depth still (200 km), the wave-speed gradient seems to occur significantly further west (near the western margin of the Eromanga Basin). Beneath most of eastern Australia, the fast seismic 'lid' does not extend to depths larger than about 100 km, and there is a pronounced low-velocity zone between about 100 and 200 km depth. The surface-wave data do not indicate a slow wave-speed channel beneath central Australia. Instead, the fast lid extends to a depth of at least 250 km, and locally to depths in excess of 350 km. The North Australian Craton is seismically well defined, but does not seem to continue into the Kimberley Block. In the shallow lithosphere, the region influenced by the Late Palaeozoic Alice Springs Orogeny (Amadeus Basin and Musgrave Block) is marked by significantly slower seismic-wave propagation than the Proterozoic cratons to the north and south. Analysis of data from individual stations has produced a set of shear-velocity profiles for the crust and uppermost mantle beneath SKIPPY stations; these provide a useful complement to the information at depth from the surface-wave studies. The crust-mantle boundary is deep (38-44 km) and mostly transitional in character along the axis of the eastern fold belt, but is relatively sharp and shallower (30-36 km) near the boundary between Precambrian and Phanerozoic outcrop. Shear-wave splitting results from the portable stations are beginning to reveal a complex pattern of seismic anisotropy that requires further analysis of both body and surface wave data.

INTRODUCTION

The surficial geology of the Australian continent is composed of an assemblage of crustal blocks that can be broadly grouped into the Precambrian western and central cratons and the Phanerozoic eastern province (Figure 1A). Structural differences in the mantle beneath the Precambrian shield and eastern Australia have previously been inferred from limited observations of surface wave dispersion [Muirhead and Drummond, 1991; Denham, 1991] and teleseismic travel-time residuals [Drummond et al., 1989]. These results have been confirmed and extended in recent studies using deployments of portable broad-band instruments across the whole continent [the SKIPPY experiment, van der Hilst et al., 1994], and it is now possible to derive three-dimensional models of the P and S wave speeds beneath the Australasian region to depths of 400 km or more.

The extensive earthquake activity in the seismic belt that runs through Indonesia, New Guinea and its offshore islands, Vanuatu, Fiji, and the Tonga-Kermadec zone provides a wide range of natural sources which can be used to constrain seismic structure in the lithosphere, asthenosphere, and the transition zone beneath. There is a rather limited number of high-quality, permanent seismological stations in Australia, so that high-resolution studies of seismic structure need these observatories to be supplemented by the installation of portable recorders (Figure 1B).

Building on the experience from array deployments in northern Australia, the Research School of Earth Sciences (RSES) started in 1993 a nation-wide observational seismology project, the SKIPPY experiment, which uses a sequence of array deployments to synthesize a continental-scale array of broad-band instruments. Major objectives of this project are to exploit different classes of seismological data for the delineation of the three-dimensional structure of the Australian lithosphere and underlying mantle, by tomographic inversions for lateral variations in seismic-wave speed and site-specific studies to extract crustal information through the construction of receiver functions, and seismic anisotropy from analysis of shear-wave splitting [van der Hilst et al., 1994; Kennett and van der Hilst, 1996].

For instruments in northern Australia, refracted arrivals for both P and S waves from the upper mantle can be used to determine upper-mantle structure. At greater distances from the sources, multiple S arrivals and surface waves provide a powerful probe for three-dimensional structure using information on many crossing propagation paths. For this class of arrival, the energy is largely directed horizontally and the wave forms are most sensitive to S-wave velocities. The information from regional earthquakes can be supplemented with the arrival times for teleseismic waves in delay-time tomography for P arrivals, where the paths through the mantle are relatively steep. In addition, information from distant events can be used to constrain crustal and upper mantle structure through the analysis of converted waves and reverberations.

The combination of many different classes of seismological information has begun to reveal the complex nature of lithospheric and mantle structure, and sheds light on the way in which the Australian continent may have been assembled. In this paper, we briefly review results of the array experiments in northern Australia, from both short-period and broad-band instruments, and present the

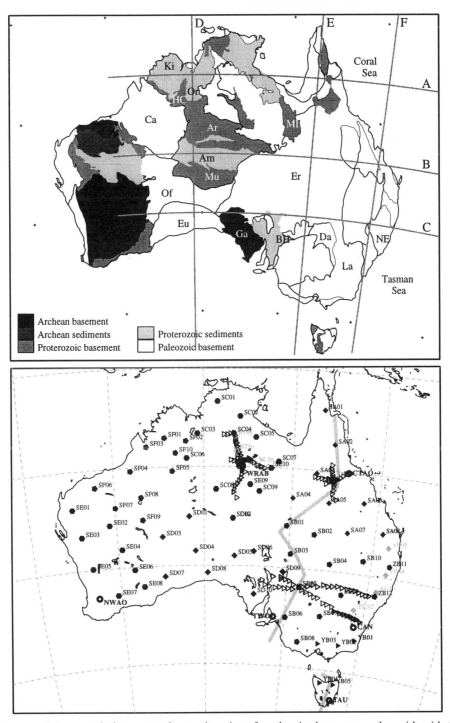

Figure 1: a) Map of the crustal elements and tectonic units referred to in the text, together with with the lines of the sections in Figure 6. (Key: Ar - Arunta Block, Am - Amadeus Basin, BH - Broken Hill Block, Ca - Canning Basin, Da - Darling Basin, Er - Eromanga Basin, Eu - Eucla Basin, Ga - Gawler Block, HC - Halls Creek Belt, Ki - Kimberley Block, La - Lachlan Fold Belt, MI - Mt Isa Block, Mu - Musgrave Block, NE - New England Fold Belt, Or - Ord Basin, Of - Officer Basin). b) Configuration of portable seismic recording stations 1985-1996. Short-period stations are indicated by open triangles. Portable broad-band stations are marked by solid symbols. Permanent stations with high fidelity recording are indicated by a double circle and station name. The approximate location of the boundary ('Tasman line'), between Precambrian outcrop in western and central Australia and the Phanerozoic east, is indicated by a shaded line.

latest results from the SKIPPY project for the continent as a whole.

P AND S VELOCITY PROFILES IN THE UPPER MANTLE

The events to the north of Australia provide a convenient set of energy sources for studies of the upper mantle, and it has proved possible to constrain the major features of the mantle velocity profile as well as the attenuation distribution with depth for both P and S waves.

Short-period studies

Much of the information on mantle structure has come from deployments of portable instruments with short-period seismometers. Hales et al. [1980] carried out a travel-time analysis for Indonesian earthquakes that occurred at a variety of depths and were recorded at a number of portable stations in northern Australia. The resulting model is rather complex, with many small discontinuities and low velocity zones, which may reflect the mapping of three-dimensional structure into a one-dimensional profile. A subsequent re-interpretation by Leven [1985], using comparisons between observed and synthetic seismograms, leads to somewhat simplified structure, but retains a prominent velocity contrast near 210 km depth.

In the period 1985-1987, RSES carried out a sequence of experiments using short-period vertical seismometers in northern Australia (Figure 1B) to record the natural seismicity in the Indonesia/New Guinea region. Many of the results were summarised by Dey et al. [1993], who presented composite record sections of upper-mantle arrivals that show significant variation in P-wave velocity structure between paths for events along the Flores arc, studied by Bowman and Kennett [1990], and paths to events in New Guinea. The P-velocity structure is well constrained from above the base of the lithosphere near 210 km down to below the 410 km discontinuity. The interpretation confirms the need for a P-velocity contrast near 210 km depth. For S waves, the corresponding record sections show a clear arrival associated with the lithosphere which cannot easily be traced beyond 2000 km but no branches associated with greater depth.

Broad-band studies

Modern broad-band seismometers provide a faithful rendition of ground motion over a wide range of frequency, and allow the full exploitation of both P and S body waves. A broad-band sensor has been operated by RSES at the Warramunga array in Northern Australia since late in 1988 (station code WRA); this facility was upgraded in 1994 by IRIS at a nearby site (WRAB). Over a period of years it has been possible to build up record sections covering the range of interest for the upper mantle by using events in the Indonesia/New Guinea earthquake belt. The records from the permanent station have been augmented by portable broadband stations deployed at distances of up to 300 km from WRA. The surface conditions in this region are such that good results can be obtained for SV waves on radial component records after rotation to the great circle path: the high surface velocities lead to little contamination by converted P waves. This represents a considerable benefit over previous S-wave studies of the upper mantle, which have been restricted to SH waves [Gudmundsson et al., 1994].

Additional information on mantle structure is provided by the SKIPPY experiment [van der Hilst et al., 1994] in which portable broad-band instruments were emplaced at a suitable distance range to complement the observations at the Warramunga array. Figure 2 shows a composite record section covering the P and S wave components returned from the upper mantle for events in New Guinea recorded at SKIPPY stations in Queensland (northeastern Australia). This section has been constructed from unfiltered vertical-component records from five shallow events and clearly displays the benefit of broadband recording. The onset of the S waves shows high-frequency behaviour (greater than 1 Hz) out to 2000 km, but beyond this distance the S-wave arrivals have a significantly lower frequency (0.2 Hz at 3000 km), and this is also seen for later arrivals at shorter distances. For P waves, the loss of higher frequencies is less pronounced.

For similar data at WRA, the change in frequency content for S waves returned from greater depth has been analysed by Gudmundsson et al. [1994] to determine the attenuation structure with depth under northern Australia. The slope of the spectral ratio between P and S wave arrivals on the same record has been used to determine the differential attenuation between P and S. This differential information can be interpreted with a knowledge of the velocity structure, and requires strong attenuation of S waves in the asthenosphere between 210 km and 410 km. In a parallel analysis, Kennett et al. [1994] have used the composite record sections, together with the earlier information from the short period studies, to build velocity profiles for P and S. These velocity models have been refined by comparison of observed and synthetic seismograms including the influence of attenuation (Figure 3) and provide a good basis from which to look at three-dimensional structure. By determining the P and S velocity profiles from the same events, the P/S velocity ratio can be well constrained, which is particularly useful for studies of mantle composition. The depth variation of the P/S velocity ratio is in good general agreement with the results for the shield areas of north America obtained by combining the P-velocity profile of LeFevre and Helmberger [1989] with the S-wave structure of Grand and Helmberger [1984].

The broad band studies indicated variations in timing of

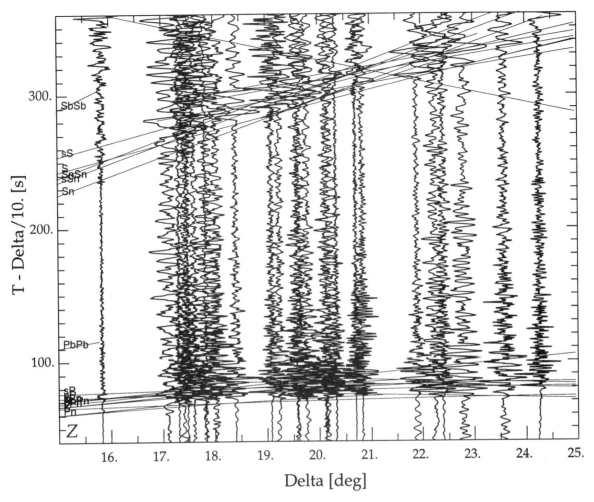

Figure 2: A composite record section of unfiltered broad-band seismograms for five New Guinea events recorded at SKIPPY stations in Queensland. The section is constructed from the vertical component for each path and covers the P and SV waves returned from the upper mantle. The travel-time curves are for an event at 25 km depth in the *ak135* reference model (Kennett et al., 1995)

events at similar distances but different azimuths from WRA, and provided the incentive for more detailed analysis of three-dimensional structure of the continent.

THREE-DIMENSIONAL STRUCTURE

The favourable position of the Australian continent relative to world seismicity can be exploited in a number of ways to obtain information on the three-dimensional seismic structure in the mantle. This can be done by means of tomographic techniques that rely on the number of high-quality crossing paths produced by many source-receiver pairs, or by site-specific studies that aim to define the subsurface structure beneath individual stations by minimizing the influence of structure elsewhere. These applications provide complementary information about Earth structure, which are being merged to obtain a more complete description of the Australian mantle.

In addition, the configuration of the seismicity around the Australian continent is also ideal for tomographic methods, since with a suitable distribution of receivers a dense and even data coverage of the continent can be achieved. RSES has just completed a three and a half year field program to install some 60 portable broadband stations across the entire Australian continent (see Figure 1B). This project used a set of up to 12 portable instruments, which occupied sites for at least 5 months at a time before being moved to a new location, so that a full continental array could be synthesized [van der Hilst et al., 1994; Kennett and van der Hilst, 1996]. The mobility of the arrays have led to the name SKIPPY (nickname for kangaroo) for the whole project. The deployments commenced in May 1993 with 8

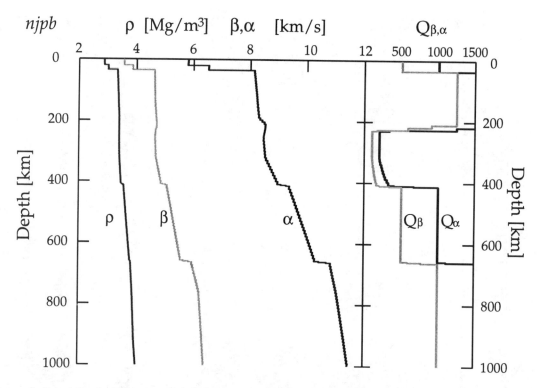

Figure 3: P and S velocity and attenuation structure for the upper mantle beneath northern Australia determined from a combination of short-period and broad-band observations.

stations in Queensland, and the coverage of the whole continent was completed at the end of September 1996. The data set from the SKIPPY stations were supplemented by records of suitable events from the permanent broadband stations. In addition to providing data from sites not occupied by SKIPPY stations, the records from the observatory instruments are important in constraining upper-mantle structure from fundamental-mode surface waves, since at frequencies less than 10 mHz they are often of better quality than the records of the portable stations.

Surface-wave studies

The records from the broad-band stations have been used in a number of different studies. A major objective is the delineation of lithospheric and mantle structure using waveform tomography for the shear-wave and surface-wave portions of the seismogram. So far the analysis has been based on the partitioned wave-form inversion technique introduced by Nolet [1990]. A nonlinear optimization is used to find a stratified model which gives the best fit to an observed seismogram, which should represent the average structure along the great circle between source and receiver. The assemblage of path averages are then used in a linear inversion to recover the three-dimensional shear-wave structure. The tomographic inversion of the linear constraints on the three-dimensional velocity structure uses a block structure of approximately equal area in latitude and longitude coupled into a set of mostly triangular basis functions in the vertical direction. Both model norm and gradient damping are used to achieve a balance between data fit and smoothness of the model. The details of the partitioned wave-form inversion method are given by Nolet [1990], and its application to Australian data is presented by Zielhuis and van der Hilst [1996].

The non-linear wave-form technique assumes the independent propagation of individual modes along a great-circle path. The neglect of inter-mode coupling may prevent reliable imaging of deep-mantle structure, and we therefore restrict the present discussion to the upper mantle. Since we do not account for out-of-plane scattering, we do not make use of high-frequency fundamental-mode surface waves, which are sensitive to large wave-speed variations in the shallow lithosphere (such as variations in crustal thickness).

The effective use of the non-linear inversion scheme for the individual paths requires synthetic seismograms that are close to the observed wave forms, and in turn this requires a reference model that is close to the averaged 1D velocity model for the wave path. Care has to be taken to account for the best averaged crustal model for the path in a region which covers both continental and oceanic lithosphere. A partial compensation for crustal thickness variations is made by using different mode files for source, propagation path, and receiver structure [Zielhuis and Nolet, 1994; Kennett, 1995]. The influence of heterogeneity is rather different for the fundamental and higher modes. For instance, at frequencies higher than about 30 mHz, the fundamental modes are sensitive to variations in crustal structure in general and the ocean-continent transition in particular. The higher modes are more sensitive to structure at larger depth, and are less influenced by scattering due to strong heterogeneity in the shallow lithosphere, and can thus be interpreted to higher frequency. The wave-form matching was, therefore, restricted to frequencies of up to 25 mHz for the fundamental mode, and 50 mHz for a time window covering the range of phase velocities expected for the higher modes. The use of higher modes, which are well excited by the deep earthquakes in the region (Tonga, New Hebrides, Java), improves resolution of structure in depth. However, the influence of the heterogeneity in structure means that it is not always possible to fit both the fundamental-mode and higher-mode windows with the same structure. In future studies, this linearization problem will be remedied by using wave-speed profiles and associated mode files of reference models that are constructed specifically for the region.

Zielhuis and van der Hilst [1996] discussed results of the application of this partitioned wave-form inversion scheme to the data from stations in eastern Australia. This study used arrays SK1 (SA01-SA08), SK2 (SB01-SB10, ZB11, ZB12) and BAS (YB01-YB05); the station locations are illustrated in Figure 1B, and the paths from events to these stations in Figure 4a. The data used in this study are consistent with a pronounced change from slow shear-wave propagation beneath the Coral and Tasman Seas and the coastal regions of continental Australia to very fast wave propagation beneath central Australia. The thickness of the high wave-speed lid increases from about 80-100 km in easternmost Australia to at least 250 km beneath the central shields. The transition is rather sharp in north Queensland but is manifest as a multiple boundary further south. Based on these data, Zielhuis and van der Hilst argued that the the conventional Tasman line (Figure 1B) marking the limits of Precambrian outcrop [see e.g. Veevers, 1984], may roughly coincide with a contrast in seismic-wave speed in the north, but that high wave-speed lithosphere extends east of the Precambrian basement further to the south. However, the data used in this earlier study does not constrain the southernmost part of the region around the Tasman Line, nor does it reliably image possible lateral variations in wave speed in the upper mantle beneath central Australia.

The model presented here is based on wave-form data from the permanent observatories in the region and from 45 SKIPPY stations covering the eastern and central parts of the Australian continent. Arrays SK1, SK2 and BAS, which were used in the previous work, are supplemented with data from the arrays SK3 (SC01-SC10) and SK4 (SD01-SD10). The station locations are indicated in Figure 1B, and the improved path coverage available with the data from the additional stations is shown in Figure 4B. Low-noise records were selected for regional earthquakes for which a Centroid Moment Tensor (CMT) is available from Harvard University. The total number of seismograms used in the construction of the three-dimensional shear velocity mode is approximately 1350, from about 360 earthquakes. Good wave-path coverage is achieved for eastern and central Australia as shown in Figure 4; this enabled the extension of the results for eastern Australia by Zielhuis and van der Hilst [1996].

Results of the inversion for three-dimensional variations in shear-wave speed are presented in Figure 5 for depths of 80, 140, 200 and 300 km. For wavelengths of 1000 km and larger, the observations are in good agreement with inferences from global inversion of Rayleigh-wave phase velocities [e.g., Trampert and Woodhouse, 1995; Ekstrom et al., 1997], but the data coverage provided by the SKIPPY arrays allows the study of continental structure in much more detail. The resolution length is about 300 km for most of the area discussed here.

A striking feature in the images of mantle structure to about 200 km depth is the large difference in shear-wave speed between the part of Australia east of 140°E, and the rest of the continent (Figures 5A,B). The wave-speed gradients are predominantly in west-east direction, and exhibit peak-to-peak amplitude variations of up to 10 per cent relative to the reference model used for the display. The patterns generally match geological trends observed at the surface. Beneath easternmost Australia and the adjacent oceanic regions, the shear-wave speed is relatively low, whereas fast wave propagation marks the central and western part of the continent. The boundary between these regions is well defined at 80 and 140 km depth. The transition from low to high velocities seems to be rather sharp in northern Queensland (near 15°S, 142°E) and lies close to the Tasman Line, the presumed eastern edge of the Precambrian shields based on geological outcrop, gravity and magnetic data (Figure 1B). Further to the south the transition from slower to faster wave speeds appears to be somewhat more complex, and may occur over at least two zones of rapid increase in wave speed. The southern part of the Eromanga Basin, and the region associated with the Lachlan Fold Belt, are characterized by wave speeds that are intermediate to that in the coastal regions and the central shields (Figure 5B). The nature of these wave-speed transitions and, in particular, the complex structure near the Broken Hill Block (32°S, 140°E) will be discussed in more detail in a separate paper [Kennett and van der Hilst, in preparation]. We remark that south of 30°S, the velocity contrast associated with the boundary between Phanerozoic and Proterozoic basement is significantly better constrained than in the study by Zielhuis and van der Hilst [1996].

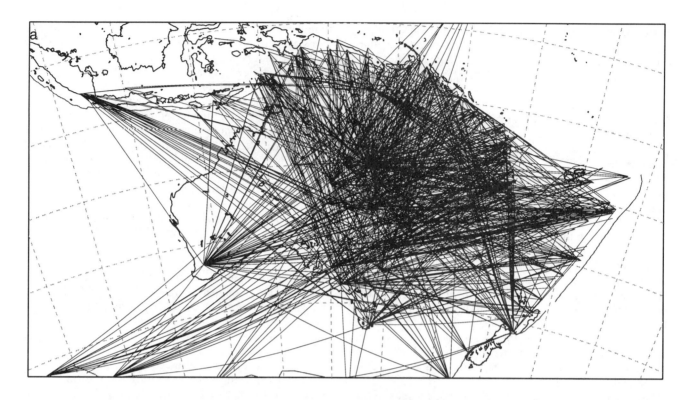

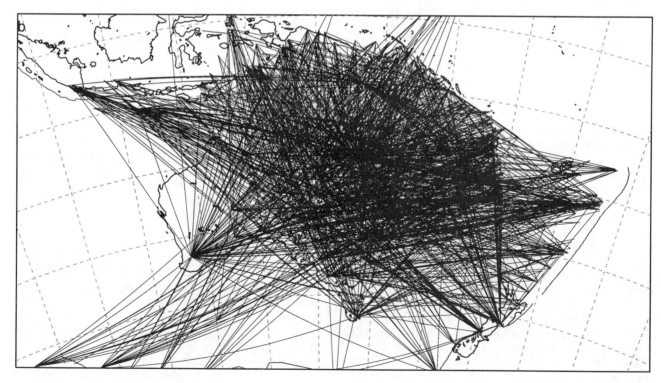

Figure 4: Wave-path coverage available for the SKIPPY experiment in eastern and central Australia. (a) wave paths exploited by Zielhuis and van der Hilst (1996); (b) wave-path coverage used in this study.

Along the eastern seaboard of the Australian continent, there is a pronounced low-velocity zone between 100 and 200 km in depth (Figure 5); the minimum wave speed seems to occur at a depth of 140 km. This low-velocity zone was recognised in earlier surface-wave dispersion studies (see Muirhead and Drummond [1991] for a review), but a striking feature in our 3-D models is the level of lateral variation in the character of the zone. The low-velocity zone is interrupted near 30°S, 150°E beneath the New England Fold Belt and the higher velocities in this region seem to be almost continuous from the surface to the transition zone. The main zones of lower velocities correlate well with recent volcanism and high heat flow. The low-speed anomalies appear to extend to greater depth beneath the Coral Sea and beneath the southeastern corner of Australia. Zielhuis and van der Hilst [1996] discuss the upper-mantle structure beneath easternmost Australia in more detail.

In the upper 200 km beneath central Australia, shear-wave speed is high, but the images suggest significant lateral variations that begin to delineate the major crustal elements. The zone of high wave speed in the central north (Figures 5A-C) coincides with the lateral extent of the North Australian Craton, and the high wave speeds in the central south coincide with the Gawler Craton and the Eucla Block (Figures 5A, B). To approximately 80 km depth, the region associated with the Late Paleozoic Alice Springs Orogeny - in particular the Amadeus Basin and the Musgrave Block - is characterized by slower shear-wave propagation than in the adjacent shields (Figure 5A). The images suggest that the North Australia Craton extends northward into western Papua New Guinea and eastern Indonesia (from Timor to Irian Jaya). Interestingly, the Kimberley Block, near 15°S, 125°E, is delineated by slower wave propagation than the shield further to the east, which suggests that the Kimberley Basin is not underlain by the same basement as the Proterozoic North Australia Craton. The change to slower wave propagation occurs across the Ord Basin and the Halls Creek Shear Zone. The high wave speeds mapped in the southwest may well be related to the Archean cratons, but they are not yet well localized owing to insufficient path coverage.

The amplitude of wave-speed variability is diminished below 200 km depth, but significant contrasts remain. At 200 km depth (Figure 5C) the images reveal again a difference between the wave speeds in eastern and central Australia. The eastern edge of the high wave speeds that characterize the regions of Proterozoic basement seems to coincide with the southeastern margin of the Mt. Isa Inlier (20°S, 140°E) and the western boundary of the Eromanga Basin, and to extend approximately along the 140°E meridian along the eastern side of the Gawler Craton (Figure 1A).

At depths larger than 300 km (Figure 5D), the general trend in the anomalies is markedly different from that in the shallower upper mantle (Figures 5A-C). The predominant orientation of the gradients is no longer west-east, although even at these large depths the strongest low wave-speed anomalies are located in the east, that is, beneath the Coral Sea and, in particular, the Tasman Sea. Between 300 and 400 km depth, there are prominent high wave speeds beneath the eastern part of the continent, in a broad band extending from 20°S, 143°E, southwards to around 35°S. We have previously noted the high velocities beneath the New England Fold Belt near 30°S, 152°E. These anomalies (discussed in more detail by Zielhuis and van der Hilst [1996]) are robust features of the solution, and are confirmed by a simple differential technique applied to a small number of high-quality Rayleigh-wave data [Passier et al., 1997]. The high wave speeds beneath the western parts of the Musgrave Block and the Officer Basin may be real, but we can not be confident in the mapping of this structure until data from the SKIPPY arrays in West Australia have been analyzed. For the deeper part of the upper mantle, the images display several 'streaks' that may result from a combination of uneven sampling, effects of source mislocation, or from the neglect of inter-mode coupling for the higher-mode packet. This will be subject to further study.

Vertical sections through the three-dimensional shear wave-speed model for the upper mantle (Figure 6) provide further insight into the lateral variation in the character of the lithosphere; in particular, they demonstrate the dramatic differences in upper mantle structure between the Phanerozoic and Precambrian parts of the continent. For the purpose of the present paper, we loosely associate the "lithosphere" with a zone of elevated shear velocity compared with the asthenosphere beneath. The locations of the cross sections are given in Figure 1A.

A section from the Coral Sea across the eastern margin of the Australian continent into the Proterozoic shield (Figure 6a) displays the large contrast in lithosphere thickness between the oceanic and continental shield regions. Beneath the Coral Sea a negative gradient in shear velocity occurs at a depth of about 50 km, beneath the coastal region of easternmost Australia the lithosphere thickness is about 100 km, and beneath the central Proterozoic cratons reduced shear wave speeds do not occur until a depth of about 250 km (Figure 6a). The virtual absence of a high wave-speed lid near 145°E coincides with the region of Neogene volcanism in the Queensland volcanic province, which may suggest that part of the continental lithosphere has been eroded by thermal processes.

The east-west cross section at 24°S (Figure 6b) reveals a similar contrast in lithosphere thickness between eastern Australia and the central shields. The high wave speeds associated with the seismological lithosphere are detected to depths exceeding 300 km beneath West Australia (130°E). The image clearly reveals the low-velocity zone extending beneath eastern Australia (east of 140°E), that separates thin high wave-speed lid from the deeper zone of high wave-speed anomalies between 300 and 400 km depth. This cross

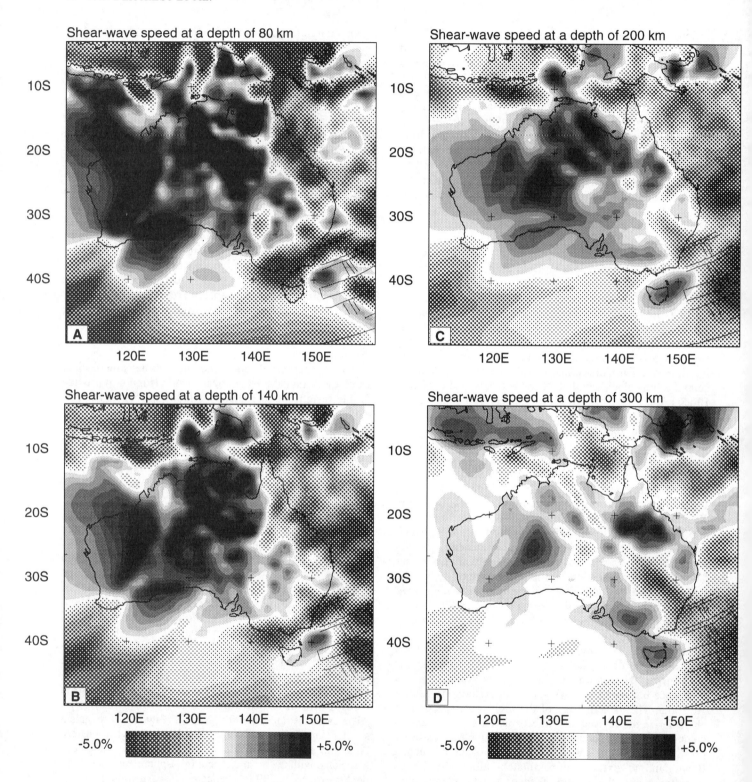

Figure 5: Three-dimensional shear-wave-speed model derived from partitioned wave-form inversion from the SKIPPY experiment in eastern and central Australia. Map views at (a) 80 km, (b) 140 km, (c) 200 km, and (d) 300 km depth. Based on the path coverage (Figure 4B) the wave-speed anomalies are reliable images east of about 125°E.

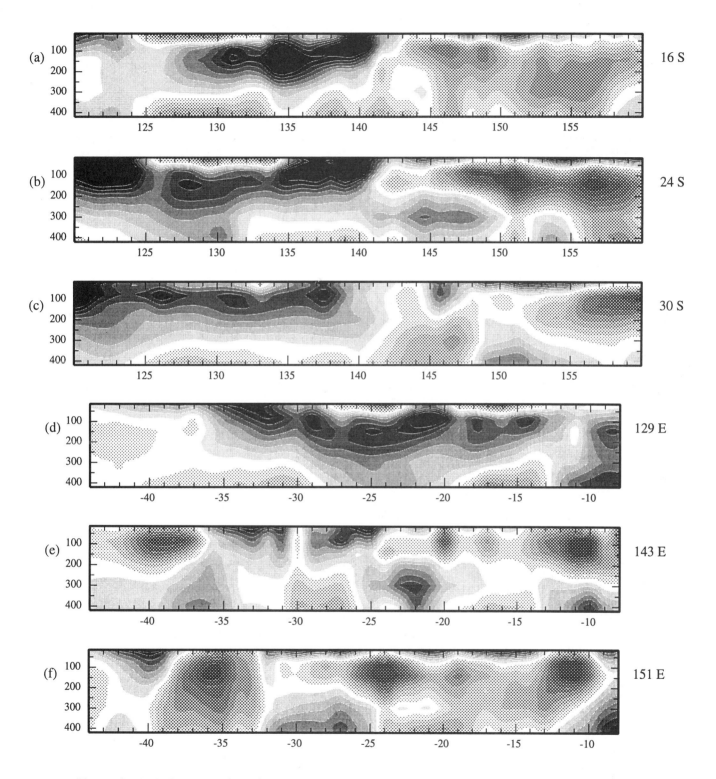

Figure 6: Vertical cross-sections through the three-dimensional shear-wave speed derived from partitioned wave-form inversion from the SKIPPY experiment in eastern and central Australia. Cross sections at (a) 16°S, (b) 24°S, and (c) 30°S, and north-south sections at (d) 129°E, (e) 143°E, and (f) 151°E.

section intersects the volcanic province at approximately 150°E.

The seismic structure shallower than 80 km may be contaminated by variations in crustal structure that are not adequately accounted for by our approach, but the cross sections clearly reveal a pronounced wave-speed anomaly in the region associated with the Alice Springs Orogeny and the eastern part of the Canning Basin (125°-135°E). At 30°S the model reveals the large lateral variations in wave speed near the Broken Hill Block, i.e. across the southern part of the Tasman Line. At this latitude, the seismically defined lithosphere is about 200 to 250 km thick beneath most of the Proterozoic shields (between 125° and 140°E), and possibly even thicker towards the Archean cratons further west. The latter observation is, however, tentative because data coverage is not sufficient to constrain the structures in the southwestern part of the continent.

The map views (Figures 5A-C) and the vertical sections (Figures 6a-c) clearly delineate the transition from thin lithosphere beneath the Phanerozoic of eastern Australia to the thick Proterozoic shields of central Australia. Locally, in particular south of 25°S, the transition is complex and may comprise multiple boundaries and basement types

The north-south cross sections (Figures 6d-f) further illustrate the pronounced lateral variation in the character of the high wave-speed lid beneath the Australian continent. The thickest lid is located beneath the western part of the Officer Basin, at about 25°S, 129°E (Figure 6d), that is near the boundary of the Proterozoic and Archean shields. The observation of such a thick high wave-speed lid in the western part of the continent, and the absence of a strong negative velocity gradient in shear velocity beneath the lid, are in good agreement with the path average shear-wave profile by Gaherty and Jordan [1995]. At 129°E (Figure 6d), the sudden increase in wave speed at about 35°S marks the ocean-continent transition in the Great Australian Bight, and the deep high wave speeds north of 10°S are probably related to subduction beneath eastern Indonesia. The cross section at 143°E (Figure 6e) displays the intermediate thickness of the lithosphere in the the region where the transition from the coastal region to the central shields is gradual. The rapid lateral variations near 30°S coincide with the boundary between the Murray and Eromanga Basins (i.e., parts of the Broken Hill Block, the Lachlan Fold Belt, and the Darling Basin). This section also illustrates the deep anomaly near 22°S in Queensland. Figure 6f displays the deep low wave-speed anomaly beneath the Tasman Sea region, the low-velocity zone beneath easternmost Australia, and the deep high wave-speed anomaly beneath the New England Fold Belt.

These significant results have been derived from the analysis of data from only part of the SKIPPY project. More detail on lithospheric and mantle structure beneath the western part of Australian region will be revealed as the data from the last two SKIPPY deployments are incorporated into the inversion for three-dimensional structure.

P-wave tomography

An additional source of information on three-dimensional structure comes from P-wave tomography using the residuals of observed arrival times of P phases compared with the predictions from a suitable reference model. The residuals reflect the integrated influence of wave-speed variations in the Earth's interior along the propagation path from source to receiver. With sufficient crossing paths, the travel time information can be used to reconstruct the three-dimensional variations in seismic wave speed.

In order to produce a homogeneous set of travel time residuals, the arrival times determined from the SKIPPY records are incorporated in the data processing scheme of the U.S. Geological Survey's National Earthquake Information Center (NEIC). In this way, the SKIPPY data help constrain the hypocenters of earthquakes in the Australasian region, and in return are made to be consistent with the global data set assembled by Engdahl et al. [1997].

Travel-time residuals from the SKIPPY project were first used by Widiyantoro and van der Hilst [1996] in their tomographic study of the complex subduction beneath the Indonesian region. Their three-dimensional mantle model was generated by embedding a high-resolution representation of regional structure in a somewhat lower-resolution cellular model for the rest of the mantle. By this means, contamination of the regional structure by features outside the region of interest can be minimised. The inversions for the Indonesian region incorporated phase data from the SKIPPY records in eastern Australia, along with data from a global distribution of permanent stations, and reveal the thick lithosphere under northern part of central Australia.

With more phase picks now available from later SKIPPY deployments (SK1, SK2, SK3, and BAS), in addition to data from the network of short-period permanent stations, the P-wave tomography has been extended to cover most of the continent, with resolution in the eastern part significantly better than in the west. Figure 7 displays the lateral variation in P-wave speed relative to the *ak135* reference model at a depth of approximately 150 km beneath the Australian region. The image reveals the lateral contrast between high and low wave speeds in northern Queensland, which is prominent in the shear-wave structure inferred from the wave-form analysis described above. The resolution of structure in oceanic regions is, of course, poor because of lack of station coverage.

Much of the information used to construct the P-wave velocity model comes from arrivals travelling rather steeply in the mantle, so that vertical resolution is limited. Nevertheless, the strong contrast in the seismic signature of the lithosphere between western and central Australia and the eastern seaboard are clearly revealed. As in the shear-wave images, the zone of higher wave speeds extends further east than the conventional Tasman Line, but thicker lithosphere lies mostly to the west of 140°E. There are

Depth slice at 150 km

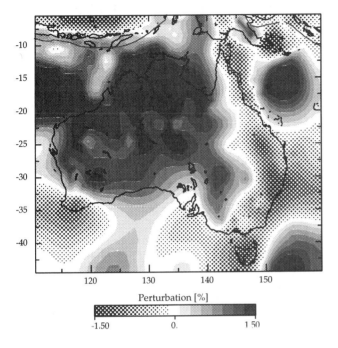

Figure 7: Cross-section through the 3D model of the P-wave velocities in the Australian region for the depth interval from 160-220 km.

some significant differences in the P and S wave-speed models, for instance beneath the New England Fold Belt. These differences may largely be due to variations in data coverage and resolution, but upon further analysis they may also yield new information about the physical nature of the anomalies, as for example whether their origin is thermal or compositional.

Receiver based studies

Along with their use in tomographic studies, the even distribution of portable broad-band stations across Australia can also be exploited to provide a range of information on the structure beneath each of the stations. Here we discuss only the results of receiver-function analyses; measurements and implications of shear-wave splitting are discussed elsewhere in this issue [Clitheroe and van der Hilst, 1997].

For each of the broad-band stations in the SKIPPY deployments, data from distant seismic events are being processed to extract a receiver function which is sensitive to the structure in the crust and uppermost mantle and, in particular, the character of major seismic interfaces such as the Moho. The analysis uses the P-wave onset and its immediate coda. The radial and vertical components of the record share the same source time function, so that the dependence on the radiation from the source can be largely eliminated by deconvolving the wave-form segment of the radial horizontal component, lying along the great-circle path from the source, with the corresponding segment of the vertical component [Langston, 1979]. The deconvolved radial receiver function then emphasises the influence of near-receiver structure, and can be inverted in the time domain for a 1-D shear wave velocity model of the crust and uppermost mantle [see e.g. Ammon et al.,1990]. A major contribution to the receiver function comes from conversions between P and S waves. The timing and amplitude of such arrivals provide constraints on the properties of major interfaces such as the crust-mantle boundary and internal boundaries within the crust. For example, a sharp crust/mantle boundary produces a converted phase with about 5 s separation from P.

The inversion of the receiver functions to recover crustal and uppermost mantle structure is widely recognised to be sensitive to the starting model if a conventional linearization scheme is employed [see e.g. Ammon et al., 1990].. However, such difficulties can be overcome by employing an inversion scheme based on a Genetic Algorithm [Shibutani et al., 1996]. This approach makes use of a "cloud" or "population" of models to minimise the dependence on a starting model; a set of "biological" analogues are used to produce new generations of models from previous generations, with preferential development of models with a good fit between observed and theoretical receiver functions. The approach provides a good sampling of the model space, and enables the estimation of the shear-wave speed distribution in the crust, along with an indication of the ratio between P and S wave speeds. Many models with an acceptable fit to data are generated during the inversion, and a stable crustal model is produced by employing a weighted average of the best 1000 models encountered in the development of the genetic algorithm. The weighting is based on the inverse of the misfit for each model, so that the best fitting models have the greatest influence.

At each station, a stacked receiver function was constructed from teleseismic observations. For each event, the receiver function was low-pass filtered to eliminate frequencies above 1 Hz in order to minimize the influence of small-scale heterogeneities. Subsequently, the receiver functions from 6 - 18 teleseismic events at each station were stacked together for a set of ranges of back azimuths. The weighting used in the stacking emphasizes receiver functions with higher signal-to-noise ratios, and those whose back azimuth and incident angle were closer to the mean for the set of events.

The radial component of the stacked receiver function was then inverted for a 1-D velocity model beneath each station. In the Genetic Algorithm inversion, the crust and uppermost mantle down to 60 km were modeled with six major layers: a sediment layer, basement layer, upper crust, middle crust, lower crust and uppermost mantle. The model parameters in each layer are the thickness, the S-wave velocity at the upper boundary, the S wave velocity at the lower boundary, and the velocity ratio between P and S waves (V_P/V_S). The S-wave velocity for each layer is constructed by linearly connecting the values at the upper and lower boundaries, to give a sequence of constant velocity-gradient segments separated by velocity discontinuities.

The results of the receiver function inversion for the stations SB06 and SA07 are shown in Figure 8. All 10,000 models searched in the Genetic Algorithm inversion for S velocity are shown as the light gray shaded area, and superimposed on this the best 1,000 models are shown with darker gray tones, where the darkness is proportional to the logarithm of the number of the models. The best fitting model for each site is shown as a black line. However a more useful and stable result is provided by the averaged model generated by weighting the best 1,000 models by the inverse of their misfit values. This averaged model is shown in white for S velocity and in a solid line for the V_P/V_S ratio. The lower panel in Figure 8 compares the wave forms of the stacked receiver function to synthetics calculated for the averaged models. The fit to the major phases is good, and this will be true of most of the best 1,000 models. The zone of darker gray in the upper part of Figure 8 can therefore be thought of as an indicator of the constraint on the crust and upper mantle structures. Both SB06 and SA07 have surficial sediment with very low velocities, so that P-to-S converted phases and reverberations (1 - 3 s) originating in the sediment layer are larger than the direct P phase, which arrives at zero time.

The analysis for crustal structure has been completed for all the SKIPPY stations in the SK1 and SK2 arrays in eastern Australia and is in progress for the permanent stations and the later SKIPPY arrays. The parametrization of the model at each station is via a sequence of velocity gradients and discontinuities; sensitivity analysis indicates that the dependence on phase conversions enhances the resolution of boundaries at the expense of the gradients. Nevertheless, all the models have been derived by the same procedure, and we can make direct comparisons and extract a wide range of information on crustal structure and uppermost mantle structure across eastern Australia. In the few places where direct comparisons can be made, e.g. ZB12, there is a very good concordance between the character of the S-velocity structure from the receiver function inversion and the previous P-velocity model from refraction studies.

Figures 9, 10 and 11 summarise the properties of the crustal models at each of the SKIPPY stations in eastern Australia in terms of the major subdivisions of the crust: the upper crust (Figure 9), middle/lower crust (Figure 10) and the crust-mantle boundary and uppermost mantle velocities (Figure 11). In each map we indicate the thickness of the layer, the corresponding crustal velocities, and the character of the boundary. A sharp boundary is associated with a clear converted phase in the receiver function wave form, whereas the expression of a transitional zone is more subtle. The nature of the boundaries is classified into the categories: SHARP (< 1 km), THIN (< 4 km), TRANSITIONAL or INTERMEDIATE (< 10 km), and BROAD (> 10 km). The depth of the crustal boundaries was estimated at each station; for a transitional zone the lower boundary was selected.

The information from the receiver-function inversion provides a major supplement to the previous results from isolated refraction and reflection experiments. For the construction of the crustal thickness map in Figure 11 the earlier information [Collins, 1991] has been combined with the present results. The crust-mantle boundary is deep (38 - 44 km) and mostly transitional in character along the axis of the fold belt zone in the east (from stations: ZB12, SB09 to SA05, SA03). A relatively sharp Moho is found at a shallower depth (30 - 36 km) at the western edge of the study area, close to the boundary between Phanerozoic and Precambrian exposure.

A major advantage of the receiver-function approach is that it provides good constraints on shear velocities in the crust and uppermost mantle, which are not well resolved by the partitioned wave-form inversion of the S wave and surface-wave portions of the seismogram. The 3-D models are most reliable from 60 km down, whereas the receiver-function analysis is most effective above 60 km. There is a good agreement between the results from the wave-form tomography and the receiver function analysis at 60 km depth.

CONCLUSION

Over the last decade, there has been a significant increase in knowledge of the P and S wave velocities in the mantle, particularly beneath northern Australia. The current generation of three-dimensional studies based on the use of portable seismometers have the potential to increase dramatically the level of understanding of mantle structure beneath the Australian region, in particular the contrast between eastern and western Australia.

Acknowledgments. The study of mantle structure beneath Australia has involved many members of the Seismology Group at the Research School of Earth Sciences, both in the field and in subsequent analysis. We would like to thank Doug Christie, John Grant, Armando Arcidiaco, Tony Percival, Gus Angus and Jan Hulse for their efforts in the field in often trying and uncomfortable conditions. Alet Zielhuis, Roger Bowman, Phil Cummins, Oli Gudmundsson, Cheng Tong, Geoff Clitheroe and Jan Weekes have all contributed to the analysis of data from the portable stations.

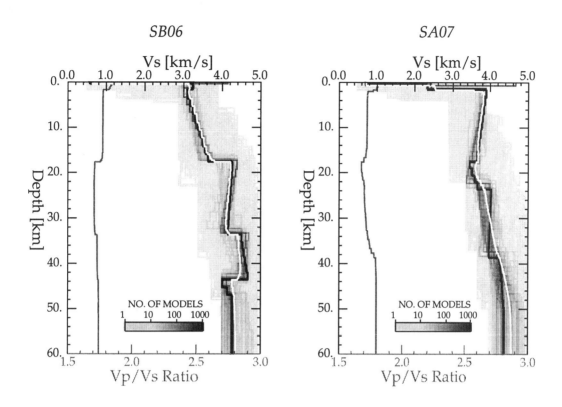

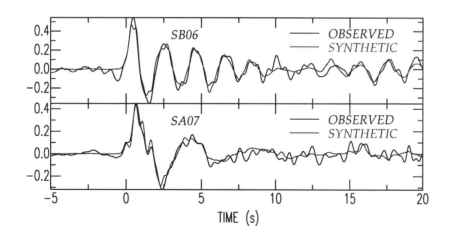

Figure 8: Genetic algorithm inversion of a receiver function to determine S-wave structure and the V_P/V_S ratio. All 10,000 models searched in the GA inversion are shown as the light gray shaded area. The best 1,000 models are shown as darker gray shaded area. The darkness is logarithmically proportional to the number of the models as shown by the scale bar. The best model and the averaged model are shown by the black solid line and the white solid line respectively. For the V_P/V_S ratio, the solid line indicates the averaged model.

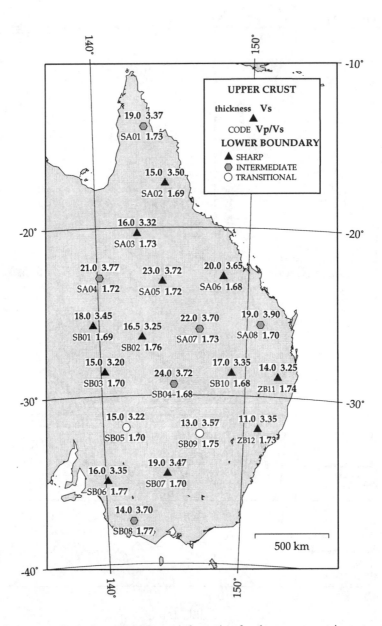

Figure 9: Summary of velocity and thickness information for the upper crust in eastern Australia.

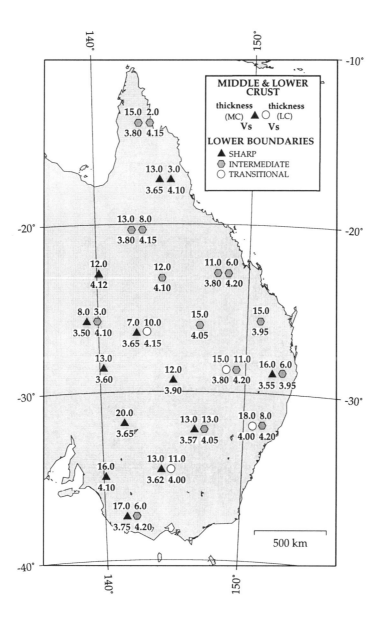

Figure 10: Summary of velocity and thickness information for the middle and lower crust in eastern Australia.

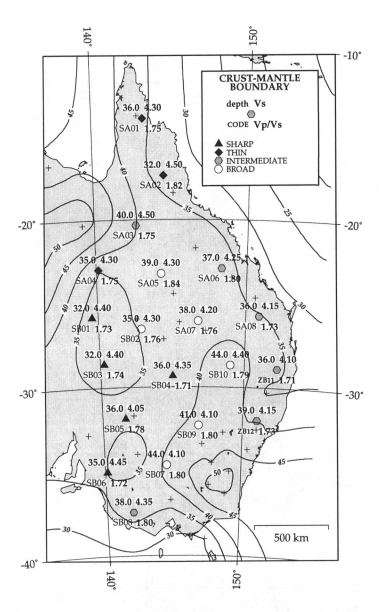

Figure 11: The depth of the crust-mantle boundary beneath eastern Australia, and the S-wave velocity and V_P/V_S ratio at the top of the mantle. The nature of the crust-mantle boundary is classified into four categories: sharp ($\leq$ 1 km), thin ($\leq$ 4 km), intermediate ($\leq$ 10 km), and broad (> 10 km). The crosses indicate the locations for which the crustal structure has been obtained previously from refraction surveys [Collins, 1991]. These data were incorporated into the contouring.

REFERENCES

Ammon, C. J., G. E. Randall, and G. Zandt, On the nonuniqueness of receiver function inversions, *J. Geophys. Res.*, 95, 15303-15318, 1990.

Bowman, J. R., and B. L. N. Kennett, An investigation of the upper mantle beneath northwestern Australia using a hybrid seismic array, *Geophys. J. Int.*, 101, 411-424, 1990.

Clitheroe, G., and R. D. van der Hilst, Complex anisotropy in the Australian lithosphere from shear-wave splitting in broad-band SKS records, *this volume*, 1997.

Collins, C. D. N., The nature of the crust-mantle boundary under Australia from seismic evidence, in *The Australian Lithosphere*, edited by B. J. Drummond, *Geol. Soc. Aust. Spec. Publ.*, 17, pp. 67-80, 1991.

Denham, D., Shear wave crustal models for the Australian continent, in *The Australian Lithosphere*, edited by B. J. Drummond, *Geol. Soc. Aust. Spec. Publ.*, 17, pp. 59-66, 1991.

Dey, S. C., B. L. N. Kennett, J. R. Bowman, and A. Goody, Variations in upper mantle structure under northern Australia, *Geophys. J. Int.*, 114, 304-310, 1993.

Drummond, B. J., K. J. Muirhead, P. Wellman, and C. Wright, A teleseismic travel-time residual map of the Australian continent, *BMR J. Austral. Geol. Geophys.*, 11, 101-105, 1989.

Ekstrom, G., J. Tromp, and E. Larson, Measurements and global models of surface wave propagation, *J. Geophys. Res.*, 102, 8137-8157, 1997.

Gaherty, J., and T. H. Jordan, Lehmann discontinuity as the base of an anisotropic layer beneath continents, *Science*, 268, 1468-1471, 1995.

Grand, S., and D. V. Helmberger, Upper mantle shear structure of North America, *Geophys. J. Roy. Astr. Soc.*, 76, 399-438, 1984.

Gudmundsson, O., B. L. N. Kennett, and A. Goody, Broadband observations of upper mantle seismic phases in northern Australia and the attenuation structure in the upper mantle, *Phys. Earth Planet. Inter.*, 84, 207-226, 1994.

Hales, A. L., K. J. Muirhead, and J. W. Rynn, A compressional velocity distribution for the upper mantle, *Tectonophys.*, 63, 309-348, 1980.

Kennett, B. L. N., Approximations for surface wave propagation in laterally varying media, *Geophys. J. Int.*, 122, 470-478, 1995.

Kennett, B. L. N., and R. D. van der Hilst, Using a synthetic continental array to study the Earth's interior, *J. Phys. Earth*, 44, 669-674, 1996.

Kennett, B. L. N., O. Gudmundsson, and C. Tong, The upper-mantle S and P velocity structure beneath northern Australia from broad-band observations, *Phys. Earth Planet. Inter.*, 86, 85-98, 1994.

Kennett, B. L. N., E. R. Engdahl, and R. Buland, Constraints on seismic velocities in the Earth from travel times, *Geophys. J. Int.*, 122, 108-124, 1995.

Langston, C. A., Structure under Mount Rainier, Washington, inferred from teleseismic body waves, *J. Geophys. Res.*, 84, 4749-4762, 1979.

LeFevre, L. V., and D. V. Helmberger, Upper mantle P velocity structure of the Australian shield, *J. Geophys. Res.*, 94, 17749-17765, 1989.

Leven, J. H., The application of synthetic seismograms in the interpretation of the upper mantle P-wave velocity structure in northern Australia, *Phys. Earth Planet. Int.*, 38, 9-27, 1985.

Muirhead, K. J., and B. J. Drummond, The seismic structure of the lithosphere under Australia and its implications for continental plate tectonics, in *The Australian Lithosphere*, edited by B. J. Drummond, *Geol. Soc. Aust. Spec. Publ.*, 17, pp. 23-40, 1991.

Nolet, G., Partitioned wave-form inversion and two-dimensional structure under the Network of Autonomously Recording Seismographs, *J. Geophys. Res.*, 95, 8499-8512, 1990.

Passier, M. L., R. D. van der Hilst, and R. Snieder, Surface wave wave-form inversion for local shear-wave velocities under eastern Australia, *Geophys. Res. Lett.*, in press, 1997.

Shibutani, T., M. Sambridge, and B. L. N. Kennett, Genetic algorithm inversion for receiver functions with application to crust and uppermost mantle structure beneath Eastern Australia, *Geophys. Res. Lett.*, 23, 1829-1832, 1996.

Trampert, J., and J. H. Woodhouse, Global phase velocity maps of Love and Rayleigh waves between 40 and 150 seconds, *Geophys. J. Int.*, 122, 675-690, 1995.

Van der Hilst, R. D., B. L. N. Kennett, D. Christie, and J. Grant, Project SKIPPY explores the mantle and lithosphere beneath Australia, *EOS*, 75, 177, 180-181, 1994.

Veevers, J. J., *Phanerozoic Earth History of Australia*, 418 pp., Oxford University Press, Oxford, UK, 1984.

Widiyantoro, S., and R. D. van der Hilst, The slab of subducted lithosphere beneath the Sunda Arc, Indonesia, *Science*, 271, 1566-1570, 1996.

Zielhuis, A., and G. Nolet, Shear-wave velocity variations in the upper mantke beneath central Europe, *Geophys. J. Int.*, 117, 695-715, 1994.

Zielhuis, A., and R. D. van der Hilst, Mantle structure beneath the eastern Australian region from Partitioned Waveform Inversion, *Geophys. J. Int.*, 127, 1-16, 1996.

B. L. N. Kennett, Research School of Earth Sciences, The Australian National University, Canberra, ACT 0200, Australia

T. Shibutani, Research Centre for Earthquake Prediction, Disaster Prevention Research Institute, Kyoto University, Kyoto 611, Japan

R. D. van der Hilst, Department of Earth, Atmospheric, and Planetary Sciences, Massachusetts Institute of Technology, Cambridge, MA 02139, USA

Mapping of Geophysical Domains in the Australian Continental Crust Using Gravity and Magnetic Anomalies

Peter Wellman

Australian Geological Survey Organisation, Canberra, Australia

Using images of gravity and magnetic anomalies, the upper crust of the Australian Continent can be subdivided into geophysical domains, of size about 500x1000 km, which differ in anomaly trend, anomaly texture, and average upper crustal density and magnetisation. Where basement is exposed, these domains correspond with geologically mapped cratons and blocks, within which there has been a common cratonisation history. At domain boundaries, the trends of the younger domain parallel to the boundary, truncate trends of the older domain, and indicate the relative age of the trends. Major anomalies occur at domain boundaries owing to differences in the average physical properties of the crust of the domains, and to geological processes at the boundaries. The domains can be grouped into mega-elements – large contiguous areas with a common geological history subsequent to formation, that are surrounded by major boundaries. The relation of these domains to variation in seismic structure and heat flow is best shown in eastern Australia, because of the greater amount of geophysical work. The domains in the eastern third of Australia contain basement rocks of Phanerozoic and Proterozoic age, and are of two types: 1. Domains of low density, low magnetisation of the upper crust, high heat flow, and thin crust. The P-wave seismic velocity has an average below 6.1 km s^{-1} in the upper crust, increases slowly in the lower crust, and has a sharp velocity increase to the mantle. Basement rocks are mainly sedimentary and non-magnetic sediment-derived granites, and are commonly overlain by platform sediments. These domains appear to be new crust formed by either crustal extension, or the addition of an accretionary wedge at a subduction zone. 2. Domains with high density and magnetisation of the upper crust, low heat flow, and thick crust. The P-wave seismic velocity increases nearly linearly with depth in the crust, with average velocities above 6.1 km s^{-1} in the upper crust. Upper crustal rock types include both volcanics and 'I'-type granites. They generally have had a long period of igneous and sedimentary activity, and form narrow, complex fold belts. Most of them are crust accreted to Australia as a major cratonised continental fragment; some are intra-continental volcanic rifts.

INTRODUCTION

Prior to about 1970 the Australian continental crust was subdivided solely by using observed geology, with age control from palaeontology and isotopic dating of upper crustal processes [GSA, 1971; Page et al., 1984]. Subsequently additional control on the age of the basement blocks was provided by interpreting Sm/Nd data to derive the time at which crustal material was derived from the mantle [McCulloch, 1987]. The advantages of using observational geology and dating are that the subdivisions of the continent are closely tied to dated geological history. The disadvantage is that the technique can be applied directly only over the approximately 25% of the continental

where deformed rock (basement) is exposed, and not over the 75% of Australia where basement is overlain by later platform sediments (cover) through which there are few wells with basement samples.

The Australian basement rocks can be independently subdivided using gravity and magnetic anomalies, as these generally reflect basement geology, except where cover is very thick. The advantage of these data-sets is that they extend over all the continental area. The difficulty with interpreting them is how to relate the patterns of anomalies closely with the basement geological history. Earlier interpretations used a complete coverage of gravity data, and an incomplete coverage of magnetic anomalies, displayed as contour maps [Wellman, 1976, 1978, 1988]. The principles and general conclusions of this early work have not been substantially changed by the more recent interpretations. Recent re-assessment of these data has resulted in a more sophisticated interpretation, by using a complete coverage of both gravity and magnetic anomalies on land, expression as both contours and images, a better understanding of outcrop geology and its constraints, and a large body of published overseas and local interpretation studies. A perspective of this later interpretation of the Australian continent biased towards its geological meaning is given by Shaw et al. [1995a, 1995b, 1996] and Myers et al. [1996]. This paper discusses more fully the geophysical patterns of anomalies on which the interpretation model is based.

GRAVITY AND MAGNETIC DATA

Surface gravity observations consist of a complete coverage of the land area at a spacing equivalent to a grid of between 4 and 11 km spacing, and a coverage of most of the continental shelf by subparallel ship traverses with spacing between 20 and 50 km [Anfiloff et al., 1976]. The surveys are mainly by the Australian Government, but include some land data from state governments, exploration companies and universities. The surface data have been integrated to the same datum and scale [Anfiloff et al., 1976], and the integrated data are available as a list of observations, as contour maps [BMR, 1976; Wellman and Murray, 1979a, 1979b; Morse et al., 1991], and as a computer generated image [Morse et al., 1992]. Free-air anomalies derived from satellite altimetry over the marine area have been integrated with the Bouguer anomalies over the land area to provide a coverage of gravity anomalies over all the Australian region [Murray et al., in press].

The magnetic anomalies over Australia are defined by numerous discrete aeromagnetic surveys, each composed of numerous, close-spaced, parallel traverses. The surveys now cover almost all the land area, and large parts of the continental margin. The surveys were mainly conducted by the Australian Government, but also by state governments and exploration companies. Flight-line spacing is generally in the range 0.4-3 km in areas of thin cover, but are up to 6 km in some areas of thick cover [Hone et al., 1997]. Data from individual surveys have been gridded, and the grids have been joined and levelled. A representation of the resultant combined grid for Australia is available as an image [Tarlowski et al., 1996]. In the following text it has been assumed that remanence is in general minor, and that relatively high magnetic features are caused by relatively high susceptibility in the underlying basement.

DEFINITION OF GEOPHYSICAL DOMAINS

The subdivision of Australia outlined in this paper is based on the concept of a geophysical domain (subsequently called a 'domain'), which is a contiguous area that has unifying geophysical features - such as trends which are subparallel but not necessarily straight, and reasonably constant.average upper-crustal density and magnetisation. The domains correspond with cratons and blocks, and generally have a size of about 1000 km by 500 km. They are inferred to correspond to areas of crust which have a similar mode of formation, resulting in density and magnetisation having a similar average and variation, and to be cratonised by a single event resulting in a similar pattern of geological and geophysical features, importantly giving subparallel trends. Where basement is exposed, domain boundaries approximate major geologically defined province boundaries. Domains can be subdivided into smaller units with more uniform properties — the bands of Wellman [1995], and crustal elements of Shaw et al. [1996].

At a boundary between geophysical domains, geological processes of the younger domain have modified a band of the crust along the margin of the older domain (reworked zone). The modification is due to processes such as metamorphism, deformation and intrusion. The domain boundary can be defined either as the extent of the older domain rocks into rocks of the younger domain, or the extent of the younger domain rocks into the older domain. The boundary adopted in this paper is the extent of the older domain rocks, as this corresponds with the major boundary in upper-crustal geophysical properties.

MAPPING OF GEOPHYSICAL DOMAINS

For the purpose of this paper, we are interested in the shorter-wavelengths (less than 400 km) of the gravity anomalies due to horizontal density changes in the upper crust and their isostatic compensation. An image of these anomalies is given in Figure 1. The gravity map shows two types of gravity feature of importance to this paper. Boundaries between domains involve large changes in average crustal density, so the wavelength of anomalies is relatively long owing to the long wavelength of the crustal change, and isostatic compensation constrained by lithospheric flexure. This type of anomaly is dominant in images of anomalies of wavelength less than 400 km

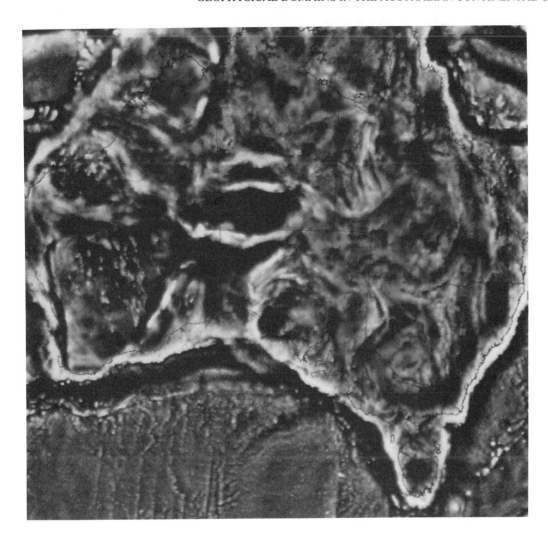

Figure 1. Gravity anomalies. Residuals from 400 x 400 km average values. The higher-amplitude longer-wavelength features are associated with changes in mean crustal density at geophysical domain boundaries. Processed from a grid derived from data of the Australian National Gravity Database held by the Australian Geological Survey Organisation.

(Figure 1). Gravity trends within domains, the second type, are caused by density anomalies in the upper crust due to features such as major granites or anticlines; these have wavelengths of less than 100 km. They are the short-wavelength low-amplitude features of Figure 1. Wavelengths smaller than 40 km are generally not mapped over Australia, because the common gravity station spacing is 11 km.

Magnetic anomalies define features with three ranges of wavelengths that are important to this study. Domains differ in average magnetisation of the upper crust, resulting in abrupt changes in average magnetic value at domain boundaries. These anomalies have a wavelength controlled by the width of the domains: 500-1000 km. They are the longer wavelength features shown on Figure 2. Major geological features within the domain (granites, anticlines etc) commonly have wavelengths in the range 10 to 50 km. The amplitude of these anomalies reflects the range in magnetisation of the upper crustal rocks of the domain, and the trend of the anomalies reflects the trend of these major features. These are the minor features in Figure 2. In areas where cover is thin or absent, the anomalies of wavelengths less than 6 km reflect faults and bedding trends of basement rocks, thereby giving an independent estimate of the structural trend and magnetisation of the area. Display of these anomalies is enhanced by derivatives of the anomalies, but they can be seen only at image scales of 1:1M or larger.

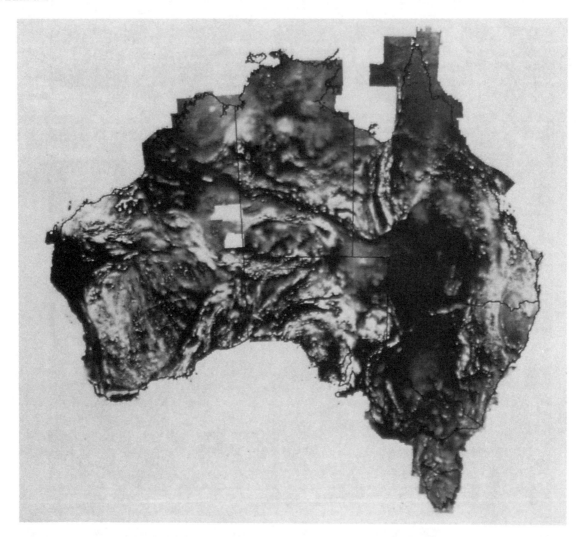

Figure 2. Magnetic anomalies. The major changes in mean value, trend and amplitude occur at the boundaries between geophysical domains. Processed from a grid of the Australian National Magnetic Database held by the Australian Geological Survey Organisation.

Domain boundaries can be mapped using gravity and/or magnetic anomalies by four main techniques. The usefulness of the techniques varies between boundaries. The first three methods are considered the more reliable.

Trends.

A domain boundary is in many cases marked by a change in dominant trend direction. Where there is no change in trend direction, the trends in both domains are parallel to the domain boundary. Where there is a change, trends of the older domain are oblique to the domain boundary, and trends of the younger domain are parallel to the domain boundary, truncating trends of the older domain [Provodnikov, 1975; Wellman, 1976]. However, because of reworking of the margin of the older domain, the extent of the younger trends in the older domain gives only the extent of reworking, and the true domain boundary is on the younger domain side of the extent of the trends characterising the older domain. The trends in each domain generally reflect the structural grain impressed during cratonisation.

Average Physical Properties.

At many domain boundaries there is a change in the average crustal magnetisation, and average upper crustal density. Abrupt changes in average crustal magnetisation are shown by changes across the domain boundary in both the average value of magnetic anomalies, and the amplitude of 6-50 km wavelength anomalies. Changes in upper crustal density are manifest as a sub-parallel, adjacent positive and negative anomaly pair along the boundary;

these have a greater than average amplitude and a wavelength of 200-400 km [Gibb and Thomas, 1976; Wellman, 1978]. The gravity high is over the denser crust and the gravity low over the less-dense crust. The major gravity anomalies near the boundary are due to the horizontal change in crustal density at the boundary, and the horizontal change in density near the base of the crust, due to regional isostatic compensation. The isostatic compensation is thought to be achieved by lithospheric flexure [Karner and Watts, 1983; Pilkington, 1990].

Geological Processes at the Domain Margin.

Two types of major geophysical feature reflect rock bodies formed during processes creating the domain boundary [Klasner and King, 1986; Wellman, 1988; Whitaker, 1994]. The first type is emplacement of igneous rocks, or mid-crustal to mantle rocks, along the margin of the younger domain (generally giving local gravity highs and intense local magnetic highs). The second type is modification of the margin of the older domain by deformation, metamorphism or igneous activity, as shown by magnetisation of uniform low value, but cut by narrow highs due to shears.

Character of Domain.

Domain boundaries can be mapped using the change in the character of the gravity or magnetic anomaly pattern, reflecting differences in the geological history of the upper crust of the two domains. The difficulty with this method is to apply it objectively. This method has been used in particular in northern Australia where other methods have not proved useful [Shaw et al., 1995a, 1996].

EXAMPLES OF DOMAIN BOUNDARIES

The following examples cover a range of ages and margin types, and they illustrate the range of geophysical responses at the domain boundaries. The examples do not include cryptic boundaries. The locations of the domain boundaries A-F are shown in Figure 3.

A, B. Yilgarn/Albany-Fraser Domain Boundary.

In southern Western Australia, a major boundary separates the Yilgarn Craton of Archaean age to the north and northwest, from the Albany-Fraser Province of Proterozoic age to the south and southeast [Beeson et al., 1988] (Figure 3). The geophysical anomalies at this boundary are particularly clear and instructive, because the geological relationships are shown by basement outcrop, and because the geophysical anomalies are of high amplitude. On the Yilgarn Craton proper, the gravity and magnetic anomalies trend north to north-northwest with characteristic sinuous positive anomalies over greenstone belts, and minor small

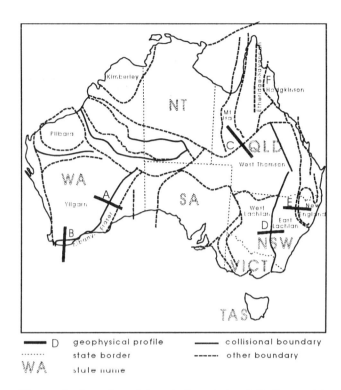

Figure 3. Location of domain boundaries A-F discussed in the text. Figure 4 gives geophysical profiles along lines A-E.

anomalies over gneiss. Along the margin of the Yilgarn Craton is a 10 to 50 km-wide reworked zone caused by deformation and metamorphism events of Albany-Fraser age. This reworking has produced generally flat and negative magnetic anomalies as a result of low magnetisation (Figure 2), and flat gravity values of short-wavelength due to only small upper crustal density contrasts. Slightly magnetic shears trend parallel to the boundary. Magnetic anomalies show that the greenstone belts are deeply eroded and sheared out along this reworked zone. Rocks of the Albany-Fraser Province strike east and northeast, truncating the north to north-northwest trends of the Yilgarn Craton. Along the length of this margin of the Albany-Fraser Province is a 10-30 km wide belt (Figure 2) containing highly-magnetic mafic and felsic granulites that are thought to be mid-crustal material thrust to the surface along the domain boundary.

A gravity dipole (adjacent high and low anomaly) traces the length of the domain boundary, but the dipole changes its polarity part way along. Seismic data show that the Yilgarn Craton has two types of crustal structure: a relatively low-velocity and low-density upper crust over most of the craton, and a higher velocity, denser upper crust in the southwest forming the area of the Western Gneiss Terrane [Dentith et al., 1996]. Where the Yilgarn Craton has a relative low upper-crustal density, it corresponds to a gravity low over the margin of the craton (Figure 4A);

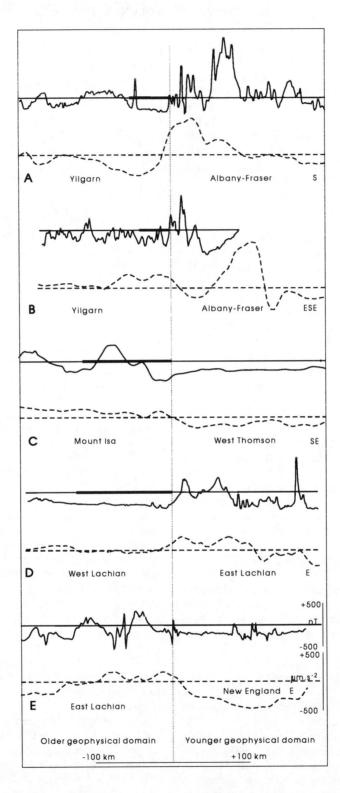

Figure 4. Gravity and magnetic anomalies associated with geophysical domain boundaries. Gravity anomalies shown as dashed lines, magnetic anomalies shown as continuous lines. The location of the lines is given in Figure 3. The extent of a band of deformed crust along the margin of the older geophysical domain is shown as a thick line; it often corresponds with low crustal magnetisation. In section E the boundary between the geophysical domains is marked by a serpentinite.

where it has a relatively high-upper crustal density (in the southwest), there is a gravity high over the margin of the craton (Figure 4B). Hence the gravity dipole anomaly over the domain boundary is largely due to the upper-crustal density difference across the domain boundary, and is not due to the high-density granulites that occur along the whole length of the Albany-Fraser Province boundary.

C. Mount Isa/West Thomson Domain Boundary.

In central western Queensland is a northeast trending domain boundary separating the Paleo-Mesoproterozoic rocks of the Mount Isa Domain from the Early Palaeozoic rocks of the West Thomson Domain (Figure 3). The boundary lies under sedimentary cover. The Mount Isa Domain corresponds to north-northwest to north-northeast trending rocks, and shows strong elongate gravity and magnetic anomalies [Wellman, 1992a]. The rocks are thought to have formed in an elongate crustal rift, 150-250 km wide and over 800 km long. The 120-km wide southeastern margin of this province was modified at the time of formation of the adjacent West Thomson Orogen by dextral shears parallel to the domain margins [Wellman, 1988] and possibly also by intrusion and metamorphism. The West Thompson Orogen has relatively low density and low magnetisation (Figure 4C). Within the West Thompson Orogen there are weak northeast trends subparallel to the domain margin, and these truncate the Mount Isa Domain trends. Seismic refraction work shows that the upper crust of the Mount Isa Domain has a relatively high seismic velocity relative to the West Thompson Orogen [Collins, 1988; Goncharov et al., 1996], consistent with the relative densities indicated by the sense of the gravity gradient over the domain boundary.

D. West Lachlan/East Lachlan Domain Boundary.

In central New South Wales there is a boundary between two Early Paleozoic domains [Wellman, 1995] (Figure 3). In the west is the West Lachlan Orogen with arcuate gravity and magnetic trends that are east to northeast where they are truncated near the domain boundary (Figures 1, 2). The margin of the West Lachlan Orogen is a metamorphosed and deformed zone about 100 km wide, with flat, slightly negative magnetic anomalies, cut by narrow slightly magnetic shear zones. In the south this band corresponds with the Wagga Metamorphic Belt. The domain boundary is marked by a major fault - the Gilmore Suture. East of the north-trending domain boundary is the East Lachlan Orogen, with high-amplitude gravity anomalies and high-amplitude positive magnetic anomalies trending parallel to the domain margin (Figures 1, 2). A gravity dipole anomaly follows the domain boundary, with a relative gravity low over the West Lachlan Orogen, and a gravity high over the East Lachlan Orogen (Figure 4D). Seismic refraction results show higher-velocity upper crust east of the boundary, consistent with the polarity of the gravity anomaly.

E. New England Accretionary Boundary.

In northeastern New South Wales there is a north-striking boundary between the East Lachlan Orogen, cratonised in the Ordovician and Silurian, and the New England Orogen to the east, which is mainly an accretionary prism formed against the East Lachlan Orogen in the Silurian to Early Carboniferous. The East Lachlan Orogen has high density, high magnetisation, and high upper-crustal seismic velocity [Collins, 1988], whereas the New England Orogen has low density, low magnetisation, and low upper-crustal seismic velocity [Finlayson and Collins, 1993]. Serpentinite emplaced along the boundary between the two crustal types, causes a narrow, elongate, magnetic high (Figure 4E).

F. Coen/Hodgkinson Domain Boundary.

In north Queensland there is a north-trending domain boundary between the Paleo-Mesoproterozoic rocks to the west of the Etheridge and Savannah Provinces, and the Early Paleozoic Hodgkinson Province to the east (Figure 3). Over the western half of Cape York Peninsula the basement is under cover, but in the south it is overlain by 1550 Ma Croydon volcanic group, so it must be Mesoproterozoic or older. Its magnetisation is very low. About 100-200 km west of the domain boundary, there is a sharp increase in both the average and the range of magnetisation, but no increase in gravity anomaly. This increase is attributed to rocks of this zone being deformed, intruded and metamorphosed during the Mesoproterozoic in the south, and during the Silurian-Devonian in the north, with these two reworkings increasing average magnetisation due to an increase in metamorphic grade [Wellman, 1992b]. East of the domain boundary are Ordovician-Devonian rocks of the Hodgkinson Province. These have low density and low magnetisation, with trends parallel to the domain boundary. No section is shown in Figure 4 because of the relative flatness of the gravity and magnetic profiles, and the difficulty in defining the position of the domain boundary owing to subsequent rifting and formation of the deep Lakefield Basin. This domain boundary is unusual in the low original magnetisation of the rocks on both sides. It shows that the margin of the older domain can in some cases have an increase in magnetisation owing to reworking.

Summary.

Using the above boundaries and others, the gravity and magnetic anomalies at domain boundaries are thought to have the following features.
1. The paired gravity high and low (dipole anomaly) is mainly caused by the boundary between crusts with

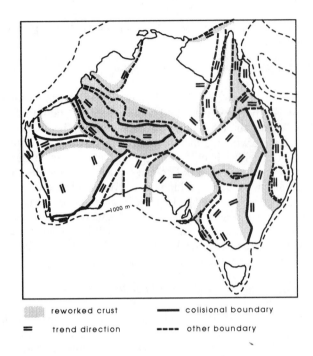

Figure 5. Boundaries between the major geophysical domains of Australia.

because the rocks are unsuitable for its recognition. These reworked bands are generally 50-150 km wide, and for each boundary they have relatively uniform width. In central Australia there is a wide area of possible reworked crust [Shaw et al., 1995a].

The domain boundaries are subdivided on the basis of geological and geophysical information into two types: those thought to be of collisional origin, and others. Boundaries of probable collisional origin are relatively few in number, and widely scattered. At collisional boundaries the younger domain generally has relatively high density and magnetisation, and its trends generally truncate the trend of the older domain. In many cases they contain a wide band of more magnetic crust along their margin with the older domain.

MEGA-ELEMENT MAP

Shaw et al. [1995a, b, 1996] and Myers et al. [1996] have grouped these geophysical domains into mega-elements - seven contiguous large cratonic regions, within which subsequent to formation there is a similar geological history. The mega-elements are separated by major boundaries. Figure 6 shows an updated version of this subdivision. The WA and SA mega-elements had an independent existence before joining. The NA and CA mega-elements together had an independent existence before

different density profiles, in regional isostatic equilibrium [Gibb and Thomas, 1976; Thomas and Gibb, 1977]. It is due only in small part to local high-density masses emplaced at the margin of the younger domain.

2. The relative average magnetic anomaly of adjacent domains is generally in the same sense as the relative upper crustal density, and relative upper-crustal P-wave seismic velocity.

3. At most boundaries the margin of the older domain is modified by geological processes active in the younger domain at the time of cratonisation of the younger domain (reworked band). Where the older domain has a high and variable magnetisation, this modification commonly results in a reduction of magnetisation to a uniform low value. Where the older domain has a very low magnetisation this modification can result in a slight increase in magnetisation.

SUBDIVISION OF AUSTRALIA INTO GEOPHYSICAL DOMAINS

Figure 5 summarises an interpretation of the Australian basement based on consideration of gravity and magnetic anomalies [Wellman, 1995; Shaw et al., 1995a]. The map shows the distribution of the major domain boundaries, the extent of reworked crust, and a very basic summary of the distribution of trend directions. Most boundaries have an associated discrete band of reworked crust along the domain boundary, and where this is not identified, it is generally

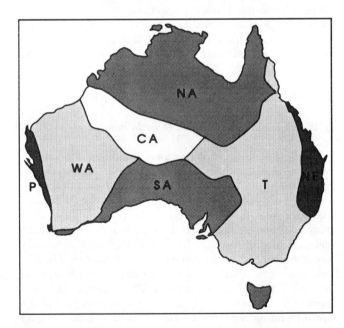

Figure 6. Australian mega-elements, representing continent-scale groups of geophysical domains (modified from Shaw et al., 1996). The crustal elements names are: CA, central Australia; NE, New England; NA, northern Australia; P, Pinjarra; WA, western Australia; SA, south Australia; T, Tasman.

joining with the combined WA-SA mega-elements. The CA mega-element is distinctive in being a wide band of long-lived interaction between mega-elements NA and the combined SA-WA.

REGIONAL ELECTRICAL CONDUCTIVITY ANOMALIES

There is a rough correspondence between some of the major regional electrical conductivity anomalies (the Canning Basin, Carpentaria and Southwest Queensland anomalies of Wang et al. [1997]) and the position of geophysical domain boundaries with the larger gravity and magnetic anomalies.

THE EASTERN AUSTRALIAN CRUST: RELATIONS BETWEEN DENSITY, MAGNETISATION, SEISMIC VELOCITY AND HEAT FLOW

A major, rather obvious, sub-division of the eastern Australian crust is into areas with low magnetisation and high magnetisation [Wellman, 1995]. The amount of magnetisation is apparent in Figure 2 as the level of magnetic anomalies, and is also apparent as the amplitude of shorter wavelength anomalies. The boundaries between crust of high and low magnetisation correspond with the domain boundaries, so the domains can be classified into high or low magnetisation types (Figures 5, 7).

Gravity anomalies over domain boundaries have a gravity low on the margin of domains with low magnetisation, and a gravity high over the margin of domains with high magnetisation (Figure 1). This is consistent with the domains of low magnetisation having low upper-crustal density and the domains of high-magnetisation having high upper-crustal density.

These two domain types have different crustal velocity profiles, as measured by P-wave seismic refraction studies [Collins, 1988; Goncharov et al., 1996] (Figure 7). In comparing these profiles, the sedimentary uppermost part of the crustal profiles is not considered further. Figure 8 shows the difference in crustal profile for Phanerozoic crust. In domains of high magnetisation (Figure 8A) the P-wave velocity is above 6.1 km s^{-1} in the upper crust, the velocity increases approximately linearly with depth, the crust is more than 40 km thick, and there is no significant velocity change at the Moho. In domains of low magnetisation (Figure 8B) the P-wave velocity is below 6.1 km s^{-1} in the upper crust, crustal velocity increases slowly in the upper crust and rapidly in the lower crust, the crust is less than 40 km thick, and the Moho is relatively well defined with a large change in velocity of 1-1.5 km s^{-1}. Figure 7 shows, in map form, that the domains with relatively high magnetisation have an average upper-crustal velocity of over 6.1 km s^{-1}, and domains with a relatively low magnetisation have an average upper-crustal velocity below 6.1 km s^{-1}. Seismic velocity correlates positively with rock

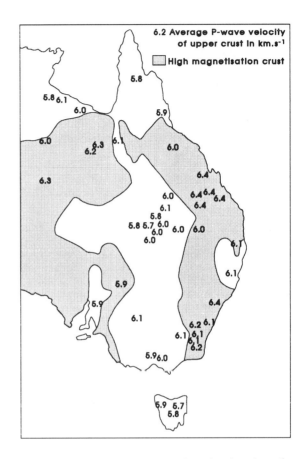

Figure 7. Eastern Australia, showing the domains of more and less magnetic crust, and the average upper-crustal P-wave seismic velocity. Domains of magnetic upper crust have a high average upper crustal P-wave seismic velocity (after Wellman, 1995).

density, so these seismic refraction results are consistent with the domains of low magnetisation having low upper-crustal density and the domains of high magnetisation having high upper-crustal density.

In eastern Australia, heat flow and crustal temperatures are measured by both widely spaced traditional measurements of temperature and thermal conductivity [Cull, 1982], and by very numerous bottom-hole temperatures of bore-holes in sedimentary basins [Cull and Conley, 1983; Somerville et al., 1994]. The area of high heat flow correlates with the area of sedimentary basins, but the presence of the basins is not thought to have affected the heat flow measurements, because the area of high heat flow is extensive and does not correlate with the localised areas of water discharge [Polak and Horsfall, 1979; Middleton, 1979; Cull and Conley, 1983; Pitt, 1986]. Figure 9 shows that the area of high-magnetisation crust correlates roughly with the area of low heat flow (less than 70 mW.m^{-3}, and temperatures at 5 km depth less than 175°), and the area of low-magnetisation crust correlates roughly with the area of high heat flow

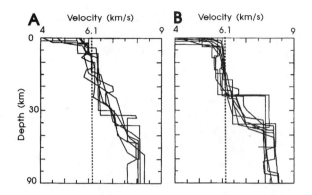

Figure 8. Seismic P-wave velocity of Phanerozoic crust of eastern Australia (after Wellman, 1995). For domains of high magnetisation (A) the upper crustal velocity is generally more than 6.1 km s^{-1}, velocity increases with depth approximately linearly, and the Moho is not a discrete velocity increase. For domains of low magnetisation (B), the upper crustal velocity is generally less than 6.1 km s^{-1}, and the Moho is a discrete velocity increase.

(greater than 70 mW m^{-3}, and temperatures at 5 km depth greater than 175°).

It is shown above that areas of low magnetisation correlate with low upper-crustal velocity and thin crust; hence in eastern Australia heat flow correlates inversely with the P-wave seismic velocity of the upper crust, and inversely with crustal thickness. This relationship for eastern Australia is consistent with results of studies of seismic velocities, heat production, and heat flow, in Europe and the former USSR. Laboratory studies of basement rock samples of Europe show that seismic velocity correlates inversely with heat production [Cermak et al., 1990]. As heat production correlates with heat flow, these data suggests that upper-crustal seismic velocity correlates inversely with heat flow. Field measurements in Europe [Cermak et al., 1991], and the former USSR [Provodnikov, 1975] show an inverse correlation of heat flow and crustal thickness (Figure 10). Crustal thickness correlates positively with upper-crustal mean density and velocity; hence these field data suggest that the upper-crustal seismic velocity correlates inversely with heat flow.

In eastern Australia the mean heat flow and mean crustal thickness are estimated to be 75 mW m^{-3} and 37 km in the low-magnetisation province, and 65 mW m^{-3} and 45 km for the high-magnetisation province (Figures 8, 9). These values imply a negative correlation between heat flow and crustal thickness, parallel to that found for European and former USSR data, but offset from it by about 10 mW m^{-3} (Figure 10). The discrepancy may be partly due to: 1) a higher mantle component of heat flow; and partly to 2) the Australian seismic results generally being determined later, with better control on the velocity-depth relationship, and deeper Moho.

COMPARISON WITH OTHER CRUSTAL MODELS

From the above discussion, we conclude that the main variation of crustal properties is due to variation in crustal types at any one time. These conclusions are inconsistent with the results of some other workers.

Drummond and Collins [1986] thought that the Proterozoic crust was thicker than Phanerozoic crust as a result of magmatic underplating at or near the base of the crust accumulating with time. The present study shows that the variation of crustal thickness and type that formed during one eon is as great, or greater than, that between eons.

Cull [1991] attributed the long-wavelength variation of magnetic anomaly over Australia to variation in heat flow, causing variations in depth to the Curie point, and hence in bulk magnetisation of crust. He dismissed the cause of variation in magnetisation as being due to variation in

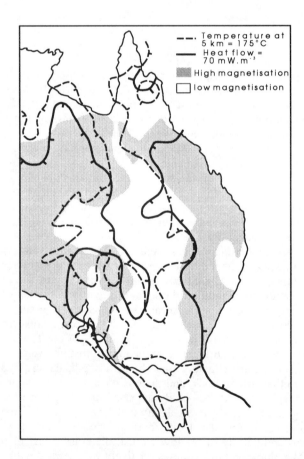

Figure 9. Relation between crustal magnetisation and heat flow. Domains of high magnetisation have low heat flow. Heat flow is indicated by contours derived from conventional measurements [Cull and Conley, 1983], and extrapolated temperature at 5 km derived from bottom hole temperatures of drill holes [Somerville et al., 1994].

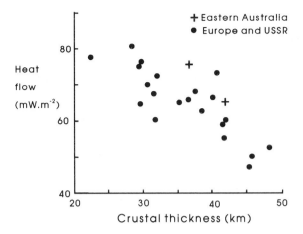

Figure 10. Inverse correlation between crustal thickness and heat flow. Data for Europe from Cermak et al. [1991], for USSR from Pavlenkova [1996], and for Australia from Figures 8 and 9.

magnetite content of the upper crust. However in this paper the magnetisation changes are thought to be mainly concentrated in the upper crust, because 1) magnetisation shows abrupt (not gradual) changes in level over domain boundaries; and 2) the short-wavelength anomalies, that must be upper-crustal, show smaller amplitudes over crust with low magnetic level, consistent with this crust having a low range of magnetisations. It is the conclusion of this study that heat flow correlates inversely with average crustal magnetisation, but does not control it.

SUMMARY

Using gravity and magnetic anomalies the Australian crust can be subdivided into numerous 'geophysical' domains within which there is a similar crustal magnetisation, crustal density, anomaly trends, and texture. These domains are interpreted to be areas of crust with similar formation history, that were cratonised during one period. Where basement is exposed, these domains correlate with geologically defined major tectonic units with similar geological history.

The boundaries between these domains have prominent gravity and magnetic anomalies due to both the difference in density and magnetisation across the boundary, and to geological processes near the boundary. These processes appear to have modified the margin of the older domain at the time of carbonisation of the younger domain, and along some margins have thrust up material to form the margin of the younger domain.

In eastern Australia the domains are of two types:

a. Domains where the upper crust has relatively low magnetisation, low density, low P-wave velocity, and relatively high heat flow. These domains have a basement mainly of sediments and granites. They are thought to have formed within the continent by crustal extension, or by addition to the continent of an accretionary wedge.

b. Domains where the upper crust has high magnetisation, high density, high seismic P-wave velocity, and relatively low heat flow. These domains have a considerable volcanic component. They are thought to have been accreted crustal fragments, or to be volcanic rifts.

Acknowledgments. I am grateful for discussion and help from John Creasy, Tim Mackey, Peter Milligan, and Russell Shaw. This paper is published by permission of the Executive Director, Australian Geological Survey Organisation.

REFERENCES

Anfiloff, W., B. C. Barlow, A. S. Murray, D. Denham, and R. Sandford, Compilation and production of the 1976 1:5,000,000 gravity map of Australia, *BMR J. Aust. Geol. Geophys.*, *1*, 273-276, 1976.

Beeson, J., C., P. Delor, and L. B. Harris, A structural and metamorphic traverse across the Albany Mobile Belt, Western Australia, *Precambrian Res.*, *40/41*, 117-136, 1988.

BMR, Gravity map of Australia, scale 1:5,000,000, Bureau of Mineral Resources, Canberra, Australia, 1976.

Cermak, V., L. Bodri, L. Rybach, and G. Guntebarth, Relationship between seismic velocity and heat production: comparison of two sets of data and test of validity, *Earth Planet. Sci. Lett.*, *99*, 48-57, 1990.

Cermak, V., M. Kral, M. Kresl, J. Kubik, and J. Safanda, Heat flow, regional geophysics and lithosphere structure in Czechoslovakia and adjacent parts of central Europe, in *Terrestrial Heat Flow and Lithospheric Structure,* edited by V. Cermak and L. Rybach, pp. 133-165, Springer-Verlag, Berlin, Germany, 1991.

Collins, C. D. N., Seismic velocities in the crust and upper mantle in Australia, *Bur. Miner. Resour. Aust. Rep.*, *277*, 1988.

Cull, J. P., An appraisal of Australian heat flow data, *BMR J. Aust. Geol. Geophys.*, 7, 11-21, 1982.

Cull, J. P., Heat flow and regional geophysics in Australia, in *Terrestrial Heat Flow and the Lithosphere Structure,* edited by V. Cermak and L. Rybach, pp. 486-500, Springer-Verlag, Berlin, Germany,1991.

Cull, J. P., and D. Conley, Geothermal gradients and heat flow in Australian sedimentary basins, *BMR J. Aust. Geol. Geophys.*, 8, 329-337, 1983.

Dentith, M. C., V. F. Dent, and B. J. Drummond, Crustal structure in the southwest Yilgarn Craton, Western Australia – an explanation for the Southwest Seismic Zone? *Geophysics Down Under, Newsletter of the Specialist Group on Solid Earth Geophysics,* Geological Society of Australia, Sydney, 23, 4-7, 1996.

Drummond, B. J., and C. D. N. Collins, Seismic evidence for underplating of the lower continental crust of Australia, *Earth Planet. Sci. Letters*, *79*, 361-372, 1986.

Finlayson, D. M., First arrival data from the Carpentaria Regional Upper Mantle Project (CRUMP), *J. Geol. Soc. Aust.* *15*, 33-50, 1968.

Finlayson, D. M., and C. D. N. Collins, Lithospheric velocity structures under the southern New England Orogen: evidence for underplating at the Tasman Sea margin., *Aust. J. Earth Sci., 40*, 141-153, 1993.

Gibb, R. A., and M. D. Thomas, Gravity signature of fossil plate boundaries in the Canadian Shield, *Nature,* 262, 199-200, 1976.

Goncharov, A. G., C. D. N. Collins, B. R. Goleby, and B. J. Drummond, Anomalous structure of the Mount Isa Inlier crust and crust-mantle transition zone in the global context, from refraction/wide-angle seismic experiments, *Geol. Soc. Aust. Abstr., 41*, 159, 1996.

GSA, Tectonic map of Australia and Papua New Guinea, scale 1:5,000,000, Geological Society of Australia, Sydney, Australia, 1971.

Hone, I. G., P. R. Milligan, J. Mitchell, and K. R. Horsfall, Australian continent-wide airborne geophysical databases, *AGSO J. Aust. Geol. Geophys., 17(2)*, 11-21, 1997.

Karner, G. D., and A. B. Watts, Gravity anomalies and flexure of the lithosphere at mountain ranges, *J. Geophys. Res., 88*, 10449-10477, 1983.

Klasner, J. S., and E. R. King, Precambrian basement geology of North and South Dakota, *Can. J. Earth Sci., 23*, 1083-1102, 1986.

McCulloch, M. T., Sm-Nd isotopic constraints on the evolution of the Precambrian crust in the Australian continent, *Am. Geophys. Union Geodynamic Series, 17*, 115-130, 1987.

Middleton, M. F., Heat flow in the Moomba, Big Lake and Toolachee Gas Fields of the Cooper Basin, and implications for hydrocarbon maturation, *Bull. Aust. Soc. Explor. Geophys., 10*, 149-155, 1979.

Morse, M. P., A. S. Murray, and J. W. Williams, Bouguer gravity anomalies, 1:1,000,000 scale maps, 50 sheets, Bureau of Mineral Resources, Canberra, Australia, 1991.

Morse, M. P., A. S. Murray, and J. W. Williams, Gravity anomaly map of Australia, scale 1:5,000,000 , pixel map series *21/A/52, 21/A/52-1*, Bureau of Mineral Resources, Canberra, Australia, 1992.

Murray, A. S., M. P. Morse, P. R. Milligan, T. Mackey, Gravity anomaly map of the Australian region, (Second Edition), scale 1:5,000,000, Australian Geological Survey Organisation, Canberra, in press, 1997.

Myers, J. S., R. D. Shaw, and I. M. Tyler, Tectonic evolution of Proterozoic Australia, *Tectonics, 15*, 1431-1446, 1996.

Page, R. W., M. T. McCulloch, and L. P. Black, Isotopic record of major Precambrian events in Australia, *Proc. 27th Int. Geol. Congr., 5*, 25-72, 1984.

Pavlenkova, N. I., Crust and upper mantle structure in northern Eurasia from seismic data, *Advances in Geophysics, 37*, 1-333, 1996.

Pilkington, M., Lithospheric flexure and gravity anomalies at Proterozoic plate boundaries in the Canadian Shield, *Tectonophysics, 176*, 277-290, 1990.

Pitt, G. M., Geothermal gradients, geothermal histories and the timing of thermal maturation in the Eromanga-Cooper Basins, in *Contributions to the Geology and Hydrocarbon Potential of the Eromanga Basin*, edited by D. I. Gravestock, P. S. Moore, and G.M. Pitt, pp. 323-351, Special Publication 12, Geological Society of Australia, 1986.

Polak, E. J., and C. L. Horsfall, Geothermal gradients in the Great Artesian Basin, Australia, *Aust. Soc. Explor. Geophys.Bull., 10*, 144-148, 1979.

Provodnikov, L. Y., The basement of platformean regions of Siberia, *Acad. Sci. USSR, Siberian Branch Trans., 194*, 1975.

Shaw, R. D., P. Wellman, P. Gunn, A. Whitaker, and W. D. Palfreyman, Australian crustal elements, based on distribution of geophysical domains, 1:5 000 000 scale map, Australian Geological Survey Organisation, Canberra, 1995a.

Shaw, R. D., P. Wellman, P. Gunn, A. J. Whitaker, C. Z. Tarlowski, and M. P. Morse, Australian crustal elements map; a geophysical model for the tectonic framework of the continent, *AGSO Res. Newsletter, 23*, 1-3, 1995b.

Shaw, R. D., P. Wellman, P. Gunn, A. J. Whitaker, C. Z. Tarlowski, and M. P. Morse, Guide to using the Australian crustal elements map, *Aust. Geol. Surv. Org. Rec., 1996/30*, 1996.

Somerville, M., D. Wyborn, P. Chopra, S. Rahman, D. Estrella, and T. van der Meulen, Hot dry rocks feasibility study, *Rep. ERDC 94/243*, Energy Research and Development Corporation, Canberra, Australia, 1994.

Symonds, P. A., C. D. N. Collins, and J. Bradshaw, Deep structure of the Browse Basin: implications for basin development and petroleum exploration, in *The Sedimentary Basins of Western Australia, Proc. Petrol. Explor. Soc. Aust. Symp., Perth,* edited by P. C. Purcell and R. R. Purcell, pp. 315-331, 1994.

Tarlowski, C. Z., P. R. Milligan, and T. Mackey, Magnetic anomaly map of Australia, scale 1:5,000,000 , Aust. Geol. Surv. Org., Canberra, Australia, 1996.

Thomas, M. P., and R. A. Gibb, Gravity anomalies and deep structure of the Cape Smith Fold Belt, northern Ungava, Quebec, *Geology, 5,* 169-172, 1977.

Wang, L. J., F. E. M. Lilley, and F. H. Chamalaun, Large scale electrical conductivity structure of Australia from magnetometer arrays, *Explor. Geophys., 28*, 150-155, 1997.

Wellman, P., Gravity trends and the growth of Australia: a tentative correlation. *J. Geol. Soc. Aust., 23*, 11-14, 1976.

Wellman, P., Gravity evidence for abrupt changes in mean crustal density at the junction of Australian crustal blocks, *BMR J. Aust. Geol. Geophys., 3*, 153-162, 1978.

Wellman, P., Development of the Australian Proterozoic crust as inferred from gravity and magnetic anomalies, *Precamb. Res., 40/41*, 89-100, 1988.

Wellman, P., Structure of the Mount Isa region inferred from gravity and magnetic anomalies, in *Detailed Studies of the Mount Isa Inlier*, edited by A. J. Stewart and D. H. Blake, pp. 15-27, *Aust. Geol. Surv. Org. Bull. 243*, 1992a.

Wellman, P., A geological interpretation of the regional gravity and magnetic features of north Queensland, Explor. Geophys., 23, 423-428, 1992b.

Wellman, P., Tasman Orogenic System: a model for its subdivision and growth history based on gravity and magnetic anomalies, *Econ. Geol., 90*, 1430-1442, 1995.

Wellman, P., and A. S. Murray, Free air gravity anomalies, Australia, scale 1:10,000,00, in *BMR Earth Science Atlas*, Bureau of Mineral Resources, Canberra, Australia, 1979a.

Wellman, P., and A. S. Murray, Bouguer gravity anomalies,

Australia, scale 1:10,000,000, in *BMR Earth Science Atlas*, Bureau of Mineral Resources, Canberra, Australia, 1979b.

Whitaker, A. J., 1994, Integrated geological and geophysical mapping of southwestern Western Australia, *AGSO J. Aust. Geol. Geophys.*, *15*, 313-328, 1994.

Wright, C., B. R. Goleby, C. D. N. Collins, R. J. Korsch, T. Barton, S. A. Greenhalgh, and S. Sugiharto, Deep seismic profiling in central Australia, *Tectonophysics*, *173*, 247-256, 1990.

Peter Wellman, Australian Geological Survey Organisation, P.O. Box 378, Canberra ACT 2601 Australia. (Phone: 616 2499653; Fax: 06 2499983; email: pwellman@agso.gov.au)

Complex Anisotropy in the Australian Lithosphere from Shear-wave Splitting in Broad-band SKS Records

Geoff Clitheroe

Research School of Earth Sciences, Australian National University, Canberra, ACT, Australia

Rob van der Hilst

Research School of Earth Sciences, Australian National University, Canberra, ACT, Australia, and Department of Earth, Atmospheric, and Planetary Sciences, Massachusetts Institute of Technology, Cambridge MA, U.S.A.

Shear wave splitting of seismic core phases (such as SKS and SKKS) reveals complex azimuthal anisotropy in continental Australia. Using broad-band seismograms recorded at stations deployed for the SKIPPY project we are able to supplement the previously spatially sparse shear-wave splitting measurements from permanent stations. This enables us to study Australian continental anisotropy on an unprecedented scale and to investigate the variation of shear-wave splitting across the continent. Using the broad band width of the seismograms, we demonstrate that differing SKS splitting phenomena are manifested at different frequencies. At a frequency of about 1 Hz, the time difference between fast and slow SKS waves is 0.3-0.6 s, and the polarization directions parallel pre-existing crustal fabric rather than present-day compressional stress axes. Over a broader frequency band, the polarization directions are inconsistent with present-day plate motion direction but can possibly be explained by predominant mineral orientation in the sub-crustal lithosphere.

INTRODUCTION

Shear waves passing through anisotropic material (characterised by direction-dependent elastic properties) split into orthogonally polarised waves with different wave speeds [Crampin, 1985]. The splitting, or birefringence, of core phases such as SKS is now almost routinely used to investigate seismic anisotropy in the Earth's mantle and its relationship to tectonic deformation [Silver and Chan, 1991; Vinnik et al., 1992].

After phase conversion at the core-mantle boundary (CMB) the shear waves propagate nearly vertically to the Earth's surface, forming isolated arrivals beyond 80_, making them ideal for the study of tectonically stable regions. In an isotropic, spherically symmetric Earth, SKS would be observed only on the radial component of the shear-wave field. Significant transverse energy, combined with non-linear particle motion, indicates shear-wave splitting due to anisotropy. Owing to the nearly vertical path to the receiver, any anisotropy inferred from shear-wave splitting is well resolved laterally, but there is little constraint on its depth. The potential source zone is approximately 3000 km, from the CMB to the surface of the Earth, although it is now generally accepted that the strain-induced Lattice Preferred Orientation (LPO) of anisotropic minerals in the upper mantle is the main source region [Mainprice and Silver, 1993]. For an extensive

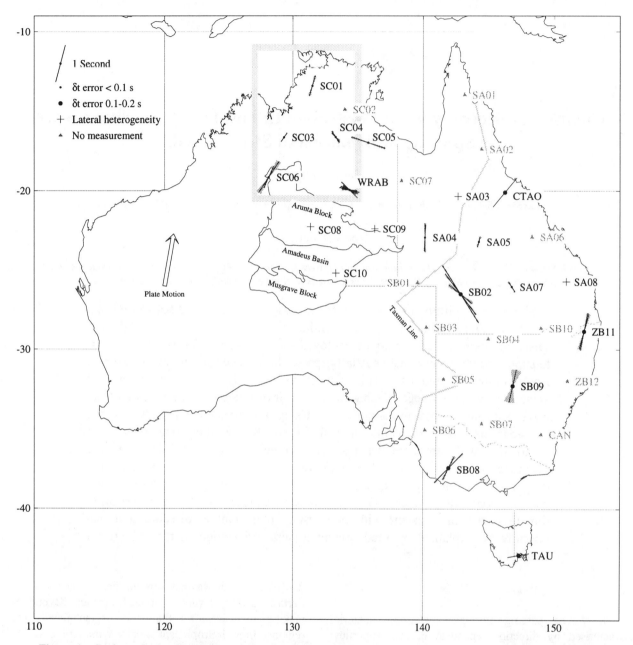

Figure 1. Study region and station locations. Also shown is the present-day motion of the Australian continent relative to Eurasia (DeMets et al, 1990), open arrow, and the outlines of geological features discussed in the text. The measurements at CTAO and TAU are after Vinnik et al, (1992).

review of investigations of anisotropy based on shear wave splitting we refer to Silver [1996].

We investigated azimuthal anisotropy for the central and eastern part of the Australian continent using SKS, SKKS and pPKS phases recorded at permanent stations and at arrays of portable broad-band seismometers of the SKIPPY project [Van der Hilst et al., 1994] (Figure 1). The SKIPPY project was designed to exploit regional seismicity [Van der Hilst et al., 1994; Zielhuis and Van der Hilst, 1996] but also provides data for the construction of an anisotropy map of the Australian continent. The data redundancy to constrain the measurements is limited as a result of the relatively short operation period of each array, but we believe that the spatial coherency of polarization directions provides significant new information on the origin of splitting. In addition, shear-wave splitting manifests itself at different frequencies, depending on site location and seismic phase used. Splitting observed at relatively high frequencies suggests that shear-wave data also carry structural signals about anisotropy in the crust. Our results are consistent with inferences from the splitting of high-frequency shear-waves excited by (sub-) crustal micro-earthquakes [Kaneshima, 1990], or by phase conversion at the crust-mantle interface [e.g., Herquel et al., 1995] and quantitatively agree with results of petrophysical modelling [Barruol and Mainprice, 1993].

METHOD AND DATA

Shear-wave splitting is usually quantified by the fast polarization direction (ϕ) and the time delay (δt) of the slow wave. Following Silver and Chan [1991], we searched for the pair ($\phi, \delta t$) that most successfully corrects for the effects of anisotropy by minimising (i) the energy, E_t, on the transverse component and (ii) the smaller eigen value, λ_2, of the two-dimensional covariance matrix of corrected horizontal particle motion, which is equivalent to searching for the most linear particle motion. We used (i) to obtain the measurements and (ii) to assess the influence of lateral heterogeneity, if any. For each measurement confidence bounds (σ_ϕ and σ_{dt}, the 1σ uncertainties), defined by ($\phi, \delta t$), are determined from the F probability distribution where E_t/E_{tmin} is less than a critical value. Three additional diagnostics were used to assess the ability of the method to correct for anisotropy (Figure 2, D-G): (i) The radial and transverse components are corrected using the splitting parameters ($\phi, \delta t$); this provides a visual measure of the ability of the method to minimise E_t, the energy on the transverse component (Figure 2, D). (ii) A contour plot of $E_t(\phi, \delta t)$ in multiples of the 95% confidence interval; this allows a check for multiple minima and a check of the uncertainty in each measurement (Figure 2, E). (iii) A check of the particle motion for the corrected fast and slow components (Figure 2, F), this allows an assessment of the degree to which the correction produces linear particle motion (Figure 2, G).

We searched for earthquakes at distances beyond 85_ and at depths larger than 80 km with isolated impulsive SKS or SKKS phases well recorded at permanent observatories and SKIPPY stations. In the case of SKKS it is important to ensure that it is well isolated from SK_3S, otherwise interference between these phases at the CMB may violate the assumption of radial polarisation. For one event, we were able to use pPKS phases. We measured ($\phi, \delta t$) and their uncertainties for all available SKS, SKKS and pPKS phases with a significant transverse signal. Unfortunately, the short deployment period of each array,

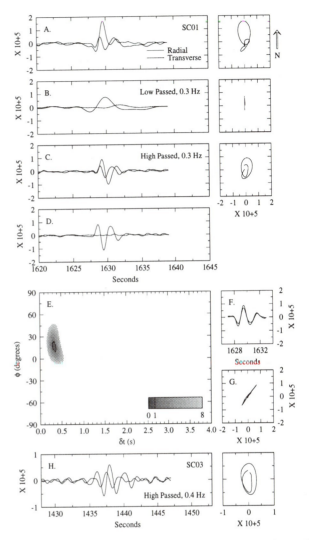

Figure 2 (A-G). SKKS splitting at different frequencies and measurement diagnostics for event 94 231, back azimuth = 160.47°, recorded at SC01.

and the need for deep, large, distant events, limits the number of suitable shear-wave phases recorded at the SKIPPY stations. In total we used 23 shear-wave phases (11 SKS, 7 SKKS, 5 pPKS) from 10 earthquakes recorded on 14 of the 30 SKIPPY stations. We used data from two events with focal depth less than 80 km (Table 1).

We could not determine the splitting parameters unambiguously for all stations. For five SKIPPY stations (SA03, SA08, SC08, SC09, SC10) and for CTAO (Charters Towers, Queensland), ϕ appeared to depend on back azimuth, and the particle motions were linear instead of elliptical. This indicates that the signal on the transverse component was not due to anisotropy but to structural heterogeneity. Of these, SC08, SC09, and SC10 were located near the Arunta and Musgrave Blocks (Figure 1) where

Table 1. List of Events Used in Study.

Date	Lat. (°N)	Long. (°W)	Depth (km)	m_b
93 221	36.436	70.711	204	5.8
93 221	36.379	70.868	215	6.2
93 247	36.429	70.812	195	5.9
93 323	54.290	-164.16	30	6.1
94 010	-13.339	-69.446	596	6.4
94 073	15.994	-92.428	165	5.8
94 119	-28.326	-63.221	571	6.3
94 130	-28.551	-63.070	603	6.4
94 143	18.174	-100.547	59	6.0
94 231	-26.653	-63.378	564	6.0

thrust faults probably dip through the entire crust [Lambeck and Burgess, 1992], and SA03 and SA08 were located in regions of complex surface faulting. For event 94 010 (see Table 1) at SB03, no significant transverse component SKS (or SKKS or pPKS) was observed even though the phase was well recorded on the radial components. This may indicate either the absence of significant anisotropy, or the alignment of event back azimuth with the fast or slow polarization direction. We found no convincing evidence for splitting at NWAO (Narrogin, Western Australia) and the splitting measurements at CAN (Canberra, ACT) are ambiguous. Reported values for CTAO and TAU are not well constrained [Vinnik et al., 1992]. The following SKIPPY stations did not record useful SKS, SKKS, or pPKS phases; SA01, SA02, SA06, SB01, SB06, SB07, SB10, SC02, SC07. At 12 SKIPPY stations and WRAB splitting measurements were made.

Shear-wave splitting is often determined from long-period or broad-band seismograms, with the latter low-pass filtered to suppress noise [Vinnik et al., 1992; Gledhill and Gubbins, 1996]. In the present case such filters were often found to remove the transverse shear-wave energy that was perceptible in the original records. In some cases, it was observed that application of filters that pass energy in different frequency bands revealed shear-wave splitting only at frequencies higher than 0.3 Hz. For an isolated SKKS phase, with clearly observed transverse energy in the record at SKIPPY station SC01, we can observe the following (Figure 2, A). When this data is low-passed at 0.3 Hz (Figure 2, B), there is little energy apparent on the transverse component and particle motion is linear. High passing at 0.3 Hz shows clear transverse energy (Figure 2, C), which has the same form as the unfiltered data. Resultant particle motion is clearly elliptical. Splitting measurements made (by minimising E_t) on the SKKS phase, high-passed at 0.3 Hz, yielded the following splitting parameters: $\phi = 17\_$ ($\sigma_\phi = 2$), $\delta t = 0.32$ s ($\sigma_{\delta t} = 0.02$). Using these parameters to correct the radial and transverse components we are able to asses their ability to remove the effects of shear-wave splitting. Using ϕ to rotate to a fast-slow reference frame there is little energy apparent on the corrected transverse component (Figure 2, D). The contour plot of E_t (Figure 2, E) shows a single, well constrained, localised minimum. Using δt to time shift the radial and transverse components (Figure 2, F) yields excellent phase coherence with resultant linear particle motion (Figure 2, G). This indicates that the method is able to correct well for the effects of anisotropy. At high frequencies, scattering may result in transverse energy. Consistency between measurements obtained by minimising E_t and λ_2 confirms that we are observing the effects of anisotropy. This can also be seen visually in the particle motion for the high-passed SKKS phase (Figure 2-C). It is elliptical, moves off in the fast direction, and returns, orthogonally to it, in the slow direction. Shear-wave splitting only at higher frequencies (higher than 0.3 Hz) was observed at SC01, SC03, SC04, SC06 and WRAB. In all cases the data were low-pass filtered at 1.5 Hz to suppress noise.

RESULTS AND DISCUSSION

The measurements along with their uncertainties are summarised in Figure 1. Each ($\phi, \delta t$) is plotted as a line; ϕ is the angle of the line and the length of the line is proportional to δt. Error estimates for each measurement are shown by: i) for σ_ϕ a shaded diabolo (e.g Figure 1. station SB09); in many cases this diabolo is too small to be visible; ii) for $\sigma_{\delta t}$ the size of the station point represents error; $\sigma_{\delta t}$ less than 0.1 s is shown with a small circle, $\sigma_{\delta t}$ 0.1-0.2 s with a large circle. Some measurements overlap. A cross denotes sites where lateral heterogeneity obscures any splitting due to anisotropy. Splitting was observed only at higher frequencies at SC01, SC03, SC06, SC04, and WRAB. These stations and the surrounding region within the grey rectangle are shown in detail in Figure 3. All other measurements (stations mainly in eastern Australia) are made over wider frequency ranges including frequencies below 0.3 Hz (low-passed at 1.5 Hz).

The directions of fast polarization inferred from the high-frequency records at WRAB show a remarkable alignment with macro-scale crustal fabric, in particular with pervasive faulting in the Proterozoic basement, and with the fault planes of the 1988 Tennant Creek earthquakes [Bowman, 1992] (Figure 3, inset). Also for the SKIPPY stations there is a strong correlation between the orientation of crustal features and the ϕ determined at high frequency. For instance, the polarization direction at SC01, SC03, and SC06 coincides with the NNE strike of major shear zones that separate the Kimberley block from the central Australian craton (Figure 3). These observations strongly suggest a relationship between splitting of shear waves with wave lengths λ of the order of 5 to 10 km and crustal structure.

The small magnitude of the delay times, a few tenths of a second, can indeed be explained by crustal anisotropy alone [Barruol and Mainprice, 1993; Alsina and Snieder, 1995], although it does not rule out a larger source of weak anisotropy. Petrophysical analysis demonstrated that several types of crustal rock may produce some amount of anisotropy [Barruol and Mainprice, 1993], and that geological structures, in particular steeply dipping foliation in felsic rocks and amphibolites, can produce significant shear-wave splitting (0.1-0.2 s delay per 10 km). These numbers are in quantitative agreement with our observations: a crustal thickness beneath WRAB of 50 km [Collins, 1991] and a delay $\delta t \approx 0.5$ s render a delay time of 0.1 s per 10 km of crustal rock, equivalent to ~3.5 % anisotropy, which is also consistent with magnitude of crustal anisotropy inferred for other regions [Kaneshima, 1990; Herquel et al., 1995; Gledhill and Stuart, 1996]. We argue that the anisotropy signal is due to crustal fabric caused by folding and faulting/shearing. At low effective confining pressures or at high pore pressures, such as in the upper crust, anisotropic fabric can also be caused by microcracks aligned in the direction of compressional tectonic stress [Crampin, 1985]. The inferred north-south orientation of compressional stress in this region of the Australian continent [Coblentz, 1995] does, however, not support this explanation for our observations.

Tong et al. [1994], using high-frequency SV and SH waves recorded at the Waramanga array after horizontal propagation through the lithosphere, argued that scattering due to heterogeneity may have masked any splitting due to lithospheric anisotropy. However, using waves refracted from the upper-mantle transition zone, they infer 1.4% transverse anisotropy in the asthenosphere. This is not necessarily inconsistent with our inferences, since shear waves do not carry information about this class of transverse isotropy.

In accord with previous observations of shear-wave splitting [Silver and Chan, 1991; Vinnik et al., 1992; Silver, 1996], the splitting of broad-band shear waves ($\delta t \approx$ 1.0-2.0 s) is significantly more pronounced than the signal that we attribute to crustal anisotropy. We remark that waves at frequencies of about 50 mHz, and $\lambda \approx 90$ km, are less sensitive to 10-km scale crustal fabric. Concurring with other investigators we therefore invoke a larger, deeper anisotropy source to explain the large time delays at the remaining stations. The alignment of ϕ inferred for eastern Australia (Figure 1) could be explained by either

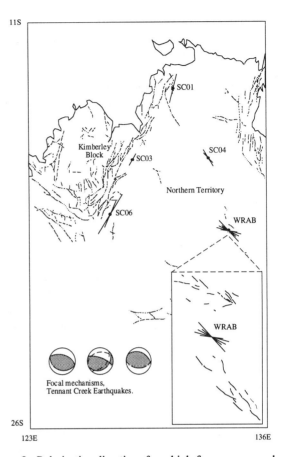

Figure 3. Polarization directions from high-frequency records of WRAB, SC01, SC03, SC04, and SC06 in relation to major crustal features (faults and thrusts). The inset shows detailed structural trends in the vicinity of WRAB and the focal mechanisms of the large Tennant Creek earthquakes of January 1988 (Bowman, 1992).

'fossil' anisotropy [Silver and Chan, 1991] or by asthenospheric flow [Vinnik, 1992]. Fossil anisotropy is supported by the tentative observation that polarization directions in eastern Australia exhibit a curvilinear trend somewhat similar to that of the "Tasman Line", the presumed, but poorly constrained [Zielhuis and Van der Hilst, 1996], eastern boundary of the Proterozoic cratons (Figure 1). Shear-wave splitting possibly indicates anisotropy frozen into the lithospheric mantle during Paleozoic terrane accretion that resulted in, from west to east, the Adelaide, Lachlan, and New England Fold Belts [Collins and Vernon, 1994]. Alternatively, the curvilinear trend of ϕ could be explained by alignment of olivine crystals by asthenospheric flow around the irregularly shaped eastern edge of the deep continental root. A similar model has been proposed to explain the spatial variation in splitting direction observed for the North American continent (Fischer, personal communication, 1996).

CONCLUSIONS

From the broad-band data we infer that shear wave splitting can be manifest at different frequencies. Higher frequency waves (λ Å 5-10 km) are polarised by anisotropic crustal fabric. In the examples given, crustal anisotropy is most likely due to structural features such as pervasive foliation, and our observations are in accord with petrophysical data. In Eastern Australia, the splitting of broad-band shear waves may be due to fossil anisotropy frozen into the lithosphere, or anisotropy induced by flow in the asthenosphere.

At this stage of data processing, we cannot yet state whether splitting occurs exclusively at either high or low frequency, or whether the splitting parameters are a more continuous function of frequency. The former could point to the existence of two anisotropic layers, which can be investigated using the two-layer approach of Silver and Savage [1994]. The latter would point to a more complex source of anisotropy, which calls for rather different analysis tools that exploit the broad-band nature of splitting, and perhaps provide a stochastic characterisation of the anisotropic medium [Gaherty et al., 1996]. We aim to investigate the effect of depth- and scale-dependent anisotropy on shear-wave splitting with wave-form modeling. Upon completion of the remaining three deployments of SKIPPY, we may also be able to infer the typical length scale of LPO and thus constrain stochastic anisotropy models proposed for the Australian lithosphere [Gaherty and Jordan, 1995].

REFERENCES

Alsina, D., and R. Snieder, Small-scale sublithospheric continental mantle deformation: constraints from SKS splitting observations, *Geophys. J. Int.*, *123*, 431-448, 1995.

Barruol, G., and D. Mainprice, A quantitative evaluation of the contribution of crustal rocks to the shear-wave splitting of teleseismic SKS waves, *Phys. Earth Planet. Inter.*, *78*, 281-300, 1993.

Bowman, J. R., The 1988 Tennant Creek, Northern Territory, earthquakes: a synthesis, *Australian, J. Earth Sci.*, *39*, 651-669, 1992.

Coblentz, D. D., M. Sandiford, R. M. Richardson, S. Zhou, and R. Hillis, The origins of the intraplate stress field in continental Australia, *Earth Planet. Sci. Lett.*, *133*, 299-309, 1995.

Collins, C. D. N., The nature of the crust-mantle boundary under Australia from seismic evidence, in *The Australian Lithosphere*, edited by B, J. Drummond, *Geol. Soc. Australia.*, Special Publ.,*17*, 67 (1991)

Collins, W. J., and R. H. Vernon, A rift-drift-delamination model of continental evolution: Palaeozoic tectonic development of eastern Australia, *Tectonophysics*, *235*, 249-275, 1994.

Crampin, S., Evaluation of anisotropy by shear-wave splitting, *Geophysics*, *50*, 142-152, 1985.

DeMets, C., R. G. Gordon, D. F. Argus, and S. Stein, Current plate motions, *Geophys. J. Int.*, *101*, 425-478, 1990.

Gaherty, J. B., and T. H. Jordan, Lehman discontinuity as the base of an anisotropic layer beneath continents, *Science*, *268*, 1468-1471, 1995.

Gaherty, J. B., R. Saltzer, and T. H. Jordan, Vertical heterogeneity in upper-mantle anisotropy from splitting of vertically propagating shear waves (abstract), *EOS Trans.AGU*, *77*, F482- F483, 1996.

Gledhill, K., and D. Gubbins, SKS splitting and the seismic anisotropy of the mantle beneath the Hikurangi subduction zone, New Zealand, *Phys. Earth Planet. Inter.*, *95*, 227-236, 1996.

Gledhill, K., and G. Stuart, Seismic anisotropy in the fore-arc region of the Hikurangi subduction zone, New Zealand, *Phys. Earth. Planet. Inter.*, *95*, 211-255, 1996.

Herquel, G., G. Wittlinger, and J. Gilbert, Anisotropy and crustal thickness of Northern Tibet, New constraints for tectonic modelling, *Geophys. Res. Lett.*, *22*, 1925-1928, 1995.

Kaneshima, S., Origin of crustal anisotropy: shear wave splitting studies in Japan, *J. Geophys. Res.*, *95*, 11121-11133, 1990.

Lambeck, K., and G. Burgess, Deep crustal structure of the Musgrave Block, central Australia: Results from teleseismic travel-time anomalies, *Aust. J. Earth Sciences*, *39*, 1-19, 1992.

Mainprice, D., and P. G. Silver, Interpretation of SKS waves using samples from the subcontinental lithosphere, *Phys. Earth Planet. Inter.*, *78*, 257-280, 1993.

Silver, P. G., Seismic anisotropy beneath the continents; probing the depths of geology, *Ann. Rev. Earth Planet. Sci.*, *24*, 385-432, 1996.

Silver, P. G., and W. W. Chan, Shear wave splitting and subcontinental deformation, *J. Geophys. Res.*, *96*, 16429-16454, 1991.

Silver, P. G., and M. K. Savage, The interpretation of shear-wave splitting in the presence of two anisotropic layers, *Geophys. J. Int.*, *119*, 949-963, 1994.

Tanimoto, T., and D. L. Anderson, Mapping convection in the mantle, *Geophys. Res. Lett.*, *11*, 287-290, 1984.

Tong, C., O. Gudmundsson, and B. L. N. Kennett, Shear wave splitting in refracted waves returned from the upper mantle transition zone beneath northern Australia, *J. Geophys. Res.*, *99*, 15783-15797, 1994.

Van der Hilst, R., B. L. N. Kennett, D. Christie, and J. Grant, Project SKIPPY explores the lithosphere and mantle beneath Australia, *Eos Trans.*, Am. Geophys.Un., *75*, 177, 1994.

Vinnik, L. P., L. I. Makayeva, A. Milev, and A. Y. Usenko, Global patterns of azimuthal anisotropy and deformations in the continental mantle, *Geophys. J. Int.*, *111*, 433-447, 1992.

Zielhuis, A., and R. D. van der Hilst, Upper-mantle shear velocity beneath eastern Australia from inversion of waveforms from SKIPPY portable arrays, *Geophys. J. Int.*, *127*, 1-17, 1996.

G. Clitheroe, Research School of Earth Sciences, Australian National University, Canberra, ACT 0200, Australia.

R. van der Hilst, Department of Earth, Atmospheric, and Planetary Sciences, Massachusetts Institute of Technology, Cambridge MA 01239, U.S.A.

A Brief Review of Differences in Lithosphere Seismic Properties Under Western and Eastern Australia Stimulated by Seismograms from the Marryat Creek Earthquakes of 1986

Barry J. Drummond

Australian Geodynamics Cooperative Research Centre and Australian Geological Survey Organisation, Canberra, Australia

Previous studies highlighted differences in the speeds at which seismic waves travel through the uppermost mantle under the shield regions of central and western Australia and the Phanerozoic regions of eastern Australia. These differences are demonstrated by observations of an earthquake and one of its aftershocks that occurred at Marryat Creek in central Australia in 1986. Wave-form data are available from the seismological observatories at Charters Towers (CTAO), on Phanerozoic basement to the northeast of the earthquakes, and Narrogin (NWAO), on the Precambrian basement to the southwest. Calculated radiation patterns for published focal mechanisms suggest that, for a laterally homogeneous upper mantle, relatively greater amounts of shear-wave energy might be expected at Charters Towers than at Narrogin. However, the opposite is observed. The recorded wave forms therefore imply much poorer S-wave propagation to Charters Towers than to Narrogin. The surface-wave coda at Charters Towers is smaller in amplitude and shorter in duration than that recorded at Narrogin. Also, the travel-times of P waves to stations in eastern Australia are greater than those to shield stations. For reasonable upper mantle models, the body-wave data would be sensitive to the velocity structure between ~100 and ~200 km depth. The body wave forms and travel times are consistent with a low-velocity zone between ~120 km and ~190 km depth in published models for eastern Australia, for which V_p/V_s ratios imply a low shear modulus. Previously published models for the shield do not have a P-wave low-velocity zone at these depths, and none is required by the data from the Marryat Creek earthquake. Travel times of Marryat Creek earthquake P-wave energy within the shield are consistent with models based on data recorded mostly within the southern half of Australia, and comparisons with times from models based on data recorded mostly in northern Australia imply a north to south zonation of the upper mantle within the shield, as well as from east to west between non-shield and shield upper mantle. The zonation could be reflecting different mantle compositions. However, paths from Marryat Creek to Charters Towers bottom under areas of high heat flow, whereas those to Narrogin bottom under areas of low heat flow. Published geothermal gradients for Australia suggest higher upper-mantle temperatures under eastern Australia than under the shield. Thermal effects are therefore suspected as a major cause of the apparent differences in upper-mantle seismic-wave propagation, particularly between eastern Australia and the shield.

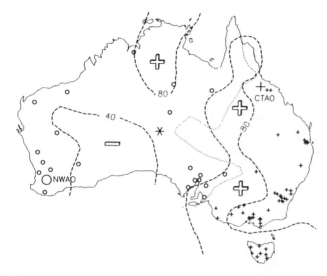

Figure 1. Seismograph stations which recorded the Marryat Creek earthquakes. The large circle (o) shows the location of the Narrogin station (NWAO); Charters Towers (CTAO) is shown with a large plus symbol (+). The location of the earthquakes is shown with an asterisk (*). Dashed lines are contours of heat flow from Cull [1982]; contour values are in mWm^{-2}. Large open plus and minus symbols highlight highs and lows in regional heat flow. The dotted line marks the Tasman Line - the interpreted boundary between lithosphere with Precambrian basement in the west and Phanerozoic basement in the east. This version of the Tasman Line is approximately that of Muirhead and Drummond [1991]. Seismograph stations with a path from the earthquake which is predominantly under the shield are shown with circles; stations with a path predominantly under Phanerozoic basement are shown with crosses.

INTRODUCTION

The Australian continent can be classified broadly into an older central and western part, where basement is Precambrian in age (the "shield"), and an eastern part, in which basement is Phanerozoic. The boundary between the shield and the eastern part is called the Tasman Line (Figure 1). Present day surface heat flow is highest just to the east of the Tasman Line, and in central northern Australia where it exceeds 80 mWm^{-2} in places (Figure 1), and lowest in the Archaean southwest part of Australia (~40 mWm^{-2}, Figure 1) [Cull, 1982].

Previous studies and reviews have concluded that the upper mantle under the shield has higher seismic velocities than that under eastern Australia, particularly in the sub-crustal lithosphere. The evidence comes from body-wave studies (e. g., see the review of early studies by Muirhead and Drummond, 1991; Bowman and Kennett, 1993; Dey et al., 1993; Kennett et al., 1994; Muirhead et al., 1977]; surface-wave studies [Goncz and Cleary, 1976; Zielhuis and van der Hilst, 1996; van der Hilst and Kennett, 1997]; and teleseismic travel-time residuals [Drummond et al., 1989].

Many of these studies, particularly those using body waves, were mostly or entirely within either the shield, particularly in northern Australia, or east of the Tasman Line. Differences between the shield and eastern Australia are therefore generally inferred from differences in the resulting models, rather than from differences evident in any single data set. The teleseismic travel time residual map of Drummond et al. [1989] represented a comparative study of the shield and eastern Australia, but the differences identified in the residuals were qualitative only, because no quantitative analysis of mantle structure ensued. The more recent SKIPPY experiment of Zielhuis and van der Hilst [1996] and van der Hilst and Kennett [1997] will provide the most comprehensive and comparative study of the continental lithosphere, particularly its shear-wave velocity structure, but results available at the time of writing [van der Hilst and Kennett, 1997] are robust only in the eastern two-thirds of Australia, and those from the western third are still preliminary.

This paper provides a simple demonstration of the differences in seismic properties in the upper mantle between the shield and eastern Australia. It compares recordings of an earthquake and one of its aftershocks at Marryat Creek in central Australia in 1986, at seismic stations both east and west of the Tasman Line. Although travel times from the main event were used in combination with those from other events in a previous study of upper mantle structure under the shield regions of Australia [Bowman and Kennett, 1993], the wave forms of the recordings have not been considered before. When taken together with the travel-time data, the wave forms provide a qualitative but clear insight into the scale of the differences in the upper mantle between the shield and eastern Australia. The degree of misfit of the travel time data from the main earthquake with published models for the shield then leads to a brief discussion on the likely further zonation of lithospheric structure within the shield. Likely causes of the regional variations in lithospheric structure are then discussed.

THE MARRYAT CREEK EARTHQUAKES

The Marryat Creek earthquakes occurred in the remote Proterozoic Musgrave Block in central Australia. Surface faulting associated with the main event formed an inverted 'L' shaped fracture, with the arms of the 'L' extending west and south from an apex in the northeast of the area of surface rupture. The southwest side of the fault was uplifted relative to the northeast side [McCue et al., 1987]. Times, locations and estimated focal depths of the events are given in Table 1. The location of the main event is plotted in Figure 1. McCue et al. [1987] and the International Seismological Centre (ISC) estimated the magnitude of the main event to be M_s~5.8 (Table 1).

Focal mechanisms were calculated by McCue et al. [1987] using P phases at regional observatories, and by

Table 1. Event Times, Locations and Depths for the March 1986 Marryat Creek Earthquake and its July aftershock.

Source	Time (h:m:s)	Lat (°S)	Long (°E)	Depth (km)	Mb	Ms
Main Event Date: March 30, 1986						
(a)	08:54	-26.22	132.82			~5.8
(b)	08:53:52.0	-26.30	132.77	5	5.7	5.8
(c)	08:53:53.20	-26.23	132.70	10	5.8	5.8
(d)	08:53:52.0	-26.21	132.82	0-3		
Aftershock Date July 11, 1986						
(a)	07:17:59.1	-26.10	132.77	10	5.5	5.1
(b)	07:17:58.49	-26.22	132.84	10	5.6	5.3

*Sources are:
(a) McCue et al. [1987];
(b) International Seismological Centre Bulletins;
(c) National Earthquake Information Centre (NEIC) NEIC earthquake data CD-ROMs. (Note: NEIC location from the NEIC earthquake data CD-ROM differs slightly from the NEIC value reported in the ISC Bulletins);
(d) values used in this study; see text for details.

Fredrich et al. [1988] using far-field body wave-form inversion (30°-90° for P and 30°-70° for SH). Fredrich et al. [1988] estimated a centroid value of the focal depth of between 0 and 3 km; the ISC solution has it at 5 km, and the aftershock deeper.

The location of the event is important for calculating epicentral distances used later in this paper. The ISC location falls outside the area of surface rupture (to the south); the NEIC location is to the west. Fredrich et al. [1988] attribute their preferred location to the ISC, but their latitude coordinate differs from that in the ISC Bulletin. The focal mechanism of Frederich et al. [1988] suggests thrusting from the southwest; this is consistent with dips on the scarps associated with surface rupture [McCue et al., 1987]. The epicentre coordinates used in this study (Table 1) were calculated by projecting the eastern and northern faults to depth, using the dips on the faults reported by McCue et al. [1987], and calculating the point where the intersection of the two fault planes reached the centroid of the focal depths estimated by Fredrich et al. [1988]. This provides epicentral coordinates within ~1 km of those of McCue et al. [1987]. The origin time of the ISC was used in this study because McCue et al. [1987] gave only an approximate origin time.

WAVE FORMS

Radiation Patterns

Digital wave-form data were recorded at Charters Towers in the northeast and at Narrogin in the southwest of Australia. These stations are almost on a great circle with the event. Their epicentral distances (Charters Towers CTAO D=13.8°, or 1532 km; Narrogin NWAO D=15.1°, or 1682 km) are such that for reasonable upper mantle models they should have recorded the same upper mantle phases.

Predicted radiation patterns (first motions) for P-, SV- and SH-waves for the focal mechanism solution of Fredrich et al. [1987] (Strike = 148°±20°; Dip = 35°±20°; Rake = 80°±25°) are shown in Figure 2. The position of Charters Towers is shown with a cross on the projection for SH; Narrogin is shown with a circle. The radiation patterns show slightly higher amplitude P radiation towards Narrogin, but greater amplitude SV radiation towards Charters Towers. Charters Towers and Narrogin lie close to a nodal plane for SH. P and SV radiation will be resolved in the vertical and radial components of the seismograph recordings, whereas SH will be resolved in the transverse component.

Short-Period Data

Short-period data for the main event recorded at Narrogin were corrupted by spikes and loss of signal. The available short-period data indicate that the coda recorded for the main event has a similar appearance to the aftershock (see also below for the long-period data). Vertical component wave-form data from the aftershock are shown in Figure 3. In order to facilitate comparisons of the relative amounts of P-wave and S-wave energy at the two stations, the amplitudes of the seismograms have been adjusted to balance the amount of energy in the P-wave coda in two-minute long windows starting at the first arrival. These windows are shown for each trace by the rectangular boxes with their left hand ends at the first, arrowed, P-wave arrival. An alternative approach might have been to adjust the amplitudes of the first arrivals to be the same, but the traces are not at exactly the same epicentral distances, and amplitudes of phases can vary along any travel-time branch,

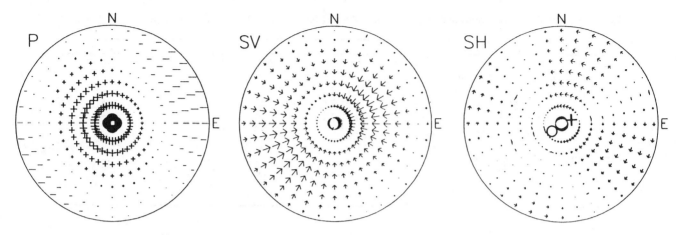

Figure 2. Predicted radiation patterns for P, SV and SH for the fault-plane solution of Fredrich et al. [1988] for the main event. In the P-wave projection, compressions are shown with plus symbols and dilatations with minus symbols. For SV and SH, the direction of particle motion is shown with arrows. The size of all symbols is proportional the amplitude of the energy radiated. The locations of Charters Towers and Narrogin are shown with a o and +, respectively, on the plot for SH.

so using the energy in a longer window was considered a more robust approach. This approach also has the fortuitous effect of creating P-wave first arrivals at Narrogin that are greater than those at Charters Towers, as predicted above by the radiation patterns. Given the analysis of radiation patterns above, if the Earth had a laterally uniform structure, larger amounts of S-wave energy should be seen in the seismogram from Charters Towers.

The seismogram from Narrogin (NWAO) (lower trace) shows a clear P-wave arrival (arrow labeled 'P'), followed several minutes later by an impulsive S-wave (arrow labeled 'S'). Surface waves near the right-hand end of the trace have exceeded the dynamic range of the instrument. The seismogram for Charters towers (CTAO) is shown as the upper trace. It also has an impulsive first P-wave arrival (arrowed, 'P'), but the S-wave arrivals several minutes later are much more emergent and have lower amplitudes than for Narrogin (arrowed, 'S'). Hence, whereas the path to Narrogin has transmitted considerable S-wave energy, relatively less S-wave energy has been recorded at Charters Towers.

Long-Period Data

Long-period data from both Charters Towers and Narrogin are plotted in Figure 4. Part (a) shows the vertical

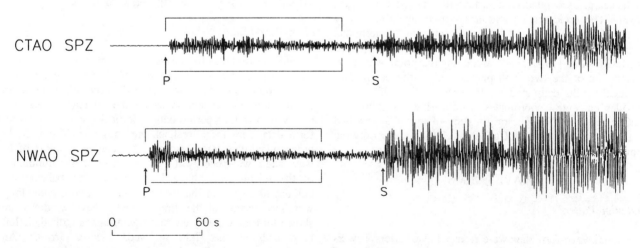

Figure 3. Short-period seismograms for the July aftershock from Charters Towers CTAO (upper trace) and Narrogin NWAO (lower trace). The arrows labeled 'P' indicate the first P-wave arrivals; those labeled 'S' indicate the interpreted first S-wave arrivals. The rectangles around the first two minutes of P-wave coda show the data used to balance the energy in the two traces (see text for discussion).

component recordings of the main event and the aftershock at Charters Towers (CTAO). The amplitudes of both traces have been scaled to show the remarkable similarity of the wave forms for the main event and the aftershock. Three-component data were available for both the main event and the aftershock. They are shown for both Charters Towers (CTAO) and Narrogin (NWAO) in Figure 4(b). The horizontal (E-W and N-S) seismograms for both events were first resolved into radial and transverse components. In order to get an indication of the differences in wave forms and energy recorded at the two stations, the amplitudes of the vertical-component seismograms for the main event at each station were adjusted to make the energy in the first two minutes of P-wave coda the same at each station, as for the short-period data. The amplitudes of the other two components for each station were then adjusted by the same factors. The amplitudes of the seismograms for the aftershock were adjusted in the same way to achieve scaling of Narrogin seismograms relative to Charters Towers. The aftershock was then scaled relative to the main event by making the maximum peak-to-trough amplitudes of the vertical-component seismograms at Narrogin the same for both events and scaling the other traces for the aftershock by the same factors.

The P- and S-wave arrival times from the short-period seismograms are shown as arrows pointing to the vertical component. Both P- and S-waves amplitudes are small on these seismograms. The bulk of the energy would be surface waves.

The long-period data recorded at Narrogin are very different from those at Charters Towers. The surface-wave coda has a considerably longer duration, and although the vertical traces from the main event and aftershock have the same general form, they are not as alike as those recorded at Charters Towers. In Figure 4(b), the seismograms from Narrogin for the radial and transverse components for the aftershock have been inverted; this makes them look more similar to those of the main event. (P. J. Gregson, Observer in Charge at AGSO Mundaring Observatory, personal communication, 1997, has advised that there is no record of either of the horizontal components being reversed between the main event and the aftershock.)

The long-period data recorded at Charters Towers and Narrogin therefore confirm the observations from the short-period data; i. e., the form of the coda recorded to the northeast in the younger eastern parts of Australia is different from that recorded along the same great-circle path to the southwest in the shield regions of the continent. The main differences are in the amount of S-wave and surface wave energy propagated, with less than expected propagated into the younger parts of eastern Australia.

P-Wave Travel Time Data

First-arrival P-wave travel times to stations across Australia for the main event from the BMR Seismological

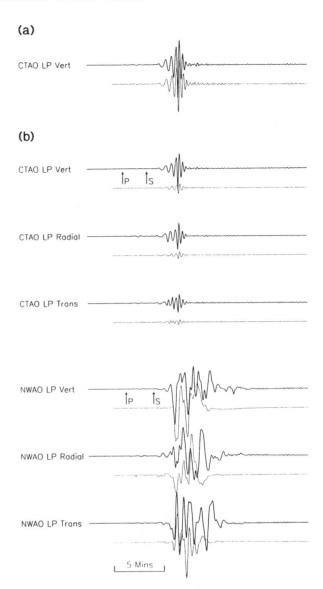

Figure 4. Long-period seismograms from Charters Towers CTAO and Narrogin NWAO. (a) vertical components for the main event and aftershock, with maximum peak to trough amplitudes scaled to be the same. (b) Three-component data for the main event and aftershock for Charters Towers (top six traces) and Narrogin (bottom 6 traces). For each station, vertical, radial and transverse components are shown. For each component for each station, the main event is shown in the upper trace, and the July aftershock is shown as the bottom trace. Arrows against the vertical components for the main event show the times of the P- and S-wave arrivals in the short period seismograms in Figure 3. See text for details of amplitude scaling.

Bulletins (K. McCue, AGSO, personal communication, 1987), are plotted in Figure 5. Times to seismograph stations located on the shield are shown as circles; these stations are also marked with circles in Figure 1. Times to

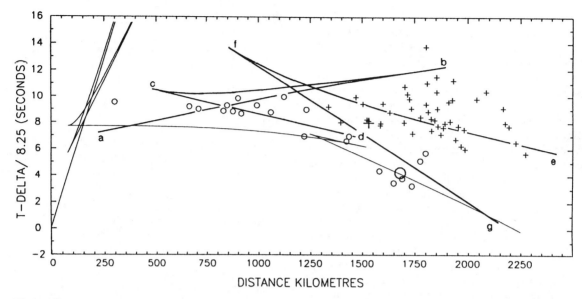

Figure 5. Travel times of the main event to stations on Precambrian basement (o) and Phanerozoic basement (+) (see Figure 1). NWAO and CTAO are shown with larger symbols. The travel time curves are for the model of Muirhead and Drummond [1991] for east of the Tasman Line (thicker line) and model SOZ-P of Bowman and Kennett [1993] for west of the Tasman Line (thinner line) (Figure 6). Phases for energy which has travelled below about 400 km have not been plotted.

stations in eastern Australia are shown with crosses; stations east of the Tasman Line are marked with crosses in Figure 1. The travel times to Narrogin (NWAO) and Charters Towers (CTAO) are shown with a larger circle and a larger cross, respectively. Bowman and Kennett [1993] estimated the reading errors of these times to be less than one second. The travel times have a reduction velocity of 8.25 km s^{-1}. One implication of the reduction velocity is that an error of ~8 km in the estimated epicentre coordinates would cause an error of up to one second in the plotted arrival times; hence the effort above to choose representative epicentral coordinates.

Seismograph stations in eastern Australia tend to be at greater epicentral distances than those farther west. However, where the distance ranges overlap between about 1400 km and 1800 km, the travel times to eastern stations are consistently later than those to stations on the shield. This is consistent with models of the upper mantle for eastern Australia and the shield. The thicker travel-time curves superimposed on Figure 5 are for the model of Muirhead and Drummond [1991] for southeast Australia (Figure 6). This model is a variation of that of Muirhead et al. [1977], which was based on body waves from explosive sources recorded across southeast Australia, predominantly east of the Tasman Line, but Muirhead and Drummond [1991] modified it for consistency with shear-wave models derived from a reinterpretation of the surface-wave dispersion curves of Goncz and Cleary [1976]. It also predicts observed travel times for P and S waves in a north-south direction through eastern-most Australia better than previous models for eastern Australia [Muirhead and Drummond, 1991]. Travel time curves of crustal phases are not shown for this model; nor are those from below about 230 km depth because the model does not extend below that depth.

Most arrivals to eastern Australia lie beyond the end of the branch *cd*, which is a forward refracted phase through a high-velocity lid at 100-120 km depth, also labeled *cd* in Figure 6. In the model of Muirhead and Drummond [1991],

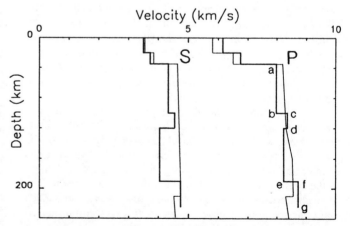

Figure 6. Plots of velocity versus depth for the P-wave and S-wave models whose calculated P-wave travel times are plotted in Figure 5. The thicker line is for eastern Australia [Muirhead and Drummond, 1991]; the thinner line is for the shield models SOZ-P (P-wave) and SOZ-S (S-wave) of Bowman and Kennett [1993].

a low-velocity layer is interpreted below 120 km depth for both P and S waves (layer *de* in Figure 6). This causes a decay of the amplitudes of arrivals along branch *cd* with increasing distance. First arrivals at greater distances are from refracted waves through the next deeper high-velocity layer (*fg*) near 190 km depth; second arrivals (*ef*) at these distances are wide-angle reflections off the layer at 190 km. Hence, many first arrivals from the Marryat Creek earthquake to stations in eastern Australia fall in the region on the travel-time plot associated with a decay in the amplitudes of a forward branch of waves refracted through high velocities between 100-120 km, and those which penetrate a low-velocity region and return as reflected waves off and refracted waves through a deeper layer with a top near 190 km depth.

In contrast, body-wave data from previous studies of the shield do not require a low-velocity zone between ~100 km depth and 200 km depth, especially for P-waves [e. g., Hales et al., 1980], and travel times from the Marryat Creek earthquake are consistent with this. They are more tightly clustered in time at any epicentral distance compared with those to the east of the Tasman Line, and generally lie within 2 s of the times predicted by the SOZ-P model of Bowman and Kennett [1993] for the shield (thin-lined travel time curves; the model is shown in Figure 6). The SOZ-P model does not contain a low-velocity zone between 120 and 190 km depth; rather, it has a high velocity gradient in the depth range where the model of Muirhead and Drummond [1991] has a low-velocity layer.

DISCUSSION

Cleary and Doyle [1962] had earlier reported poor S_n (uppermost mantle) propagation in eastern Australia. Narrogin and Charters Towers are at epicentral distances from the Marryat Creek earthquakes for which first-arrival P-wave energy at each station would be expected to have travelled at similar depths (~100 to ~200 km) through the lithosphere. For the published earthquake focal mechanisms, relatively larger amounts of S-wave energy might be expected to propagate through the upper mantle under the Palaeozoic basement to Charters Towers. However, the opposite is observed, with relatively larger amounts of both short-period and long-period S-wave energy recorded at Narrogin on shield basement in the southwest. The wave forms for surface wave data are also very different. Travel times to eastern stations are also greater than those to stations on the shield. This is consistent with the teleseismic travel-time residual results of Drummond et al. [1989]. The Marryat Creek data therefore suggest that the poor S-wave propagation reported by Cleary and Doyle [1962] extends to depths greater than just the uppermost mantle.

The SOZ-P model of Bowman and Kennett [1993] fits the P-wave travel time data from the Marryat Creek main event reasonably well for distances between 1,200 and 1,800 km. Its predicted travel times to shield stations at distances between 250 and 1,100 km (P_n, or upper-most mantle refracted/diving waves) are 1-2 s earlier than the recorded travel times, although the apparent velocity predicted by the model is similar to that indicated by the recorded arrival times. This would imply that the average crustal thickness for the shield assumed in the SOZ-P model (35 km) is less than that under Marryat Creek, and although no estimates of crustal thickness are available for the Marryat Creek region to confirm this, crustal thicknesses greater than 45 km are likely elsewhere under the Musgrave Block [Leven and Lindsay, 1995].

Other shield models [e. g. TASS1a and TASS2a of Finlayson et al., 1974; CAP8 of Hales et al., 1980; NWB-1 of Bowman and Kennett, 1990; and njpb of Kennett et al., 1994; Dey et al., 1993] either have greater travel-time misfits at distances greater than 1200 km, or more importantly do not predict the uppermost mantle velocities demonstrated by the Marryat Creek earthquake travel times at distances less than 1200 km. The SOZ-P model therefore represents a better fit to the Marryat Creek earthquake travel time data within the shield (Figure 5) than other models for the shield. Some of the Marryat Creek travel times were used by Bowman and Kennett [1993] in developing the SOZ-P model, but these data comprise only a small portion of the data used, and are therefore unlikely to have influenced their inversion to such an extent that the observation that SOZ-P represents a better fit than other models is the result of a circular argument.

The key issue is that in each of the studies that has produced an upper-mantle model for the Australian shield, the derived model fits the observed data from that study, but not necessarily the data from other studies. However, in many cases, the data come from different regions. For example, the main reason that SOZ-P represents a better fit to the Marryat Creek data is probably because Bowman and Kennett [1993] used travel times for rays which mostly crossed the central and southern part of the continent. The calculated travel time curves between 250 and 1250 km from the SOZ-P model have apparent velocities and travel times almost the same as those from the TASS2a model of Finlayson et al. [1974], which was based on explosive sources distributed along a north-south corridor through central Australia and recorded at stations with a distribution biased to the southern half of the continent, as was the SOZ-P model. Models from studies based on data propagating from the northern margins of the continent into the shield of northern Australia have greater travel times and lower apparent velocities at distances between 250 and 1250 km. This is also the case for the TASS1a model of Finlayson et al. [1974], which had purely east-west oriented ray paths across southern Australia; however, the TASS1a model is a better estimator of observed Marryat Creek travel times at distances greater than 1,300 km than all other models except SOZ-P. Models of the upper mantle based on energy sources to the north of the continent [e. g., CAP8 of

Hales et al., 1980; NWB-1 of Bowman and Kennett, 1990; and njpb of Kennett et al., 1994; Dey et al., 1993] have lower velocities from the Moho to about 100 km depth and, apart from SOZ-P, higher velocities between about 100 km and 165 km depth, than models derived from data restricted mostly to southern Australia.

This is consistent with the preliminary results from the SKIPPY experiment, which shows higher modeled P-wave velocities at 150 km depth under northern Australia compared to southern Australia [Figure 8 of van der Hilst and Kennett, 1997]. This would perhaps suggest a north to south regionalisation of lithospheric structure of the shield, as well as the more pronounced east to west regionalisation between the shield and eastern Australia which is the main subject of this paper.

Dey et al. [1993] found that in northern Australia, models satisfying data recorded to the northeast from the Papua New Guinea region, were different from those for data recorded to the northwest through the Indonesian Archipelago. Their data would have sampled the upper mantle under the most northern parts of the Australian shield. This suggests that the structure differs from east to west even within the shield, particularly in the north.

Four reasons can explain the regionalisation of upper mantle velocity values: different compositions, varying degrees of partial melt extraction, different mineralogies, and different temperatures.

Jordan [1979] examined the effects of composition on the physical properties of a number of garnet lherzolite compositions thought representative of the upper mantle. His results show normative P-wave velocities which range from 8.19 to 8.34 km s^{-1}. Although these values need to be adjusted to the pressure and temperature conditions of the upper mantle, the variability as a fraction of an average value would stay about the same, and could explain a lot of the variability in the published models. The removal of partial melts from an initially fertile mantle will alter the seismic P-wave velocity, but the effects are likely to be less than the apparent natural variability in the starting material; for example, Jordan estimates that the removal of a 20% partial melt from a pyrolite will increase the seismic velocity by less than 0.01 km s^{-1}. A test of whether natural variability is the main cause of the differences in mantle seismic models will be provided by the SKIPPY results for the entire continent, because natural variability would be expected to track the evolution and growth of the continent. For example, Shaw et al. [1996] divided the continent into basement elements, which could then be grouped into megaelements, where each mega-element is comprised of basement elements which amalgamated and then acted mostly as large individual tectonic units [Myers et al., 1996]. If each mega-element was brought to the continent with its own upper mantle root, the mantle roots might be expected to have different starting properties. The S-wave tomograms of van der Hilst and Kennett [1997] do broadly reflect the mega-element boundaries of Shaw et al., although their present P-wave model does not, at least within the shield.

Temperature can affect seismic velocity in three ways. Firstly, Hales et al. [1975] attributed a discontinuity at depths of 60 - 70 km under parts of Australia to the spinel (above) to garnet (below) phase transition. The spinel to garnet phase boundary has a positive, concave-upwards gradient in pressure-temperature space, so as a region at depth cools it can cross from the spinel to the garnet field. Seismic velocity increases with increasing garnet content [e. g., Jordan, 1979], so hot uppermost mantle may sit within the spinel field and have lower velocities than comparable cold mantle at the same depth. This would appear to be the case for some models for northern Australia, where heat flow is higher than 80 mWm^{-2} (Figure 1) and geotherms predicted by Cull [1991] lie outside the garnet stability field of O'Reilly [1984].

Secondly, temperature has an intrinsic effect on seismic velocity: increasing temperature causes a drop in velocity. Energy from the Marryat Creek earthquakes propagating to Narrogin would have bottomed in or under lithosphere which has some of the lowest surface heat flow in Australia (£ 40 mWm^{-2}). Energy propagating to Charters Towers bottomed in or under lithosphere with surface heat flow exceeding 80 mWm^{-2}. The contribution of mantle heat flux to surface heat flow is uncertain, and estimates of geothermal gradients for western, central and eastern Australia show considerable variation. However, upper mantle temperatures for eastern Australia are assumed to be higher than those in the shield, especially in western Australia [Cull, 1991]. The seismic velocity in the model of Muirhead and Drummond [1991] for eastern Australia is of the order of 0.2 - 0.3 kms^{-1} lower than that in the SOZ-P model of Bowman and Kennett [1993] from the Moho to about 190 km depth. If this difference is due only to the intrinsic effect of temperature on seismic velocity, then for a temperature derivative of velocity of

$$\left(\frac{dV}{dT}\right)_p = -4 \times 10^{-4} \text{ km s}^{-1}{}^\circ\text{C}^{-1}$$

[Anderson et al., 1972], the upper mantle under eastern Australia needs to be at least 500°C hotter than that farther west under the shield. The geotherms of Cull [1991] indicate that this might be possible, although 500°C lies at the upper end of likely temperature differences.

Finally, temperature can affect seismic velocities if the rocks are at or above the solidus by reducing the shear modulus. On the basis of electrical sounding anomalies, Lilley et al. [1981] predicted low portions of partial melt in the upper mantle in southeast Australia, although at depths of 200-300 km, which is deeper than the seismic low velocity zone of Muirhead and Drummond [1991]. No partial melting is inferred in the electrical conductivity models under the shield in central Australia. Muirhead and Drummond [1991] predicted a V_p/V_s ratio of 2.03 within the low-velocity zone under eastern Australia; this would imply a low shear modulus.

Hence the spatial correlation of poor S-wave energy propagation with high surface heat flow in eastern Australia and good S-wave energy propagation with low surface heat flow in western Australia is likely to have a causal relationship. Its effect on mineralogy, its intrinsic effect on seismic velocity, or the possible presence of partial melts in the upper mantle under eastern Australia are all potential explanations for all or some of the observed velocity differences. The relative effects of these factors cannot be determined with the available data.

Acknowledgements. Ken Muirhead helped as a sounding board for some of the ideas and the kernel of the computer code used for calculating the travel-time curves in Figure 5. Kevin McCue provided travel-time data and copies of seismograms. Brian Kennett provided the software used to calculate the radiation patterns in Figure 2, some of the wave-form data, and the details of several of his upper-mantle models. David Denham, Ken Muirhead and Clive Collins and two anonymous reviewers provided constructive comments on the manuscript. This paper is published with the permission of the Executive Director of the Australian Geological Survey Organisation (AGSO) and the Director of the Australian Geodynamics Cooperative Research Centre (AGCRC). The AGCRC is funded by the Government of the Commonwealth of Australia through its Cooperative Research Centres Program.

REFERENCES

Anderson, D. L., C. Sammis, and T. H. Jordan, Composition of the mantle and core, in *The Nature of the Solid Earth,* edited by E. C. Robertson, pp. 41-66, McGraw Hill, New York, 1972.

Bowman, J. R., and B. L. N. Kennett, An investigation of the upper mantle beneath northwestern Australia using a hybrid seismic array, *Geophys. J. Int., 101,* 411-424, 1990.

Bowman, J. R., and B. L. N. Kennett, The velocity structure of the Australian shield from seismic travel times, *Bull. Seism. Soc. Am., 83,* 25-37, 1993.

Cleary, J. R., and H. A. Doyle, Application of a seismograph network and electronic computer in near earthquake studies, *Bull. Seism. Soc. Am., 52,* 673-682, 1962.

Cull, J. P., An appraisal of Australian heat-flow data, *BMR J. Aust. Geol. Geophys., 7,* 11-21, 1982.

Cull, J. P., Geothermal gradients in Australia, in *The Australian Lithosphere,* edited by B. J. Drummond, pp. 147-156, *Spec. Pub. Geol. Soc. Aust., 17,* 1991

Dey, S. C., B. L. N. Kennett, J. R. Bowman, and A. Goody, Variations in upper mantle structure under northern Australia, *Geophys. J. Int., 114,* 304-310, 1993.

Drummond, B. J., K. J. Muirhead, C. Wright, and P. Wellman, A teleseismic travel time residual map of the Australian continent, *BMR J. Aust. Geol. Geophys., 11,* 101-105, 1989.

Finlayson, D. M., J. P. Cull, and B. J. Drummond, Upper mantle structure from the trans-Australia seismic survey (TASS) and other seismic refraction data, *J. Geol. Soc. Aust., 21,* 447-458, 1974.

Fredrich, J., R. McCaffrey, and D. Denham, Source parameters of seven large Australian earthquakes determined by body-wave inversion, *Geophys. J., 95,* 1-13, 1988.

Goncz, J. H., and J. R. Cleary, Variations in the structure of the upper mantle beneath Australia from Rayleigh wave observations, *Geophys. J. Roy. Astron. Soc., 44,* 507-516, 1976.

Hales, A. L., K. J. Muirhead, J. M. W. Rynn, and J. F. Gettrust, Upper mantle travel times in Australia - a preliminary report, *Phys. Earth Planet. Int., 11,* 109-118, 1975.

Hales, A. L., K. J. Muirhead, and J. M. W. Rynn, A compressional velocity distribution for the upper mantle. *Tectonophysics, 63,* 309-348, 1980.

Jordan, T. H., Mineralogies, densities and seismic velocities of garnet lherzolites and their geophysical implications, in *The Mantle Sample: Inclusions in kimberlites and other volcanics,* edited by F. R. Boyd and Henry O.A. Meyer, pp. 1-14, *Proc. Second Int. Kimberlite Conf., Vol. 2,* Am. Geophys. Union, Washington, 1979.

Kennett, B. L. N., O. Gudmundsson, and C. Tong, The upper-mantle S and P velocity structure beneath northern Australia from broad-band observations. *Phys. Earth Planet. Int., 86,* 85-98, 1994.

Leven, J. H., and J. F. Lindsay, A geophysical investigation of the southern margin of the Musgrave Block, South Australia, *AGSO J. Aust. Geol. Geophys., 16,* 155-161, 1995.

Lilley, F. E. M., D. V. Woods, and M. N. Sloane, Electrical conductivity profiles and implications for the absence or presence of partial melting beneath central and southeast Australia, *Phys. Earth Planet. Int., 25,* 419-428, 1981.

McCue, K., B. C. Barlow, D. Denham, T. Jones, G. Gibson, and M. Michael-Leiba, Another chip off the old Australian Block, *EOS Trans. Am. Geophys. Union, 68,* 609, 1987.

Muirhead, K. J., and B. J. Drummond, The base of the lithosphere under Australia, in *The Australian Lithosphere,* edited by B. J. Drummond, pp. 23-40, *Spec. Pub. Geol. Soc. Aust., 17,* 1991.

Muirhead, K. J., J. R. Cleary, and D. M. Finlayson, A long-range seismic profile in south-eastern Australia, *Geophys. J. Roy. Astron. Soc., 48,* 509-520, 1977.

Myers, J. S., R. D. Shaw, and I. M. Tyler, Tectonic evolution of Proterozoic Australia, *Tectonics, 15,* 1431-1446, 1996.

O'Reilly, S. Y., The mantle environment, in *Kimberlite Occurrence and Origin,* edited by J. E. Glover and P. G. Harris, pp. 63-102, *Pub. Univ. W.A. Dept. Geol. and Univ. Ext., 8,* 1984.

Shaw, R. D., P. Wellman, P. Gunn, A. J. Whittaker, C. Tarlowski, and M. Morse, Guide to using the Australian Crustal Elements Map, *Aust. Geol. Surv. Org Rec., 1996/30,* 1996.

van der Hilst, R. D., and B. L. N. Kennett, Upper mantle structure beneath Australia from portable array deployments, *This volume, in press.,* 1997.

Zielhuis, A., and R. D. van der Hilst, Upper-mantle shear velocity beneath eastern Australia from inversion of waveforms from SKIPPY portable arrays, *Geophys. J. Int., 127,* 1-16, 1996.

B. J. Drummond, Australian Geodynamics Cooperative Research Centre and Australian Geological Survey Organisation, PO Box 378, Canberra, ACT 2601, AUSTRALIA.

Lithospheric Structure in Southeast Australia: a Model Based on Gravity, Geoid and Mechanical Analyses

Y. Zhang[1], E. Scheibner[2], B. E. Hobbs[3], A. Ord[1], B. J. Drummond[4], and S. J. D. Cox[1]

Eastern Australia is considered an upper-plate passive margin with a dense oceanic lithosphere and an adjacent dense upper-mantle wedge created by igneous underplating during lithospheric thinning/extension in the Jurassic-Early Cretaceous. Mechanical modeling of such a tectonic framework suggests that gravitationally-induced body forces contribute significantly to the compressive stress field of the region. In this study, we aim to obtain further constraints on the lithospheric density/geometry structures in southeast Australia by modeling the gravity field and the geoid, and analysing the mechanical behaviour of the constrained system. Gravity and geoid anomalies are calculated for east-west, profiles across the passive margin, and are compared with observed gravity and geoid anomalies. A close match between the calculated and observed values is achieved by adjusting the geometries and densities of the layers in the studied profiles. This approach enables us to define a lithospheric structural model, with specific geometries and densities for southeast Australia, which is compatible with gravity and geoid observations, geological arguments and density distributions constrained by petrological observations. The model is characterised particularly by upper-plate passive- margin geometries, an upper-mantle wedge of similar density to the oceanic lithosphere, and a density contrast between the oceanic and continental lithosphere. Mechanical analyses of this system demonstrate that gravity relaxation-induced body forces associated with the geometry-density configurations of southeast Australia can generate horizontal east-west, compression dominating in southeast Australia, approximately normal to the passive margin. This is consistent with the bulk of the measured stress orientations in the region. The locations of the uplift predicted by the models also agree well with the geological observations and with theories based on igneous underplating and isostasy.

1 CSIRO Exploration and Mining, Nedlands, WA, Australia
2 CSIRO Exploration and Mining, North Ryde, NSW, Australia
3 CSIRO Exploration and Mining, Wembley, WA, Australia
4 Australian Geological Survey Organisation, Canberra, ACT, Australia

REGIONAL TECTONIC FRAMEWORK

The Australian continental lithosphere has been evolving as a separate major continent since the breakup of Gondwanaland in Jurassic-Cretaceous time, itself being part of the Pangean oscillation and the breakup of Pangea [Veevers, 1984; 1990]. The fragmentation of East Gondwanaland was accompanied by a series of mantle convection-related lithospheric processes such as rifting, divergent plate motions and sea-floor spreading. These processes led to the formation of passive margins surrounding the Australian continent roughly in the Jurassic-Early Cretaceous, with the eastern Australian passive margin

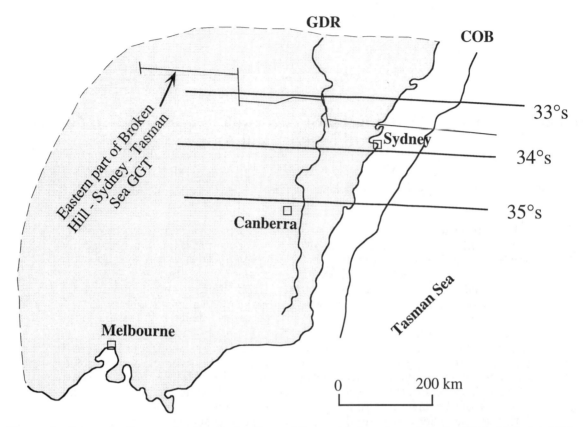

Figure 1. Schematic diagram showing the locations of the modeled sections along the 33°S, 34°S and 35°S parallels, the eastern part of the Broken Hill-Sydney-Tasman Sea Global Geoscience Transect [Scheibner et al., 1991a], the Great Dividing Range (GDR), the coast line and the continent/ocean boundary (COB) in the southeast Australian passive margin.

setting developing behind the active plate margin of the Southwest Pacific; see Johnson [1989] for a review of the history of the southeast Australian highlands.

Etheridge et al. [1989] and Lister et al. [1991] proposed a modified detachment model of continental extension and ultimate breakup (see Lister et al. [1986] for a description of the original model) to explain the formation of passive margins. This model demonstrates that asymmetrical passive margin pairs (upper- and lower-plate margins) can develop as the result of an interplay of simple and pure shearing below and above the master detachment(s); the asymmetry is reflected in the morphology, structure, uplift/subsidence history and/or thermal evolution of the passive margins [Etheridge et al., 1989; Lister and Etheridge, 1989; Scheibner, 1992]. Much of eastern Australia has an upper-plate passive-margin geometry according to this theory.

Scheibner et al. [1991a] incorporated the idea of the upper-plate passive margin in the construction of the Broken Hill-Sydney-Tasman Sea Global Geoscience Transect (GGT) (Figures 1 and 2), and further suggested a denser mantle wedge beneath the edge of extended eastern continental lithosphere, next to the oceanic lithosphere (see also Scheibner [1992] and Zhang et al. [1996]). This was based on the consideration that one important aspect of the Jurassic-Cretaceous continental extension and final breakup in eastern Australia was thinning of the crust and upper mantle towards the line of breakup. Such thinning must have been accompanied by significant igneous underplating which represents a material transfer from shallow asthenosphere to lithosphere. This igneous underplating created a new upper-mantle wedge compositionally similar to the adjacent oceanic upper mantle (Figure 2).

MOTIVATION FOR THIS STUDY

The regional modern day stress field in southeast Australia is characterised by high horizontal stresses (~50-100 MPa) oriented predominantly in an east-west, direction, as revealed both by (shallow) *in situ* stress measurements and by (deeper) earthquake focal mechanism studies [Denham et al., 1979; Enever et al., 1990; Brown and Windsor, 1990]. Locally, however, high horizontal stresses directed approximately north-south dominate, and in some localities the stress field is close to biaxial with north-south stresses approximately equal to east-west, stresses [e.g., Denham et al., 1979; Enever et al., 1990]. Moreover, the pattern of east-

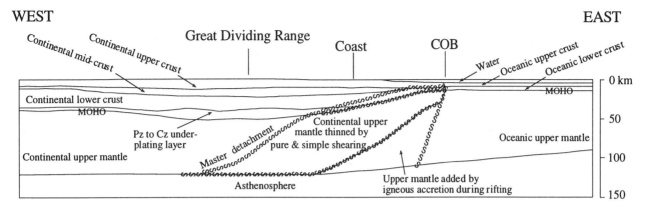

Figure 2. The crustal and lithospheric structural framework along the eastern part of the Broken Hill-Sydney-Tasman Sea Global Geoscience Transect [Scheibner et al., 1991a]. The vertical and horizontal scales are identical here (the same in Figures 8, 9, 10 and 12).

west, high horizontal stress seems to have been present in southeast Australia at least back to the Miocene (30 million years) as indicated by displacements on approximately north-south striking high-angle reverse faults revealed by fission track studies [A. Gleadow, personal communication, 1996].

There are also many instances in southeast Australia of approximately north south striking high angle reverse faults and thrusts that displace old (Palaeozoic) rocks against Miocene sediments. Examples are the Khancoban-Yellow Bog Thrust and the Tawonga Thrust [Beavis, 1960] both of which thrust granite over Tertiary gravels [Ken Sharp, personal communication from Snowy Mountain Authority records, 1996], the Adaminaby Fault (a high-angle reverse fault displacing Palaeozoic granite against Miocene sediments), the Lake George Fault and the Shoalhaven Fault (high-angle reverse faults displacing Palaeozoic sediments against Miocene sediments). Thus, it appears that approximately east-west, compression has been important in southeast Australia at least since the Miocene and there are even indications that such a stress field has been important as far west as Spencer Gulf in South Australia, where north-south striking reverse faults displace Cambrian rocks against Miocene sediments [M. Sandiford, personal communication, 1997].

The important point here is that the absolute plate-tectonic displacement vector is oriented approximately north-south in southeast Australia over the time period Miocene - present, and hence one might expect high horizontal stresses oriented north-south rather than the observed dominant east-west, trend [Zoback et al., 1989; Zoback, 1992; Denham et al., 1979]. This means that the stress pattern of southeast Australia and the observed recent intraplate deformation cannot be explained simply by modeling plate-motion related boundary forces [Cloetingh and Wortel, 1986; Coblentz et al., 1995; Sandiford et al., 1995].

Zhang et al. [1996] simulated the 2D stress distribution in southeast Australia using the tectonic framework outlined in the Broken Hill-Sydney-Tasman Sea Global Geoscience Transect [Scheibner et al., 1991a] (Figure 2). That study demonstrated that the observed high east-west, horizontal compressive stresses could be explained by lithospheric structure-related gravitational adjustments, that induce dominantly elastic stresses normal to the trend of a mantle wedge of underplated mafic material in the lithosphere.

This paper is a refinement of the earlier paper by Zhang et al. [1996]. Here we modify the model by adjusting the crust-lithosphere geometry and density distribution to fit observed gravity and geoid data. More realistic starting conditions for mechanical calculations are also adopted in this study, including initial stresses and boundary conditions. The results confirm high horizontal east-west, stresses in the bulk of the region, generated by density contrasts in the lithosphere. The generation of these stresses is also associated with displacements which are partly responsible for the formation of the Great Dividing Range and other topographical features in southeast Australia.

We have chosen to model east-west cross sections as these are approximately normal to plate motion and parallel to the anomalous east-west, stress orientation. As the variation in lithospheric geometry is negligible in the north-south direction in comparison to the east-west, direction, the geometry is essentially prismatic perpendicular to the plane of the 2D section selected. This geometry also means that there are no density-induced tectonic stresses in the north-south direction. The interest here is to test whether east-west, maximum horizontal compression is possible purely due to the geometry and density structures in the section. The discussion of the stresses orthogonal to the sections is not included in this study owing to the limitations of a 2D model.

THEORETICAL BASIS AND MODEL DESCRIPTION

Gravity Modeling

Computation of gravity anomalies was carried out using the forward modeling computer program of Roach [1993].

The program uses the two-dimensional line integral method of Talwani et al. [1959] to calculate the gravity fields resulting from geological layers or bodies with polygonal cross sections and infinite extent orthogonal to the modeled section. The shapes of the calculated gravity anomalies for a cross section are entirely controlled by the geometries and densities incorporated.

The cross sections modeled here are the simplified geological profiles along parallels of latitude across the eastern Australian passive margin. These profiles are compiled using knowledge of the regional geology and structures, and their structural features are essentially similar to the Broken Hill-Sydney-Tasman Sea Global Geoscience Transect [Scheibner et al., 1991a]. The approach adopted here is to adjust iteratively the geometries and densities of a geological profile (still constrained by the knowledge of the regional tectonic framework), so that a reasonable match between the calculated gravity anomaly curve and the observed gravity anomaly profiles is achieved. In doing so, constraints on the geometry-density structure of the profile are obtained.

The observed gravity anomalies used here are Bouguer anomalies onshore but free-air anomalies offshore. Accordingly, the real density of sea water was used for the sea-water layer in the models. The density of the sea-water layer must be replaced with the densities of ocean-floor sediments or upper-crustal rocks if comparison with the observed Bouguer gravity anomalies is attempted for the offshore areas [Symonds and Willcox, 1976; Simpson and Jachens, 1989].

The achievement of perfect local isostatic equilibrium is not aimed for here, either for gravity modeling or geoid modeling. Small departure from strict local isostatic compensation in a section is allowed by the elastic nature of the lithosphere and is accommodated by elastic stresses. There do exist examples in the Earth where an approximate regional equilibrium is possible only for an elastic plate [e.g., Vening Meinesz, 1941; Gunn, 1943]. For the Australian plate, some studies [e.g., Stephenson and Lambeck, 1985; Dooley, 1991] suggest incomplete local compensation, but others [e.g., Wellman, 1976, 1979] believe otherwise. A regional equilibrium may be possible only at very large depth [Dooley, 1991].

Geoid Modeling

A geoid anomaly is the elevation difference between the geoid and the reference ellipsoid [Ranalli, 1987, p. 189]. Modeling of geoid anomalies provides constraints on the lithospheric geometry-density structure of a geological profile. The approach is similar to gravity modeling. The normalised geoid anomalies (assuming zero anomaly for a reference location in a profile) were calculated for a profile and then compared with the observed normalised geoid anomalies. A reasonable match between the calculated and observed anomaly curves is considered, in conjunction with the results of gravity modeling, as an indication of the acceptability of proposed lithospheric geometry-density structures for the profile.

The geoid anomalies can be calculated from Haxby and Turcotte [1978] or Coblentz et al. [1994]:

$$\Delta N = -\frac{2\pi G}{g} \int_{h}^{L} \Delta\rho(y)\, y\, dy, \qquad (1)$$

where ΔN is the geoid anomaly, G is the gravitational constant, g is the gravitational acceleration, $\Delta\rho(y)$ is the density contrast relative to a reference state at depth y, h is the surface elevation and L is the maximum depth of the lithospheric column. The observed geoid anomaly profile used for comparison was interpolated from the contour geoid map published by Allman [1982].

Finite-Difference Mechanical Modeling

The methodology employed for mechanical modeling is the same as that of Zhang et al. [1996], using the finite-difference computer code FLAC (Fast Lagrangian Analysis of Continua) [Cundall and Board, 1988; Itasca, 1992]. This model simulates the mechanical behavior of a lithospheric profile using a 2-D mesh comprising quadrilateral elements. In this study, the deformation of the materials is controlled by a prescribed elastic-plastic constitutive law under the constraints of starting geometrical and mechanical configurations.

The elastic-plastic constitutive law assumes that the material deforms elastically and isotropically until a yield stress is reached [Vermeer and de Borst, 1984; Hobbs et al., 1990; Ord, 1991]. The elastic part of the deformation is determined by Hooke's law. The material yields once the material reaches the maximum yield stress (the Mohr-Coulomb yield criterion), after which irrecoverable deformation occurs with continuing load. The plastic portion of the deformation follows a non-associated flow law with constant (non-hardening) values of cohesion, friction angle, and dilation angle (Table 1). The total strain of the material is the sum of the elastic and plastic components. Although nominally two dimensional, the model simulates a plane-strain situation with out-of-plane stresses considered in the calculations.

The construction of the model requires the specification of geometries, material properties, initial stresses and boundary conditions. The geometries are based on the profiles constrained by gravity and geoid modeling, and are defined by the boundary coordinates of various layers in a profile. The material properties include density, bulk modulus, shear modulus, cohesion, friction angle and dilation angle. The specification of these parameters is based on representative values for the rocks in each layer in the modeled profile (Table 1). However, the choice of density values is based on

TABLE 1: Material properties

Layer	Rocks	Density (kg·m⁻³)	Bulk mod. (Pa)	Shear mod. (Pa)	Cohesion (Pa)	Friction angle	Dilation angle
1a	greenschist, granite & seds.	2700	3.00×10^{10}	2.25×10^{10}	2.5×10^7	20°	2°
1b	granite & tonalite gneiss	2800	3.57×10^{10}	2.46×10^{10}	2.8×10^7	25°	2°
2	tonalitic granulite & gneiss with restite	2900	4.32×10^{10}	2.85×10^{10}	3.0×10^7	30°	2°
3	mafic granulite	3120	5.13×10^{10}	3.23×10^{10}	4.0×10^7	35°	4°
4	lherzolite, garnet pyroxenite & wehrlite	3270	9.42×10^{10}	5.1×10^{10}	5.6×10^7	40°	6°
5	water	1030	1.00×10^9	1.00×10^8	2.0×10^7	0°	0°
6	sediments	2200	9.5×10^9	8.6×10^9	2.0×10^7	15°	1°
7	basalt & gabbro	2900	4.63×10^{10}	3.05×10^{10}	3.0×10^7	30°	4°
8	ultramafic-mafic pyroxenite & wehrlite	3100	7.53×10^{10}	4.52×10^{10}	4.0×10^7	35°	4°
9	ultramafic pyroxenite, ecologite & wehrlite	3300	10.61×10^{10}	5.47×10^{10}	5.6×10^7	35°	6°
10	asthenosphere	3210	10.83×10^{10}	5.0×10^{10}	4.6×10^7	40°	6°

the results from gravity and geoid modeling but constrained by our understanding of the petrology of the rocks concerned.

Initial stresses for the model must reflect stresses in a static lithosphere. McGarr [1988] proposed that in the absence of applied tectonic forces the horizontal and vertical stresses (σ_{xx} and σ_{yy} respectively) in the lithosphere are given by

$$\sigma_{xx} = \sigma_{yy} = \int \rho(y)\, g\, y\, dy, \qquad (2)$$

where y is depth, $\rho(y)$ is density as a function of depth, and g is gravity. This is contrasted with previous simple models based on Hooke's law with the assumption of zero horizontal strain and pre-stress [Jaeger and Cook, 1979], yielding

$$\sigma_{yy} = \int \rho(y)\, g\, y\, dy$$

$$\sigma_{xx} = \frac{v}{1-v}\, \sigma_{yy} \qquad (3)$$

where v is Poisson's ratio. McGarr's argument is that lithospheric materials are prestressed systems which can deform horizontally in response to applied loads or remote stresses, and therefore lithostatic stresses develop. The stresses given by (2) are consistent with the observation that the stresses within plate interiors are often horizontally compressional, rather than horizontally extensional as one would expect from (3).

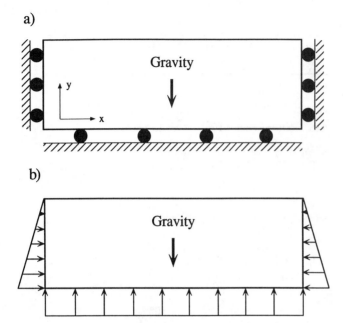

Figure 3. Boundary conditions for the mechanical models. (a) Container boundary conditions with a free surface at top. Lateral boundaries are fixed in the x (horizontal) direction but free in the y (vertical) direction; the base is fixed in the y direction but free in the x direction. (b) Stress boundary conditions with a free surface at top. Pressures are applied to the lateral boundaries and the base of the model to prevent gravitational collapse; the pressures are set up to balance the initial lithostatic stresses near these boundaries. The left boundary is horizontally fixed. Gravity is applied in both boundary conditions.

In this model, therefore, the initial stresses are set up using (2) for the entire section, precisely according to the density and geometry structures. This is more rigorous than the simplified scheme in the model of Zhang et al. [1996], where the initial stresses are equal to $\rho g y$ only at the bottom of the section and then vary linearly to zero at the surface. Such simplification overlooks the inhomogeneous density distribution in the section and creates an initial stress field slightly different from the true $\rho g y$ situation.

Two types of boundary conditions are adopted here to simulate the situation of static gravitational relaxation. The first (Figure 3a) is a container boundary condition with the influence of gravity, where the top is a free surface, while roller boundary conditions (no displacement perpendicular to the boundary) are applied to the base, left and right sides of the model. This means that there is zero horizontal bulk deformation, that is, the Jaeger constraint [Jaeger and Cook, 1979], though materials within the bulk of the model are free to deform both vertically and horizontally. The second (Figure 3b) can be described as a stress boundary condition or the McGarr constraint [McGarr, 1988]. Here the top is still a free surface, but pressures are applied to the base, left and right sides of the model to prevent gravitational collapse; the left side is horizontally fixed to provide a fixed reference boundary for the model. The pressures are precisely set up to balance the initial lithostatic stresses near these boundaries. The major difference between the two is that the stress (or McGarr) boundary condition includes remote stress by incorporating pressures, and allows bulk deformation in both the horizontal and vertical directions.

LITHOSPHERIC STRUCTURES INTERPRETED FROM GRAVITY AND GEOID MODELING

The 34°S Profile

This cross section is 700 km long and 150 km deep, extending east-west from 146° 30'E, 34°S, to 154° 4.7'N, 34°S (Figures 1 and 4, bottom). The section has significant similarities to the Broken Hill-Sydney-Tasman Sea Global Geoscience Transect [Scheibner et al., 1991a] (Figure 2) and is composed of eleven layers. The continental lithospheric part includes five layers, that is, four crustal layers on the top of the upper mantle layer. The oceanic lithosphere below the sea water layer contains four layers, including sea floor sediments, two crustal layers and oceanic upper mantle. The asthenospheric layer is located at the bottom of the section. Topographic elevation and bathymetry data are incorporated in the profile; these data define topographic features such as the Great Dividing Range (GDR), the coast line, and the shelf and sea floor geometries.

The modeled profile (Figure 4) shows a good match between calculated and observed gravity anomalies. The gravity high across the coastline mainly reflects the existence of a denser mantle wedge and thinned continental crust; the continent-ocean edge effect [e.g., Walcott, 1972] may also contribute to the gravity high, but this effect is small here because the model reaches a depth of 150 km and involves complex lithospheric structures, whereas the Walcott model is concerned with only near surface features. This contrasts with the sharp drop in gravity anomalies near the continent/ocean boundary (COB), which seems to arise mainly from a sharp increase in the depth of sea water, and also with a broad gravity low across the GDR, which is the result of thicker continental crust as a consequence of underplating and intraplating (emplacement into the crust). The gravity curves become elevated further towards the ocean (east) direction, possibly reflecting the changes of sea-water depth and sea-floor sediment thickness towards the ocean ridge. Some small (short wavelength) anomalies shown in the observed gravity profile could be the result of local density changes caused by geological bodies such as granite plutons. The modeling of such fine gravity features has not been attempted here because of the scale of the present model.

The geoid anomaly curves are geometrically much simpler than the gravity profiles. The observed and calculated curves both display a wide gentle peak roughly across the GDR–

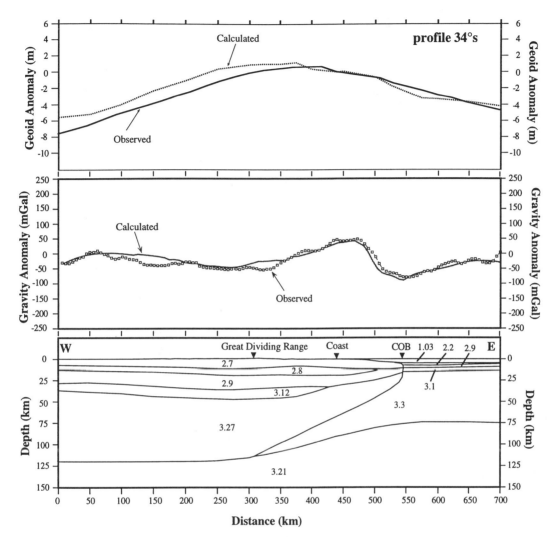

Figure 4. 2D gravity and geoid analyses for the 34°S profile (146° 30', 34°S to 154° 4.7', 34°S) across southeast Australia. The observed gravity anomaly profile is extracted from the Australian Geological Survey Organisation data base and the observed geoid anomaly profile is derived from the contour map published by Allman [1982]; these details also apply to the other models in this paper.

Coast region, below which the mantle wedge exists and the top boundary of the asthenosphere becomes shallower. Their overall shapes match reasonable well, though the convergence for the west part of the section is not as good as for the east part.

From the gravity and geoid modeling above, we have obtained a geological cross section associated with particular geometry-density structures (Figure 4), which generates gravity and geoid anomalies consistent with the observed data. This section contains a lithosphere with two major components, continental and oceanic, with variable thicknesses. The maximum thickness of the continental lithosphere is 120 km, located in the west portion of the section. This thickness decreases gradually in the region where the continental lithosphere becomes thinned and the assumed mantle wedge exists, and reaches about 80 km below the shelf near the COB (Figure 4). This is consistent with the results of seismic velocity models [Muirhead and Drummond, 1991; van der Hilst et al., 1996; B. L. N. Kennett, personal communication, 1996], which suggest that the continental lithosphere is 100-120 km thick in eastern Australia. The thickness of the oceanic lithosphere obtained here for the section is about 75 km. This is a reasonable thickness for the oceanic lithosphere of the Tasman Sea with ages between about 55.5-84 Ma, that is, magnetic anomalies 24 to 34 [Scheibner et al., 1991b], according to the concept of thickening of oceanic lithosphere with age [Forsyth, 1975]. Another important feature of the section is the geometrical configuration of the mantle wedge. In this profile, this wedge extends about 235 km horizontally

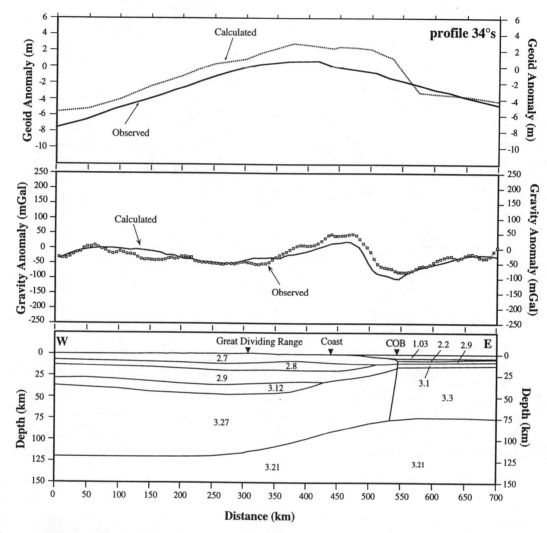

Figure 5. 2D gravity and geoid analyses for the same profile (34°S) as in Figure 4, but the mantle wedge is excluded here and replaced by a steep boundary between the continental lithosphere and the oceanic lithosphere.

between the locations of the GDR and the COB, and has a top-boundary dip angle of about 19.5°.

The density values derived from the gravity and geoid modeling are given in Table 1 (also see Figure 4). These are generally consistent with the data revealed by experimental studies [Clark, 1966; Lama and Vutukuri, 1978; Turcotte and Schubert, 1982] for the rocks or materials relevant to these layers. A significant feature of this density structure is that there exist density contrasts between neighbouring oceanic and continental crustal layers, between the oceanic and continental lithospheric upper mantle, and between the lithospheric mantle and the asthenosphere. These density contrasts are assumed to result from the development of continental and oceanic lithospheres and of partial melting in the asthenosphere, in the context of plate tectonics. They are critical to and are used in the following mechanical analyses.

A sensitivity analysis. To evaluate the sensitivity of the geometry to the results of gravity and geoid modeling, we have constructed another model for the 34°S profile, but with a much steeper contact between the continental and oceanic lithospheric mantle (Figure 5, bottom). Now the transition between the continent and ocean becomes abrupt and the oceanic mantle wedge no longer exists. The modeling results (Figure 5) show that the matches between the observed and calculated gravity-geoid anomalies become poorer for the region across the coast line, where the wedge exists in the previous model (Figure 4). In particular, the sharp drop in calculated geoid anomalies near the COB clearly contrasts with the smooth change of the observed anomalies. Such changes in the calculated gravity and geoid anomalies are obviously caused by the loss of mass due to the exclusion of the denser wedge. This mass loss cannot be realistically compensated for to recover the changes in gravity and geoid anomalies for the lithospheric column, as the crust should not be significantly thinned further and the continuity of the bottom of the lithosphere must be maintained across the COB. The results of this model seem to suggest that the inclusion of the dense mantle wedge in the model is realistic and required.

The 33°S and 35°S Profiles

We performed gravity modeling for two more profiles across the passive margin along the 33°S and 35°S latitude lines (Figures 1 and 6). The two profiles are 740 and 675 km long, respectively, with the same depth of 150 km.

The modeling results for the 33°S and 35°S profiles show that consistency between the observed and calculated gravity anomalies can be achieved in both models, based on the same density assemblies and similar geometries as used in the 34°S profile (Figure 6). This is particularly well demonstrated by the close match between the two curves for the areas across the coast line, which reflects structures including the thinned continental lithosphere, the mantle wedge, and the transition between the continental and oceanic lithospheres.

A major difference between the 33°S, 34°S and 35°S models is that the geometrical configuration of the assumed mantle wedge varies as one progresses along the passive margin (Figures 4 and 6). The wedge extends 204 km, 235 km and 316 km horizontally in the 35°S, 34°S and 33°S models, respectively. The corresponding dip of the wedge's top boundary in these three profiles is 22.3°, 19.5° and 14.6°. These measurements indicate that the mantle wedge becomes longer and less steep from south to north along the passive margin. Such a geometrical feature correlates with the increase of the distance between the GDR and the COB from 35°S to 33°S along the passive margin (see Figure 1). This supports earlier suggestions of Lister and Etheridge [1989], that the detachment geometry (i.e. dip) under the upper-plate passive margin determines the offset of the Passive Margin Mountains from the continental margin.

Summary of Lithospheric Structures

The modeling results presented above have constrained a lithospheric model for southeast Australia, which involves specific passive margin geometries and density contrasts (Figures 4 and 6, Table 1). These are characterised by: a lithosphere with a variable thickness (about 75-120 km), a thinned continental lithosphere near the COB, a dense mantle wedge with particular geometrical configuration beneath the thinned continental margin, and various density contrasts between neighboring layers (e.g., between the oceanic and continental lithosphere). This information is the basis for the mechanical analyses which follow.

STRESS AND DISPLACEMENT PREDICTED BY MECHANICAL ANALYSIS

The geometry and density structures shown in Figure 4 (the 34°S profile) have been reconstructed in the numerical model, using a carefully designed finite difference mesh of 210 × 140 elements (Figure 7). The mechanical parameters assigned to the mesh are given in Table 1, and are consistent with the data obtained experimentally for the relevant rocks or materials [Clark, 1966; Lama and Vutukuri, 1978; Turcotte and Schubert, 1982].

A general feature of the mechanical system is that densities and elastic constants increase with depth, partly reflecting the effect of pressure increase; however, the density and shear modulus of the bottom asthenosphere layer are smaller than those of the neighboring upper-mantle lithosphere (layers 9 and 10) because of partial melting [Jordan, 1981; Muirhead and Drummond, 1991]. Furthermore, the oceanic lithosphere is overall denser than the continental lithosphere [Jordan, 1981]. The modeling was carried out for "container" boundary conditions and stress boundary conditions respectively (Figure 3).

The position of the bottom boundary of the model (150 km) is not important in the solution of the mechanical model. An analysis shows that extending the lower boundary of the model to 250 km by including a deeper portion of homogeneous asthenosphere does not modify the results in any significant way.

Model with "Container" Boundary Conditions – the Jaeger Assumption

The distribution of the finite displacement vectors (Figure 8a) mainly reflects the gravitational effect of density distribution in the simulated section (see Figure 7a). In the east oceanic domain, owing to the density contrast between the oceanic lithosphere and the continental lithosphere, the materials creep overall westwards and downwards. Such material movement represents a major continent-ward gravitational push combined with subsidence on the oceanic lithosphere. Material movement contrary to this dominant pattern is observed in some other domains. In the west part of the section below the crust, the vectors point towards the east, because the lithospheric column near the west side has a thinner crust but thicker lithospheric mantle, and therefore has more mass than the central column. This material movement trend converges with the overall continent-ward push from the mantle wedge and oceanic lithosphere. Because the push from the ocean is stronger than the push from the west, all the displacement vectors change directions here to form a single movement trend in the crust, pointing to the west and expressing anomalous uplift across the GDR. The other location showing a contrary movement is the shallower region across the coast. Here high uplift at the GDR and weak constraints on the shelf from the ocean result in a creep movement towards the east. However, this creep movement

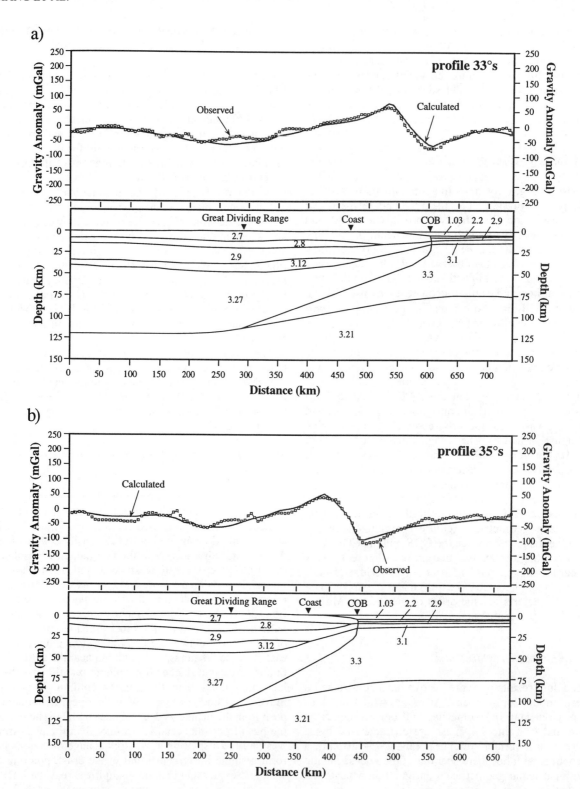

Figure 6. 2D gravity analyses for: a) the 33°S profile (146° 42', 33°S to 154° 37', 33°S); b) the 35°S profile (146° 42', 35°S to 154°, 35°S) across southeast Australia.

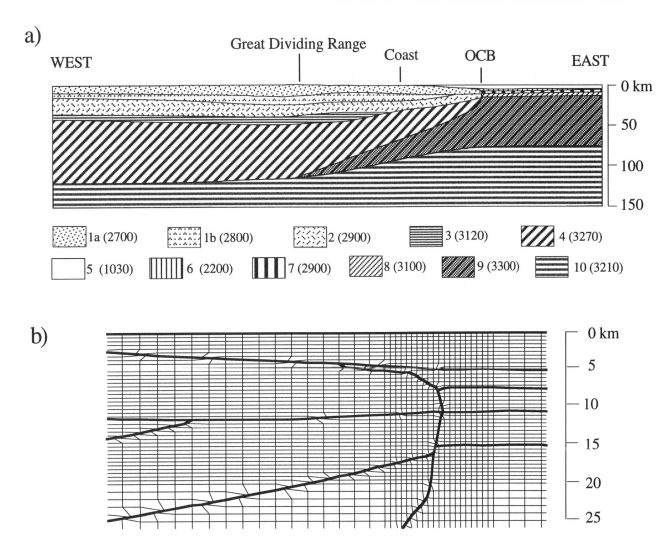

Figure 7. (a) The numerical design for the geometry-density structure in the 34°S profile (see Figure 4, bottom). Layers are labelled from 1a through 10 and numbers in brackets give the densities of the layers in kg.m^{-3}. (b) A small portion of the finite difference mesh underlying the numerical model above. Layer boundaries are highlighted by thicker lines.

terminates at the COB, where it collides with the continentward movement from the ocean.

The contours of the difference between horizontal stress (σ_{xx}) and vertical stress (σ_{yy}) are shown in Figure 8b. A negative value of ($\sigma_{xx} - \sigma_{yy}$) represents a deformation domain dominated by maximum horizontal compression, whilst a positive value indicates maximum vertical compression; σ_{xx} and σ_{yy} are both compressive with negative sign.

We find that a significant part of the continent is in a state of maximum horizontal compression (horizontal compression is greater than vertical compression), though enclosed in it are some pockets of maximum vertical compression located below topographic elevations (e.g., the GDR). Another domain exhibiting major maximum horizontal compression is around the COB. The maxima of ($\sigma_{xx} - \sigma_{yy}$) contours in these horizontal-compression dominated domains coincide with the locations where creep movements with different orientations collide (see Figure 8a). In addition to the small areas already mentioned, some other domains showing maximum vertical compression occur near the east and west side of the section and below the coast (Figure 8b). These are the places where materials generally creep away. We need to emphasise that the difference between σ_{xx} and σ_{yy} is small in this model. One of these stresses is only marginally larger than the other in the bulk of the section with the value of $|\sigma_{xx} - \sigma_{yy}|$ mostly less than 10 MPa.

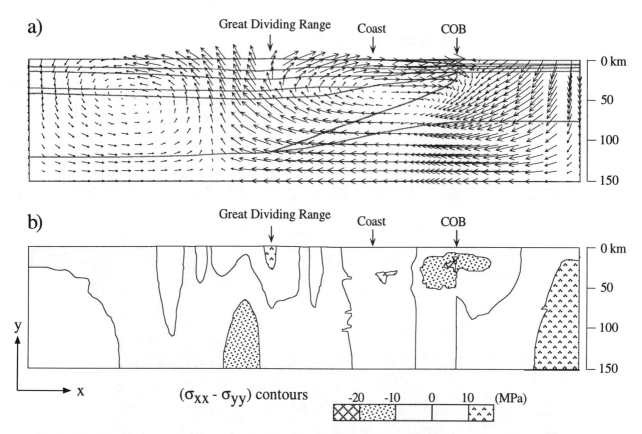

Figure 8. Displacements and stress results for the numerical model with container boundary conditions. (a) Displacement vectors; the length is proportional to the magnitude of displacement (the same in Figures 9, 10 and 12) and the maximum displacement is 13 m. (b) ($\sigma_{xx} - \sigma_{yy}$) contours (the difference between the horizontal stresses and the vertical stresses). Compressive stress is negative (also in the other Figures).

Model with Stress Boundary Conditions – the McGarr Assumption

The pattern of finite displacements (Figure 9a) for the model with stress boundary conditions displays clear differences from that described for the model with container boundary conditions (see Figure 8a). Displacement vectors suggest that the materials creep entirely towards the continent. The magnitude of displacement is also larger (the maximum is 97 m as opposed to 13 m in the "container" model). These observations imply a greater gravitational push on the continent from the oceanic lithosphere and the mantle wedge. Material movement in the vertical direction varies systematically across the section. The oceanic lithosphere is dominated by subsidence, which gradually decreases from the COB towards the coast. Then the vertical movement switches to uplift in the continent, with the highest uplift approximately located at the GDR. Material movement shows a small tendency towards subsidence near the west boundary of the section, corresponding to mass increase in the lithospheric column there.

The ($\sigma_{xx} - \sigma_{yy}$) contours for this model also show remarkable changes from the former model (Figure 9b). The bulk of the section now has negative ($\sigma_{xx} - \sigma_{yy}$) values, indicating the domination of maximum horizontal compression. Maximum vertical compression is observed only in two pockets below the GDR and the coast, and also a small area at depth near the east boundary of the section. The intensity of the horizontal compression has also increased. The ($\sigma_{xx} - \sigma_{yy}$) values are overall larger than those in the previous model (see Figure 8b), with the maxima reaching 30 MPa near the COB and the tip of the wedge.

It is clear that stress boundary conditions intensify oceanic gravitational push and enhance horizontal compression. Applying boundary supporting stress to prevent gravitational collapse of the lithospheric column (see Figure 3b) simulates the effects of remote stresses or an extended lithosphere segment (beyond the section) in the model. In fact, the boundary supporting stress on the oceanic side is larger than on the continental side, as the denser oceanic lithosphere has larger lithostatic stresses than the continental lithosphere at the same depth. Moreover, the stress boundary for the base of the section allows this boundary to move up or down in response to deformation. All these effects represent a more

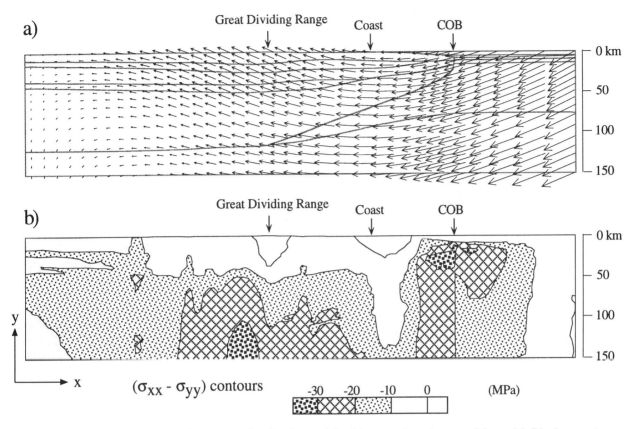

Figure 9. Displacements and stress results for the model with stress boundary conditions. (a) Displacement vectors; the maximum displacement is 96.7 m. (b) ($\sigma_{xx} - \sigma_{yy}$) contours.

realistic tectonic environment for modeling the gravitational push of the oceanic lithosphere. In contrast, with container-boundary conditions (see Figure 3a), he section is treated as a fully independent unit, where the effects of remote stresses are entirely excluded. The fixed lateral boundaries may have a "hold back" effect on creep movement towards the section interior. This weakens oceanic push and horizontal compression.

Effect of additional in-plate compressive stress. Remnant compressive stresses from earlier tectonic events may exist as additional in-plate compressive stresses and may influence the stress field in the region [Denham et al., 1979; Zhang et al., 1996]. To examine this situation, the model presented above is repeated, but a 20-MPa initial compressive horizontal stress is superimposed onto the initial lithostatic stresses.

The modeling results indicate that the displacement field for this model is very similar to that for the previous model (see Figure 9a), because the density-geometry structures and boundary conditions remain unchanged. The pattern of ($\sigma_{xx} - \sigma_{yy}$) contours (Figure 10) is also similar to the previous pattern (see Figure 9b). However, the negative values of ($\sigma_{xx} - \sigma_{yy}$) have changed by -20 MPa, with the maxima reaching -50 MPa. The interpretation is that the whole section is now in more severe horizontal compression, including the small domains below the GDR and the COB which were previously in a state of maximum vertical compression. It seems that the 20 MPa initial horizontal stress increment is largely preserved in the model, superimposed upon stress changes or adjustments resulting from the density-geometry structures and boundary conditions. This is true because the model remains elastic for these stress levels.

The finding of this model is similar to the results of Zhang et al. [1996]. If the crust or the lithosphere behaves elastically or partly elastically, any remaining stress could be preserved and influence the present stress field. An east-west compressive horizontal remnant stress could enhance the horizontal compression generated from the gravitational push related to the dense oceanic lithosphere and mantle wedge.

Influence of plate boundary forces. Although southeast Australia seems not to be influenced by subduction along the east boundary of the plate, because of the seafloor spreading activities in the Lau Basin [e.g., Sclater et al., 1972] and further south in the Havre Trough, such micro-spreading and the convergence along the New Zealand strike-slip plate

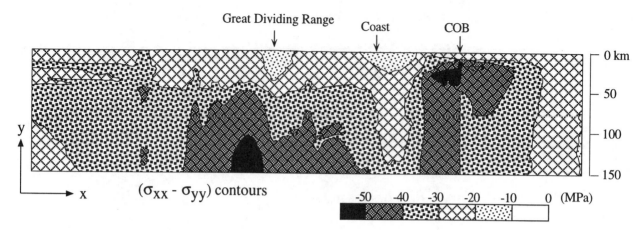

Figure 10. ($\sigma_{xx} - \sigma_{yy}$) contours for the model with stress boundary conditions and also with -20 MPa additional horizontal compressive stress.

boundary may impose a push on southeast Australia. To evaluate the effect of such a speculative plate boundary force, we have constructed a model incorporating an additional compressive supporting stress of 20 MPa on the oceanic lithospheric part of the east side of the section, otherwise identical to the model shown in Figure 9.

The increase in the boundary pressure on the oceanic lithosphere has intensified continent-ward material movement (Figure 11a) and horizontal maximum compression (Figure 11b). The amount of displacement has increased significantly as indicated by a maximum displacement of 156 m, and this increase is dominantly expressed as horizontal components.

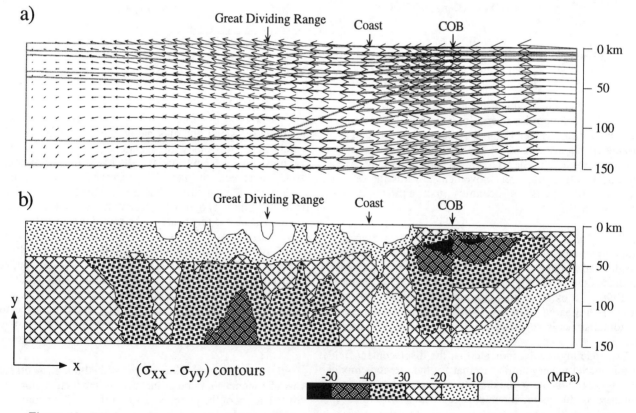

Figure 11. Displacement vectors (a) and ($\sigma_{xx} - \sigma_{yy}$) contours (b) for the model with stress boundary conditions and also with 20 MPa additional supporting stress at the eastern edge of the oceanic lithosphere; the maximum displacement is 156.3 m. See text for further explanation.

The negative values of ($\sigma_{xx} - \sigma_{yy}$) have also increased in comparison to those illustrated in Figure 9b, reflecting stronger horizontal compression. The only location showing a weak vertical maximum compression is a small area right below the GDR. An interesting feature here is that the geometry of the continental crust is clearly manifested in the stress distribution, characterised by negative ($\sigma_{xx} - \sigma_{yy}$) values between 10-20 MPa (a few pockets of smaller values exist below topographic elevations). This is because the effect of the pressure increase at the edge of the oceanic lithosphere is partitioned into the continental crust and the layers below in different manners.

DISCUSSION

Lithospheric Geometry and Density Structure

We have developed a lithospheric structure model for southeast Australia which is consistent with the gravity and geoid observations. If we assume that the gravity and geoid analyses adopted here give reasonable constraints on the density and geometry structures, the geometry and density parameters incorporated in the model of Zhang et al. [1996] need to be modified.

The lithosphere in the model of Zhang et al. [1996] has a thickness ranging from 125 km (continent) to 90 km (ocean). A 125-km thick continental lithosphere may be appropriate for southeast Australia according to the present gravity-geoid modeling and interpretation of seismic velocities [Muirhead and Drummond, 1991]. However, oceanic lithosphere with a minimum thickness of 90 km seems to be too thick, in comparison with the thickness of 75 km defined by this study. The 90 km thickness may be reasonable for the oldest parts of the oceanic lithosphere in the Tasman Sea (84 to 74 Ma [Scheibner et al., 1991b]). However, the oceanic lithosphere in our sections is younger than 74 Ma, and theoretical age-thickness relations [Forsyth, 1975] suggest a thickness around 80±5 km. A slightly thinner lithosphere might be related to heat input during the passage of the lithosphere over the Tasman hot spots [McDougall and Duncan, 1988; Sutherland, 1994] which possibly delayed the thermal maturation.

Density parameters used by Zhang et al. [1996, Figure 1a and Table 1] are such that the density contrast is about 6.12% between the oceanic and continental crust, 2.57% between the oceanic and continental lithospheric upper mantle, and -3.1 % between the asthenosphere and the lithospheric upper mantle. In contrast, these parameters derived from gravity and geoid analyses are about 4.16%, 0.92% and -2.3%, respectively (see Table 1 and Figure. 7a). Therefore, the density contrasts between the oceanic and the continental part in the model of Zhang et al. [1996] are probably too large; however, larger density contrasts between the oceanic and continental upper mantle have been used in other published lithospheric models (e.g., 3.03% in Couch

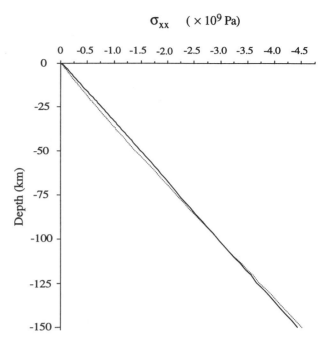

Figure 12. Variation of initial horizontal stress with depth at a location in the section modeled by Zhang et al. [1996] for two different ways to generate initial stresses. Solid line gives the initial stress in the model of Zhang et al. [1996], which assumes that the lithostatic stresses at the base of the section equal 4.5 × 10^9 Pa (close to the value of equation (2) at the base), and then vary to zero to the top linearly. The dashed line gives the initial stress precisely according to equation (2).

and Riddihough [1989]). The consequence of larger density contrasts between the ocean and the continent, similar to that of a thicker oceanic lithosphere, is that the oceanic lithospheric column in the model has excessive mass relative to the true situation.

Geodynamics

The current mechanical model predicts clearly less intense horizontal compression in comparison with the results of Zhang et al. [1996, Figure 5]. This is partly because Zhang et al. [1996] has larger density contrasts between continental and oceanic layers and a thicker oceanic lithosphere which generate stronger continent-ward push. The other important reason is that Zhang et al. [1996] adopted a simplified scheme to set up initial stresses, overlooking density inhomogeneity. This leads to larger initial horizontal stresses in shallower layers, and smaller horizontal stresses in deeper layers, than the initial stresses proposed by equation (2) (Figure 12). The effects of such slightly anomalous initial horizontal stresses are largely preserved in the final results since the lateral boundaries of the model are fully constrained in the horizontal direction under the container boundary conditions and the material is essentially elastic throughout.

104 ZHANG ET AL.

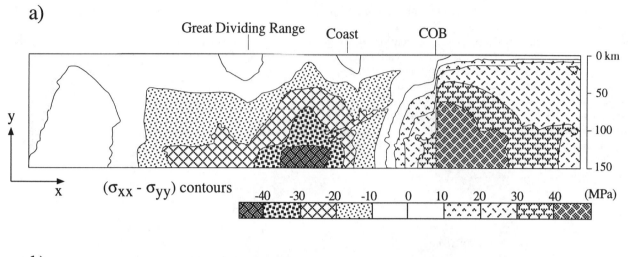

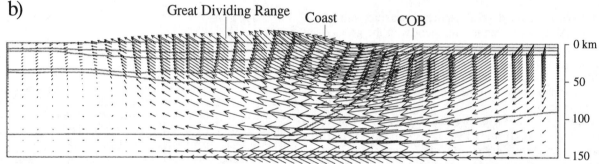

Figure 13. Reconstruction of one of Zhang et al.'s numerical models [1996, ref. Figures 3, 5b and 7a]. The only change in this model relative to the original model is that the initial stress is set up according to equation (2); the geometries, densities and container boundary conditions remain unchanged. (a) ($\sigma_{xx} - \sigma_{yy}$) contours; (b) Displacement vectors; the maximum displacement is 52.4 m.

As a result, the shallower layers show more intense horizontal compression, with the maximum vertical compression localised in the deeper layers. Modifying the model using the realistic initial stress (the same densities, geometries and boundary conditions) gives different stress results (Figure 13a); the pattern of displacement remains unchanged (Figure 13b). Now the maximum horizontal compression is weakened and localised in the continent, with the maximum vertical compression dominating the oceanic part.

The mechanical results of this study demonstrate that the maximum horizontal compression could occur in the bulk of the modeled 2D section owing to regional geometry-density structures. The projection of this into the 3D space of southeast Australia means that the regional maximum compression should be roughly horizontally east-west oriented (normal to the passive margin), essentially in line with the bulk of the measured stress orientations in the region [Denham et al., 1979; Denham, 1988; Enever et al., 1990]. The mechanism behind this is that the dense oceanic lithosphere and mantle wedge induce gravity relaxation body forces, pushing the continent and expressed as a material creep movement towards the continent (consistent with the results of Vetter and Meissner [1979]]. These features are more prominent in the models with stress boundary conditions. Maximum vertical compression can be expected in the shallower locations below topographic elevations (e.g., the GDR), and at the coast or deep locations where materials creep away, where horizontal extension is possible. Furthermore, any east-west horizontal compressive remnant stress or any compressive plate boundary force from the east boundary could further enhance the magnitude of horizontal compression stresses.

Therefore, the results of this study support the speculative model of Zhang et al. [1996] regarding the origin of regional stresses. In southeast Australia, the density-geometry related gravitational body forces (in conjunction with remnant stresses or eastern-plate boundary force, if any), and the plate motion from the southern ridge boundary northwards, could be two complementary mechanisms controlling the regional stress regime. They tend to generate compression directions almost orthogonal to each other in the horizontal plane.

Commonly, the former mechanism and the associated stress field become marginally dominant in the bulk of the region. The resultant maximum horizontal compression is mostly approximately normal to the COB. However, local topography (e.g., the GDR and the coast) and/or geological structures modify local stresses in the shallow crust near the surface. If the effect of these factors is to weaken the role of the regional body force or to create horizontal gravitational extension, the plate-motion induced stress then becomes dominant and the local maximum compression changes orientation sharply. This creates an observed stress field with variable maximum compression orientations or, in some cases, almost biaxial stresses, although the dominating orientation is approximately normal to the COB [see Zhang et al., 1996, Figure 1]. This is consistent with the observation that the maximum compression parallel to the direction of plate motion appears to be localised at the edge of the continent [Denham et al., 1979; Enever et al., 1990]. Furthermore, topographic uplift predicted by this model (the highest uplift across the GDR and subsidence at the ocean) generally coincides with the present topographic framework.

The hypothesis developed here that the density-geometry related gravitational body forces contribute to high horizontal compression at continental margins may hold in general for other continents, in particular those with upper-plate margins. The world stress map [Zoback et al., 1989; Zoback, 1992] shows that the orientations of the maximum horizontal compression at continent margins are mostly dominantly at high angles to the margins, for example, in South Africa and North and South America. Of course, the situation in North and South America differs from southeast Australia, where the direction of plate motion is normal to the COB. Any compression due to the body forces aligns with and intensifies the compression due to plate motion.

In this study, we cannot truly evaluate the magnitude-orientation relationship between the principal stresses on the horizontal plane, because our 2D model does not predict the stresses in the north-south direction which is orthogonal to our profile. Three-dimensional stresses will be investigated in our future 3D model. Furthermore, the continent-ward material creep movement predicted by this study gives only the material movement pattern associated with the modeled density-geometry structure. In real lithosphere, such movement may gradually slow down as a result of lithospheric isostatic adjustment and thermal maturity, but could survive for a long time if the strength of lithosphere prevents a strict local isostatic equilibrium. The relationship between material movement and geological time scales can be evaluated in future through an advanced viscous/brittle model.

Temporal Development of southeast Australia

The development of southeast Australia seems to be an integral part of passive margin evolution since the breakup of Gondwanaland in Jurassic-Cretaceous time, being part of Pangean oscillation after the singularity of Pangea at 230 Ma [Veevers, 1990]. The first event is continental extension, caused probably by mantle upwelling (and/or thermal activity), and the related uplift during the dispersal and breakup of Gondwanaland [Veevers, 1990], associated with Jurassic intraplate igneous activity [Scheibner, 1979]. These were followed by the formation of an extensional lithospheric detachment and the onset of the eastern passive-margin development in Jurassic-Early Cretaceous.

The displacement along the main detachment and the development of passive margins appear to follow the "lithospheric wedge delamination" models of Lister et al. [1991]. As the lower plate was pulled out by simple shearing, asthenospheric upwelling developed beneath the detachment in the toe region. This upwelling caused new intraplate igneous activity and underplating, resulting in further thermal uplift. As extension continued, the thermally weakened upper mantle and lower crust close to the detachment appear to have deformed by pure shearing. This led to thinning of the lithosphere, which was compensated by further igneous underplating.

The asymmetric breakup finally occurred in the Late Cretaceous [Jongsma and Mutter, 1978]. Seafloor spreading commenced at about 84 Ma in the Tasman Sea [Veevers et al., 1991] and continued into the Paleocene, terminating around 54 Ma. The earliest oceanic crust formed in the above process started to subside after about 30 Myr, owing to thermal decay (i.e. the crust is heavier). Similarly the upper-mantle wedge which formed due to igneous underplating under the edge of the upper-plate margin, also becomes thermally mature and therefore denser. This means that probably already in the Eocene at about 54 Ma, the mantle wedge and oceanic lithosphere started to exert the continent-ward gravity relaxation-induced compressive stress predicted by this study. It is interesting that this is when the South Australian Highlands originated [Wellman and Greenhalgh, 1988], and other morphological effects including basin formation and changes in drainage were observed [Veevers, 1984].

Such a continent-ward compression should have become intensified towards the Miocene and until the present day. This dynamic system may have continuously contributed to the distribution of uplift in the eastern highlands of southeast Australia, such as anomalous uplift observed across the GDR, and to the activation of various regional reverse faults affecting Miocene-Oligocene strata and morphological features.

CONCLUSIONS

Gravity and geoid modeling for east-west profiles across the passive margin in southeast Australia has enabled us to constrain a lithospheric-structure model for this region. The model has specific geometry and density configurations, and is compatible with regional gravity and geoid observations. The main features of this southeast Australian lithospheric structure model include: upper-plate passive-margin geometries, a lithosphere with gradually variable thickness (about 75-120 km), a thinned continental lithosphere near the continental boundary, a dense mantle wedge beneath the thinned continental margin, and density contrasts between neighboring layers, particularly between the oceanic and continental lithosphere.

Mechanical modeling of the above system demonstrates that creep movements, dominantly towards the continent, occurred as a result of gravity relaxation associated with the geometry-density configuration of southeast Australia. The rise of the Great Dividing Range is consistent with this displacement field. Such material movement creates body forces acting roughly normal to the passive margin (and the continental boundary), and these body forces can generate horizontal compression greater than that in the vertical direction in the region. This direction is roughly normal to the passive margin and consistent with the bulk of the measured maximum-stress orientations in the region. The involvement of any remnant horizontal compressive stress, or any stress related to boundary forces at the eastern plate boundary, could enhance such a stress pattern further.

It is therefore speculated that the present intraplate stress field in south-east Australia may be a combination of the stress generated gravitationally by the upper-plate passive-margin setting and the stress associated with plate rotation/motion, further superimposed upon local topographic and structural effects. This results in a highly variable complex stress regime with a slight dominance of east-west maximum compression, but some local instances of nearly north-south maximum compression.

Acknowledgements. We would like to acknowledge the Australian Geological Survey Organisation for the use of gravity data. Special thanks go to John Creasey and Peter Wellman for processing gravity data and extracting gravity anomaly profiles. The manuscript was improved from reviews by Mike Sandiford and an anonymous reviewer. Ken Baxter is thanked for his comments on an earlier version of this manuscript. Stuart Craig is thanked for useful suggestions and comments on the work and particularly pointing out the importance of initial stress assumptions. Discussions with Ken Sharp are greatly appreciated; the views expressed in this paper concerning steep reverse faults in southeast Australia are our views and are not necessarily Ken's views. This paper is published with the permission of the Director of the Australian Geodynamics Cooperative Research Centre.

REFERENCES

Allman, J. S., A geoid for south-east Asia and the Pacific, *Aust. J. Geod. Photo. Surv., 36,* 59 -63, 1982.

Beavis, F. C., The Tawonga Fault, north-east Victoria, *Proc. Roy. Soc. Victoria, 72,* 95-100, 1960.

Brown, E. T., and C. R. Windsor, Near surface *in situ* stresses in Australia and their influence on underground construction, in *Proc. IEA Seventh Australian Tunnelling Conference,* pp. 18-48, 1990.

Clark, S. P. Jr. (ed.), *Handbook of Physical Constants,* 587 pp., *Geol. Soc. Am. Memoir, 97,* New York, 1966.

Cloetingh, S., and R. Wortel, Stress in the Indo-Australian Plate, *Tectonophysics, 132,* 49-67, 1986.

Coblentz, D. D., R. M. Richardson, and M. Sandiford, On the gravitational potential of the Earth's lithosphere, *Tectonics, 13,* 929-945, 1994.

Coblentz, D. D., M. Sandiford, R. M. Richardson, S. H. Zhou, and R. Hills, The origins of the intraplate stress field in continental Australia, *Earth Planet. Sci. Lett., 133,* 299-309, 1995.

Couch, R. W., and R. P. Riddihough, The crustal structure of the western continental margin of North America, in *Geophysical Framework of the Continental United States, Memoir 172,* edited by L. C. Pakiser and W. D. Mooney, pp. 103-128, The Geological Society of America, 1989.

Cundall, P. A., and M. Board, A microcomputer program for modelling large-strain plasticity problem, in *Numerical Methods in Geomechanics, Proc. 6th Int. Conf. on Numerical Methods in Geomechanics,* edited by C. Swoboda, pp. 2101-2103, Balkema, Rotterdam, 1988.

Denham, D., Australian seismicity - the puzzle of the not so stable continent, *Seism. Res. Lett., 49,* 289-295, 1988.

Denham, D., L. G. Alexander, and G. Worotnicki, Stress in the Australian crust: evidence from earthquakes and in-situ stress measurements, *BMR J. Austr. Geol. Geophys., 4,* 289-295, 1979.

Dooley, J. C., Velocity variations and isostatic compensation in the Australian region, in *The Australian Lithosphere, Geol. Soc. Spec. Publ. 17,* edited by B.J. Drummond, pp. 41-58, 1991.

Enever, J. R., R. J. Walton, and C. R. Windsor, Stress regime in the Sydney basin and its implications for excavation design and construction, in *Proc. IEA Seventh Australian Tunnelling Conference,* pp. 49-59, 1990.

Etheridge, M. A., P. A. Symonds, and G. S. Lister, Application of the detachment model to reconstruction of conjugate passive margins, in *Extensional Tectonics and Stratigraphy of the North Atlantic Margins, Am. Assoc. Petrol. Geol., Mem. 46,* edited by A. J. Tankard, and H. R. Balkwill, pp. 23-40, 1989.

Forsyth, D. W., The early structural evolution and anisotropy of the oceanic upper mantle, *Geophys. J. R. Astr. Soc., 43,* 103-162, 1975.

Gunn, R., A quantitative study of isobaric equilibrium and gravity anomalies in the Hawaiian Islands, *Jour. Franklin Instit., 236,* 373-390, 1943.

Haxby, W. F., and D. L. Turcotte, On isostatic geoid anomalies, *J. Geophys. Res., 83,* 5473-5478, 1978.

Hobbs, B. E., H.-B. Mühlhaus, and A. Ord, Instability, softening and localisation of deformation, in *Deformation Mechanisms, Rheology and Tectonics, Geol. Soc. Spec. Pub. 54,* edited by R. J. Knipe and E. H. Rutter, pp. 143-165, 1990.

Itasca, *FLAC: Fast Lagrangian Analysis of Continua,* user manual, version 3.2, Itasca Consulting Group Inc., Minneapolis, 1992.

Jaeger, J. C., and N. G. W. Cook, *Fundamentals of Rock Mechanics*, 593 pp., Chapman and Hall, London, 1979.

Johnson, R. W. (ed.), *Intraplate Volcanism in Eastern Australia and New Zealand*, Australian Academy of Sciences, 408 pp., Press Syndicate of University of Cambridge, 1989.

Jongsma, D., and J. C. Mutter, Non-axial breaching of a rift valley: evidence from Lord Howe Rise and the south-eastern Australian margin, *Earth Planet. Sci. Lett., 39*, 226-234, 1978.

Jordan, T. H., Continents as a chemical boundary layer, *Phil. Trans. Roy. Soc. Lond., A301*, 359-373, 1981.

Lama, R. D., and V. S. Vutukuri, *Handbook on Mechanical Properties of Rocks—Testing Techniques and Results (Volume II)*, 515 pp., Trans Tech Publications, Clausthal, Germany, 1978.

Lister, G. S., and M. A. Etheridge, Detachment models for the uplift and volcanism in the eastern Highlands, in *Intraplate Volcanism in Eastern Australia and New Zealand*, edited by R. W. Johnson, pp. 297-313. Cambridge University Press, Cambridge, UK, 1989.

Lister, G. S., M. A. Etheridge, and P. A. Symonds, Detachment faulting and evolution of passive continental margins, *Geology, 14*, 246-250, 1986.

Lister, G. S., M. A. Etheridge, and P. A. Symonds, Detachment models for the formation of passive continental margins, *Tectonics, 10*, 1038-1064, 1991.

McDougall, I., and R. A. Duncan, Age progressive volcanism in the Tasmantid Seamounts, *Earth Planet. Sci. Lett., 89*, 207-220, 1988.

McGarr, A., On the state of lithospheric stress in the absence of applied tectonic forces, *J. Geophys. Res., 93*, 13609-13617, 1988.

Muirhead, K. J., and B. J. Drummond, The base of the lithosphere under Australia, in *The Australian Lithosphere, Geol. Soc. Spec. Publ., 17*, edited by B. J. Drummond, pp. 23-40, 1991.

Ord, A., Deformation of rock: a pressure-sensitive, dilatant material, *PAGEOPH, 137*, 337-366, 1991.

Ranalli, G., *Rheology of the Earth*, 365 pp., Allen and Unwin, London, UK, 1987.

Roach, M., *Model2d Version 3.0, User Manual*, 23 pp., The University of Tasmania, 1993.

Sandiford, M., D. Coblentz, and R. M. Richardson, Focusing ridge-torques during continental collision in the Indo-Australian Plate, *Geology, 22*, 831-834, 1995.

Scheibner, E., Geological significance of some lineaments in New South Wales. *LANDSAT79 Proceedings*, pp. 318-332, 1st Australian Remote Sensing Conference, Sydney, 1979.

Scheibner, E., Influence of detachment-related passive margin geometry on subsequent active margin dynamics: applied to the Tasman Fold Belt System, *Tectonophysics, 214*, 401-416, 1992.

Scheibner, E., C. McA. Powell, and R. Spencer, *Broken Hill-Sydney-Tasman Sea Transect, New South Wales, Eastern Australia (2 sheets), Global Geoscience Transect 5, Expl. Notes*, 29 pp., Inter-Union Commission on the Lithosphere and American Geophysical Union, 1991a.

Scheibner E., T. Sato, and C. Craddock, Tectonic Map of the Circum-Pacific Region, Southwest Quadrant, U.S. Geological Survey, 1991b.

Sclater, J. G., J. W. Hawkins, J. Mammerickx, and C. D. Chase, Crustal extension between the Tonga and Lau Ridges: petrologic and geophysical evidence, *Geol. Soc. Am. Bull., 83*, 505-518, 1972.

Simpson, R. W., and R. C. Jachens, Gravity methods in regional studies, in *Geophysical Framework of the Continental United States*, edited by L. C. Pakiser, and W. D. Mooney, pp. 35-102, The Geological Society of America, Memoir 172, 1989.

Stephenson, R., and K. Lambeck, Isostatic response of the lithosphere with in-plane stress: application to central Australia, *J. Geophys. Res., 90*, 8581-8588, 1985.

Sutherland, L., Tasman Sea evolution and hot spot trails, in *Evolution of the Tasman Sea Basin*, edited by G. van der Linger, K. M. Surauson, and R. Muir, pp. 35-51, A. A. Balkema, 1994.

Symonds, P. A., and J. B. Willcox, The gravity field of offshore Australia, *BMR J. Aust. Geol. Geophys., 1*, 303-314, 1976.

Talwani, M., J. L. Worzel, and M. Landisman, Rapid gravity computations for two-dimensional bodies with application to the Mendocino submarine fracture zone, *J. Geophys. Res., 64*, 49-59, 1959.

Turcotte, D. L., and G. Schubert, *Geodynamics: Applications of Continuum Physics to Geological Problems*, 450 pp., Wiley, New York, 1982.

van der Hilst, R. D, A. Zielhuis, and B. L. N. Kennett, Mapping seismic structure in the Australian lithosphere - the SKIPPY project, in *Geological Society of Australia, Abstracts No. 41*, p. 453, 13th Australian Geological Convention, Perth, 1996.

Veevers, J. J. (ed.), *Phanerozoic Earth History of Australia*, 418 pp., Oxford Univ. Press, Oxford, 1984.

Veevers, J. J., Tectonic-climatic supercycles in the billion-year plate-tectonic eon: Permian Pangean icehouse alternated with Cretaceous dispersed-continents greenhouse, *Sediment. Geol., 68*, 1-16, 1990.

Veevers, J. J., C. McA. Powell, and S. R. Roots, Review of seafloor spreading around Australia: I. Synthesis of the pattern of spreading, *Aust. J. Earth Sci., 38*, 373-389, 1991.

Vening Meinesz, F. A., Gravity over the Hawaiian archipelago and over the Madeira area, *Koninklijke Nederlandsche Akademie van Wetenschappen Proceedings, 44*, 1-12, 1941.

Vermeer, P. A., and R. de Borst, Non-associated plasticity for soils, concrete and rock, *Heron, 29*, 1-64, 1984.

Vetter, U. R., and R. O. Meissner, Rheologic properties of the lithosphere and applications to passive continental margins, *Tectonophysics, 59*, 367-380, 1979.

Walcott, R. I., Gravity, flexure, and the growth of sedimentary basins at a continental edge, *Geol. Soc. Amer. Bull., 83*, 1845-1848, 1972.

Wellman, P., Regional variation of gravity, and isostatic equilibrium of the Australian crust, *BMR J. Aust. Geol. Geophys., 1*, 297-302, 1976.

Wellman, P., On the isostatic compensation of Australian topography, *BMR J. Aust. Geol. Geophys., 4*, 373-382, 1979.

Wellman, P., and S. A. Greenhalgh, Flinders/Mount Lofty Ranges, South Australia: their uplift, erosion, and relationship to crustal structure, *Roy. Soc. South Aust. Trans., 112*, 11-19, 1988.

Zhang, Y., E. Scheibner, A. Ord, and B. E. Hobbs, Numerical modelling of crustal stresses in the eastern Australian passive margin, *Aust. J. Earth Sci.*, *43*, 161-175, 1996.

Zoback, M. L., First- and second-order patterns of stress in the lithosphere: the world stress map project, *J. Geophys. Res.*, *97*, 11703-11728, 1992.

Zoback, M. L., M. D. Zoback, J. Adams, *et al.*, Global patterns of tectonic stress. *Nature*, *341*, 291-298, 1989.

S. J. D. Cox, CSIRO Exploration and Mining, PO Box 437, Nedlands, W.A. 6009, Australia

B. J. Drummond, Australian Geological Survey Organisation, GPO Box 378, Canberra, ACT 2601, Australia

B. E. Hobbs, CSIRO Exploration and Mining, Private Bag, Wembley, W.A. 6014, Australia

A. Ord, CSIRO Exploration and Mining, PO Box 437, Nedlands, W.A. 6009, Australia

E. Scheibner, CSIRO Exploration and Mining, PO Box 136, North Ryde, NSW 2113, Australia

Y. Zhang, CSIRO Exploration and Mining, PO Box 437, Nedlands, W.A. 6009, Australia

The Mount Isa Geodynamic Transect - Crustal Implications

B. R. Goleby
Australian Geological Survey Organisation, Canberra, Australia

T. MacCready
Department of Earth Sciences, Monash University, Clayton, Victoria, Australia

B. J. Drummond and A. Goncharov
Australian Geological Survey Organisation, Canberra, Australia

The Mount Isa Inlier of northern Australia is a Proterozoic region renowned for the number of world-class ore deposits found within. Structurally, it has been sub-divided into three regions, an Eastern Fold Belt and a Western Fold Belt, separated by the Kalkadoon-Leichhardt Belt. Data from a regional 250 km deep seismic reflection transect indicates a marked difference in the structural style in the top 5-10 km between Eastern and Western Fold Belts and a major mid-crustal high-velocity anomaly. The Eastern Fold Belt contains a series of young, shallow, west-dipping features which are interpreted as reverse faults. These faults cut a highly reflective, shallow, east-dipping reflective zone that lies beneath almost the entire Eastern Fold Belt. We interpret this as a major, regional detachment surface. In contrast, the Western Fold Belt contains several steeper, highly reflective, west-dipping fault zones which suggest a 'thick-skinned' style of deformation with west-dipping or nearly vertical faults and no evidence of major shallow-dipping structures. Key boundary faults or domain faults, e. g. May Downs, Pilgrim and Fountain Range Faults, are interpreted to be near-vertical strike-slip faults within the upper crust.

INTRODUCTION

The Mount Isa Inlier in northwest Queensland, Australia (Figure 1), represents a portion of a Middle Proterozoic mobile belt over 200 km wide and 600 km long. The main north-south structural fabric of this area developed between ~1600 Ma and ~1500 Ma during east-west crustal shortening associated with the Isan Orogeny [Blake, 1987]. Recent work [O'Dea and Lister, 1997] has been effective in outlining the history of ~200 Myr of ensialic rifting that preceded this orogeny. However, little is known of the tectonic setting during the Isan Orogeny.

In 1994 the Australian Geodynamics Cooperative Research Centre (AGCRC) undertook a multidisciplinary study of the Mount Isa Inlier that included deep seismic-reflection traverses, seismic-refraction profiles, potential field studies and geological mapping [Drummond et al., 1995]. The study was concentrated in the central portion of the Inlier, trending east-west, across the strike of the major structural fabric (Figure 1). It extends to the geophysically defined limits of the mobile belt in both directions, and runs across the most intensely studied strip of the inlier [Stewart and Blake, 1992].

The Mount Isa Inlier is divided into three regions, an Eastern Fold Belt and a Western Fold Belt, separated by a block of competent basement rocks referred to as the Kalkadoon-Leichhardt Belt (Figure 1). In this study, we include both the Eastern Fold Belt proper and the Mary Kathleen Zone of Blake [1987] in our 'Eastern Fold Belt', because the early east-directed thin-skinned style of deformation can be traced across both belts.

The Eastern Fold Belt is a north-south trending region of tight folding and west-directed thrusting, developed during the early east-west shortening of the Isan Orogeny. This

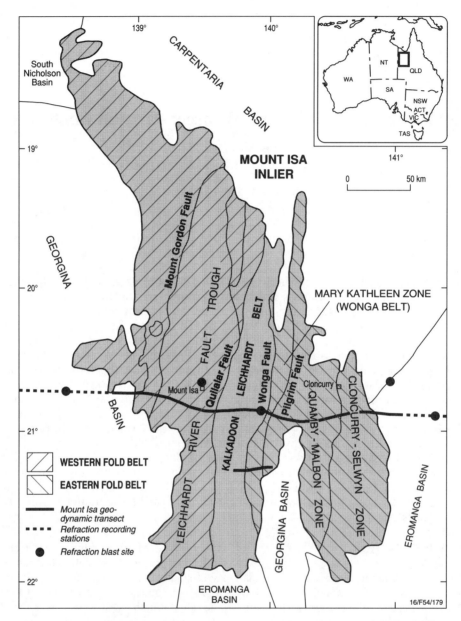

Figure 1. Locality map of the Mount Isa Inlier, subdivided into its main tectonic provinces and showing the location of both the crustal refraction profile and the deep seismic reflection traverse, Mount Isa Geodynamic Transect. Map modified from Blake [1987].

fold and thrust belt is cut by a major north-south fault, the Pilgrim Fault, which separates the Mary Kathleen Zone from the Quamby-Malbon Zone (Figure 1). To the east of the Pilgrim Fault, the Eastern Fold Belt is intruded by late voluminous granitoids of the Williams/Naraku suite. The western third of the seismic transect crosses the Western Fold Belt, a region of distinctly different character from the fold-thrust phase in the Eastern Fold Belt. This section crosses the Leichhardt River Fault Trough, an economically important region of the Mount Isa Inlier just to the south of the major Pb/Zn and Cu mineralization at Mount Isa. The Trough is intruded by granitoids of the Sybella Granite suite. The intensity of deformation decreases rapidly west of the Sybella Granite, so that deformation associated with the Isan Orogeny is essentially absent at the western end of the transect.

This paper describes a new crustal section for the Mount Isa Inlier based on the recent deep seismic reflection and refraction profiles, obtained as part of the AGCRC's multidisciplinary study of the Mount Isa region. The paper then presents the main tectonic implications of the cross section and investigates several of the main metallogenic implications of this model.

DEEP SEISMIC PROFILING TRANSECT

Two deep seismic transects were recorded. The main transect was approximately 250 km in length, crossing the Inlier just south of the main mineralized regions (Figure 1). It was recorded using an explosive energy source recorded into a 120 channel seismic array [Drummond et al., 1995]. The array interval was 40 m, with the shot interval nominally 240 m. Seismic data were recorded over a 20-s recording length at a 2-ms sample rate. A second, shorter traverse was recorded to the south, where the Wonga Belt thickens (Figure 1). Similar acquisition parameters were used on this traverse.

A conventional processing sequence was used for both traverses, with additional emphasis on determining optimum static corrections and velocity corrections. A detailed refraction static-analysis and processing methodology was required to adjust for the effects of the weathering. Near-surface velocity variations across the Inlier are great, varying from around 1000 m/s within parts of the regolith, up to around 6500 m/s in more mafic bedrock material in several areas.

The length of individual reflector segments is variable, but is usually far shorter that that encountered within sedimentary basins. However, reflector amplitude is often excellent, and the interpretation strategy is based on identifying regions of similar reflector coherency and dip and their linkage to the known surface geology. These regions of similar coherency and dip are used to define geologically distinct terrains, the boundaries of which are in many cases defined by shear zones or fault planes. In addition, any unusual or spurious reflections were investigated as being possibly from out-of-plane by their consistency with the regional geological trends. Diffractions and P-S reflections were identified using conventional diffraction curves and P-S ratios. The orientation of the seismic profile and the regional geological trends resulted in few out-of-plane events being recorded, the majority of the reflections being consistent with the geological trends.

Both unmigrated and migrated sections were produced and used in the interpretation; however most illustrations within this paper use the unmigrated seismic sections because they show details within the section more clearly. Migration of the short reflector lengths results in the typical deterioration in resolution within the migrated sections, and so the unmigrated sections are used for clarity.

The seismic-refraction/wide-angle profile was coincident with the deep seismic-reflection transect, but was extended both to the east and west, well into the adjacent Phanerozoic Basins to ensure that the neighboring crustal velocity field was sampled. The maximum shot-to-receiver offset used was approximately 390 km, which produced good crust-mantle refractions. In addition to the shots recorded as part of the refraction experiment, three additional large mine blasts were also recorded as part of the survey, extending the data coverage to approximately 500 km.

The resulting interpreted sections from both seismic surveys have effectively revealed the deep architecture of the crust, and show the structures related to the Isan Orogeny. The combination of the reflection and refraction interpretations have helped define a two-stage evolution that records a shift from thin- to thick-skinned deformation, and reveals a full crustal architecture, including major mid-crustal mafic layers.

MOUNT ISA SEISMIC REFRACTION RESULTS

The refraction/wide-angle seismic profile (Figure 1) recorded refracted events from within the Mount Isa Inlier, as well as from the crust to the east and west of the Inlier. These results indicate that the crust beneath the Mount Isa terrane is ~55 km thick, with a transition from lower crustal material to mantle material occurring over a broad zone ~15 km thick [Goncharov et al., 1997]. There is a significant lateral variation in seismic velocity distribution along the transect (Figure 2). Low-velocity layers are common in the crust and in the crust-mantle transition zone. Lateral variations in velocity within the upper 15 km of the crust can be seen, with relatively lower velocities (5.7-6.1 km/s) more common within the Western Fold Belt. These lower velocities may indicate the major influence of fault zones on the physical parameters along this part of the transect. Compared to this, the Eastern Fold Belt looks more consolidated, with typical velocities of 6.1 km/s and higher prevailing in the upper 15 km.

The most intriguing feature of the model is the presence of two anomalous high-velocity bodies (Figure 2). These bodies have seismic velocities higher than 7 km/s and are ~100 km long in cross section, ~5 km thick, and, based on the seismic velocities, probably contain mafic to ultramafic material. The shallower body occurs in the middle crust beneath the Eastern Fold Belt at a depth of 20-30 km dipping shallowly west [Goncharov et al., 1997]. The deeper body occurs beneath the Western Fold Belt and merges with the transitional lower crust (Figure 2).

Although the two high-velocity bodies are not linked, they have similar dips, and are aligned to define one large, shallowly west-dipping zone in the middle crust. This geometry implies some fundamental structural control, and such a large structure is expected to have had an important influence on the tectonic evolution on the Mount Isa Inlier.

MOUNT ISA SEISMIC REFLECTION RESULTS

Seismic data from the main traverse within the Eastern Fold Belt image a zone of prominent subhorizontal

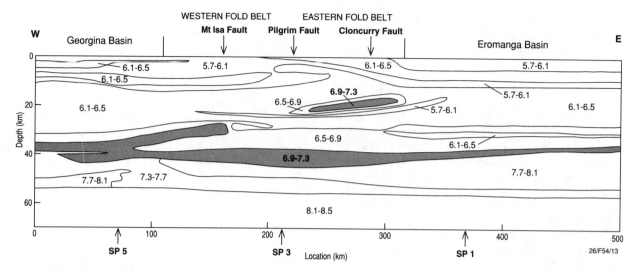

Figure 2. Velocity model from the refraction/wide-angle seismic profile showing the seismic velocity distribution throughout the crust. Several of the key surface structural features are marked.

reflections at depths ranging from 4-9 km, averaging ~6 km, cut by several steeper reflecting zones (Figure 3). Supracrustal rocks are folded at the surface into a series of major anticlines (Mittakoodi folds), including the broadly folded Duck Creek Anticline and the overturned, tightly folded Bulonga Anticline (Figure 3). Some of these folds verge west in the hanging walls of low-angle thrusts such as the Roos Mine Thrust [see also Huang, 1994] and Cloncurry Fault (Figure 4). The tight folding in this anticlinal region contrasts with the lack of strain apparent in the subhorizontal reflections imaged between 1-3 s two-way-traveltime, and we interpret this as indicating that detachment tectonics were active. Thus the Eastern Fold Belt is part of a fold and thrust belt with a thin-skinned tectonic style. We also infer that the subhorizontal reflective zone represents the top of crystalline basement, where a rheological contrast may have focussed the decoupling.

The Soldiers Cap Group, at the eastern end of the seismic transect, shows a similar deformational style (Figure 4). In this region there are also two anticlines, with the western limb of the Snake Creek Anticline overturned near the Cloncurry Fault (Figure 4). The Cloncurry Fault itself is a late reverse fault, imaged in the seismic data, that links into, or cuts, the subhorizontal detachment surface.

Near the western edge of the Eastern Fold Belt, within the Lake Mary Kathleen Zone, the seismic data images a

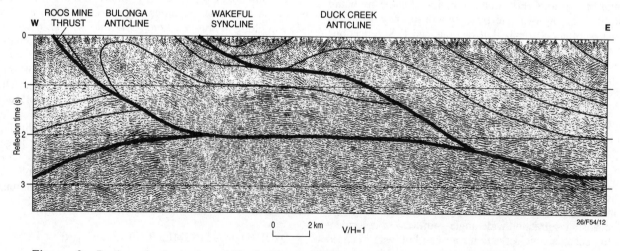

Figure 3. Portion of seismic data within the Eastern Fold Belt (Mittakoodi Fold Belt). This shows one of a series of easterly-dipping features that cut a highly reflective, shallowly east-dipping reflective zone zone (2 s -3 s TWTT) that lies beneath almost the entire Eastern Fold Belt.

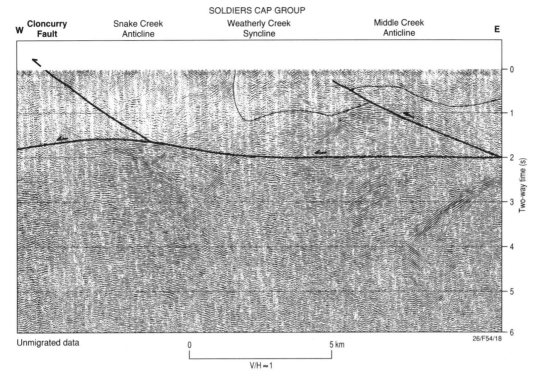

Figure 4. Interpretation in the Soldiers Cap Group (Eastern Fold Belt). The Cloncurry Fault is shown at the western part of the section.

number of early thrust faults that cut at a low angle through stratigraphy and appear to merge with a main detachment horizon.

Both surface geological mapping and seismic observation indicated that the thin-skinned system has been dissected by a number of reverse and strike-slip faults. One of the reverse faults is particularly reflective in the Marimo area (Figure 5). The enhanced reflectivity is probably due to hydrothermal alteration associated with mineralisation in this region, which is on strike with the Mt McNamara Cu-Au mine [Goleby et al., 1996]. The Eastern Fold Belt is dissected by two sets of strike-slip faults, the Pilgrim Fault, separating the Mary Kathleen Zone from the Quamby-Malbon Zone, and a set of north-east trending faults (including the Fountain Range Fault) restricted to the Mary Kathleen Zone. Both the reverse faults and later strike-slip faults cut into the basement, offsetting the highly reflective detachment. This indicates a fundamental change in tectonic style from early thin-skinned to later thick-skinned deformation.

There is little seismic evidence for thin-skinned deformation within the Western Fold Belt, apart from minor low-angle fault-related stratigraphic duplication; rather, the seismic data suggest that thick-skinned tectonics were the dominant regime. The seismic data reveal no basal detachment and suggest that controlling structures within this belt involved existing basement structures. Prominent features imaged by the seismic data include several moderate to steeply dipping fault zones (Figure 6). The strongest of these is a west-dipping zone beneath the Sybella Granite, that projects to the surface near the Adelheid Fault. Both P-wave and S-wave reflections were observed from this fault. The P-wave reflections are labeled 'Adelheid Fault'; the S-wave reflections are not labeled and lie between the Adelheid and Mount Isa Faults. The Mount Isa Fault is steeper than the Adelheid Fault and is, in places, directly imaged; in other places, it is inferred from reflector terminations. The Adelheid Fault is only gently folded, and probably developed late in the deformation (Figure 6). The enhanced reflectivity of this fault is probably due to hydrothermal alteration [Goleby et al., 1996].

The seismic section from the shorter traverse indicates a similar geometry to that interpreted within the Wonga Belt/Mary Kathleen Zone on the main seismic traverse. However, in this traverse the faults are all shallower dipping than to the north. The Pilgrim Fault, inferred as a strike slip fault on the main traverse, appears to dip to the west.

There is also a series of strong reflectors just to the east of the Pilgrim Fault, which we infer to represent the western extent of the high-velocity body seen on the main

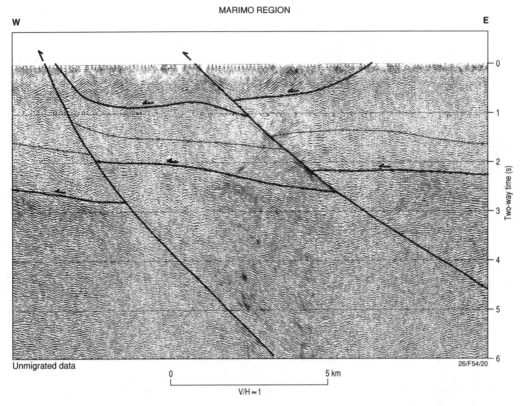

Figure 5. Interpretation in the Marimo region (Eastern Fold Belt), showing another of the series of easterly-dipping features (in this case the Marimo Fault) that cut a highly reflective zone (2 s -3 s TWTT) that lies beneath almost the entire Eastern Fold Belt.

traverse. If this is correct, then this high-velocity body is not only three-dimensional in extent but also dips slightly to the south.

CRUSTAL IMPLICATIONS

The interpreted seismic transect section, together with surface geologic mapping studies, have resulted in an 'east-west' crustal-scale cross section through the Mount Isa Inlier. No indication is given in the seismic data of the significance and role of north-south shortening during the earliest post-rifting deformation events [e.g. Bell et al., 1992]. Basement rocks are composed of gneisses deformed and metamorphosed during the Barramundi Orogeny ~1870-1850 Ma [Etheridge et al. 1987]. This deformation was accompanied by the intrusion of voluminous granitoids (Kalkadoon Granite, ~1870-1840 Ma) [Wyborn and Page, 1983] and extrusion of felsic volcanics (Leichhardt Volcanics). Five separate rift-sag cycles have been recognised in a regional stratigraphy [O'Dea and Lister, 1997] divided into two cover sequences, Cover Sequence 2 (~1790-1720 Ma) and Cover Sequence 3 (~1710-1620) [Blake, 1987].

The seismic structure of the Eastern Fold Belt described previously indicates a two-stage evolution of shortening. The first stage involved a thin-skinned style of deformation represented by low-angle detachment tectonics, including low angle faulting within the upper sequences, which is overprinted by thick-skinned, steeper faults that dissect the thin-skinned system, and then by east-west thrusting (Figure 5). Constraints on the age of this stage come from the relation of deformation to metamorphism, which indicates that this stage probably occurred prior to ~1540 Ma.. Four major zones of low-angle thrusting are crossed in the Eastern Fold Belt, and in each case thrusting developed prior to or during Isan metamorphism. The second stage involved the development of reverse faults and strike-slip faults that, in places, cut through the thin-skinned structures [MacCready et al., 1997]. Although some of the strike-slip faults (e.g. Pilgrim Fault, which offsets Cambrian rocks to the south) have multiple deformation histories, the steep faults in the Eastern Fold Belt appear to have developed prior to or during the intrusion of the Williams and Naraku Batholiths.

In contrast, the seismic structure of the Western Fold Belt indicates that deformation is dominated by basement-involved faulting, which may be linked with the thick-skinned tectonism imaged within the Eastern Fold Belt.

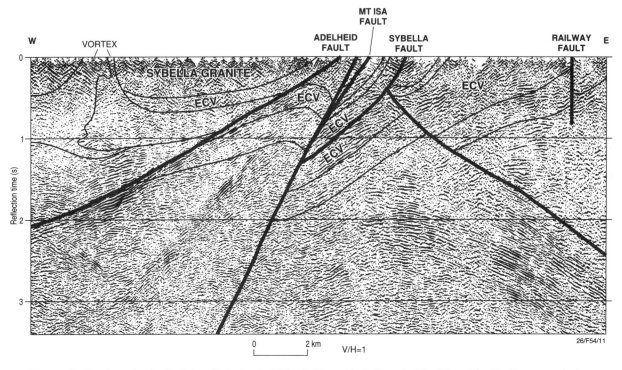

Figure 6. Portion of seismic data within the Leichhardt River Fault Trough. The Mount Isa Fault appears to be part of a family of west-dipping faults, within a sequence of faulted and folded Haslington Group units. The Adelheid Fault is imaged as a highly reflective zone that projects downwards from the surface and extends westwards beneath the Sybella Granite. This granite is imaged as a relatively thin body intruded sub-parallel to stratigraphy.

This belt exhibits an overall deformation style involving east-verging, tight folding, cut by steeper faults [MacCready et al., 1997].

The refraction section geometry suggests the presence of a fundamental crust-penetrating structure that had an important influence on the tectonic evolution of the Mount Isa Inlier (Figure 7). Although the reflection data does not clearly image this major structure, it is coincident with the top of the high-velocity linked bodies interpreted in the refraction interpretation. The emplacement options for the high-velocity bodies can be subdivided into two basic categories - magmatic and/or structural/tectonic. These define two end-members in a continuum of possible emplacement options. The linear nature of the two high-velocity layers suggests structural control; however, models of magmatic emplacement are easier to reconcile with the way the high-velocity rock is sandwiched within a generally low-velocity crust. Further research is being done on this

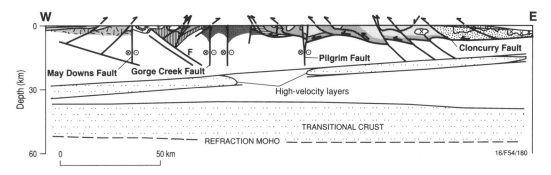

Figure 7. Interpreted crustal model for the Mount Isa Inlier, combining features from both the seismic reflection and seismic refraction interpretations as well as the surface geological information

data set to investigate relationships between the refraction velocities and the petrology of the middle and lower continental crust [Goncharov et al., 1997].

The deeper crustal structure indicates a series of east-dipping faults that all sole onto the main westerly dipping crust-penetrating surface bounding the top of the high-velocity bodies (Figure 7). In the Eastern Fold Belt, these faults link back to the surface faults discussed previously. In the Western Fold Belt, these deeper faults are antithetic to the thick-skinned faults imaged.

CRUSTAL STRUCTURE AND FLUID-FLOW MODELS

Within the Marimo area (discussed earlier), the seismic data have imaged a major fault zone whose strong reflectivity is due to hydrothermal alteration. This interpretation is supported by the presence of the Mount McNamara Cu-Au, mine where a 200-m wide, reflective zone of alteration has been mapped. Further evidence that this fault zone has acted as a fluid pathway comes from the geographical location of other faults in the region. Southwards, along the fault trace, there is a series of significant operational Cu-Au mines (e.g. Hampden, Mount Dore, Selwyn, and Osborne), suggesting that the fault system has played a major part in either the migration of fluids from their source to their final reservoir or in providing a pathway for fluids to circulate, driving mineral concentrating processes. The significance of the low-angle mid-crustal structure to these reflective zones is as yet unclear. However, they are apparently all connected, indicating the presence of a path to facilitate fluid flow from the middle crust and focus these fluids directly into late deformational structures near the surface.

The enhanced reflectivity observed on several of the later faults is inferred to be related to metasomatism and hence fluid flow. The map and seismic patterns indicate that these faults were linked in a large network, creating fluid pathways through the crust which may have been important in localising mineralisation in the region.

The Adelheid Fault is another fault whose highly reflective nature may be a result of the hydrothermal alteration. Four major Pb/Zn and Cu deposits are located just east of this fault zone, indicating the presence of a fundamental and very effective plumbing system. Fluid and mass transfer work by Heinrich et al. [1995] indicates that the significant mass transfer occurring during late to post metamorphic copper enrichment and kilometre-scale carbonate-Fe oxide alteration required a major influx of fluids from a source outside the main Isa Valley. They suggest the major major fluid influx could have come from an eroded sedimentary cover; however, model of Heinrich et al. [1995] for the Isa Valley also includes an influx of fluids from deeper within the crust. This influx is not inconsistent with that suggested by the seismic results.

We postulate that there is a fundamental relationship between the larger deposits in the Mount Isa Inlier and the more reflective faults as imaged by the seismic data. Most of the major structures associated with mineralisation that are visible in the seismic data are inclined and terminate in the mid-crust. In contrast, the major vertical structures imaged in the seismic do not host any significant mineralisation, even though they appear to penetrate through to the middle crust. This could imply that major movement on these vertical structures post-dated any significant hydrothermal activity, or their steep orientation may not have been favorable for significant fluid circulation.

Acknowledgments. This paper is published with the permission of the Executive Director, AGSO, and the Director, AGCRC. We thank the two reviewers for their useful comments and suggestions. The authors would also like to thank Chris Fitzgerald for drafting the figures.

REFERENCES

Bell, T. H., J. Reinhardt, and R. L. Hammond, Multiple foliation development during thrusting and synchronous formation of vertical shear zones, *J. Struct. Geol., 14,* 791-805, 1992.

Blake, D. H., Geology of the Mount Isa Inlier and environs, Queensland and Northern Territory, *Bur. Miner. Res. Bull., 225,* 83pp., 1987.

Drummond, B. J., T. MacCready, G. S .Lister, A. Goncharov, B. R. Goleby, R. Page, and L. A. I. Wyborn, AGCRC Mount Isa Transect, paper presented at AGCRC Mount Isa Transect Workshop, (unpublished), 50pp., 1995.

Etheridge, M. A., R. W. R. Rutland, and L. A. I. Wyborn, Orogenesis and tectonic process in the Early to Middle Proterozoic of northern Australia, in *Precambrian Lithospheric Evolution,* edited by A. Kröner, pp 131-147, American Geophysical Union Geodynamic Series, 17,. 1987

Goleby, B. R., Drummond, B. J., and MacCready, T., The Mount Isa deep seismic transect, *Aust. Geol. Surv. Org., Res. Newsl., 24,* 6-8, 1996.

Goncharov, A., B. R. Goleby, T. MacCready, and B. J. Drummond, Tectonics of the Mount Isa Inlier from seismic studies, paper presented at Geodynamics and Ore Deposits Conference, Australian Geodynamics Cooperative Research Centre, Ballarat, Victoria, 42-45, 1997.

Heinrich, C. A., J. H. C. Bain, T. P. Mernagh, and L. A. I. Wyborn, Fluid and mass transfer during metabasalt alteration and copper mineralization at Mount Isa, Australia, *Econ. Geol., 90,* 705-730, 1995.

Huang, W., Structural and stratigraphic relations on the western flank of the Mittakoodi Culmination: a case study in the Roos Mine Area south of the Corella Dam, Eastern Mount Isa Inlier, NW Queensland, *Aust. Crustal Res. Centre Tech. Publ., 21,* 28pp., 1994.

MacCready, T., B. R. Goleby, A. Goncharov, G. S. Lister, and B. J. Drummond, An evolutionary framework for the Isan Orogeny Proterozoic terranes, paper presented at Geodynamics and Ore Deposits Conference, Australian Geodynamics Cooperative Research Centre, Ballarat, Victoria, 42-45, 1997.

O'Dea, M. G., and G. S. Lister, The evolution of the Mount Isa Orogen - from start to finish, paper presented at Geodynamics and Ore Deposits Conference, Australian Geodynamics Cooperative Research Centre, Ballarat, Victoria, 34-37, 1997.

Stewart, A. J., and D. H. Blake, Detailed studies of the Mount Isa Inlier, *Aust. Geol. Surv. Org. Bull., 243,* 374pp., 1992.

Wyborn, L. A. I., and R. W. Page, The Proterozoic Kalkadoon and Ewen Batholiths, Mount Isa Inlier, Queensland: source, chemistry, age, and metamorphism, *BMR J.Aust. Geol. Geophys., 8,* 53-69, 1983.

___

B. J. Drummond, B. R. Goleby, and A. Goncharov, Australian Geological Survey Organisation, PO Box 378, Canberra, ACT, 2601, Australia.

T. MacCready, Department of Earth Sciences, Monash University, Clayton, Victoria, 3168, Australia.

Intra-Crustal "Seismic Isostasy" in the Baltic Shield and Australian Precambrian Cratons from Deep Seismic Profiles and the Kola Superdeep Bore Hole Data

A. G. Goncharov[1], M. D. Lizinsky[2], C. D. N. Collins[1], K. A. Kalnin[2], T. N. Fomin[1], B. J. Drummond[1], B. R. Goleby[1], and L. N. Platonenkova[2]

[1]Australian Geological Survey Organisation, Canberra, Australia
[2]Department of Geophysics, St.-Petersburg Mining Institute, St.-Petersburg, Russia

Low-velocity layers are important elements of seismic models of Precambrian regions. Commonly, anomalously high-velocity rocks in these regions are underlain by anomalously low-velocity rocks and vice-versa, so that a balancing of high and low velocities can be seen along any vertical profile through the crust. The best characteristic to quantify this balancing effect is average velocity (i.e. the ratio of depth to vertical travel time). The average velocity-depth distribution gives an estimate of the degree of isostatic compensation ("seismic isostasy") at any given depth. In the Baltic Shield, velocity variations in the upper crust (up to 1 km/s or almost 20% of average values) are compensated above the lower crust. Significant lateral variations in compressional-wave velocity at a mid-crustal level (20-35 km depth) underneath the Proterozoic Mount Isa Inlier (Australia) are also compensated within the crust well above the Moho. A similar conclusion can be derived from the analysis of average velocity-depth functions from other Australian Precambrian terranes. These results are consistent with the data collected by different seismic methods, including deep seismic sounding profiles, and vertical seismic profiling (VSP) in the Kola Superdeep Bore Hole (KSDBH) in the northern part of Russia. Direct pressure estimates based on the density data from the KSDBH confirm that objects with anomalous seismic velocity and density in the upper crust may be isostatically compensated well above the Moho, but the possibility of global translation of "seismic isostasy" into conventional isostasy is a subject for further studies. Intra-crustal isostasy may exclude or reduce the need for isostatic equilibrium to be achieved at the Moho or at the lithosphere-asthenosphere boundary in Precambrian crust. A concept of multi-level "seismic isostasy" below the Precambrian crust emphasises the essence of our observations: wherever a high- or low-velocity anomaly occurs in the crust, it tends to be compensated by its counterpart immediately underneath.

1. INTRODUCTION

The hypothesis of isostatic equilibrium of the Earth's crust goes back to the 19th century. Two major concepts of isostasy have been known for a long time: Airy type and Pratt type isostasy [Heiskanen and Vening-Meinesz, 1958]. In the case of the Airy hypothesis (modified by Heiskanen), the Moho depth varies, the crustal density is constant, and the depth of isostatic compensation corresponds to the maximum depth of the Moho. In the Pratt isostasy scheme (modified by Hayford), the crustal density changes, the Moho depth remains constant, and the depth of compensation corresponds to the Moho. It was recognised later that neither of the above concepts is universal and that elements of both concepts have to be taken into consideration to explain isostatic equilibrium as well as lateral variations in the density of subcrustal material [Andreyev and Klushin, 1965].

Nevertheless, most researchers agree that a major part of the Earth's crust, with the exception of presently tectonically active regions (e.g. the Crimea Mountains in Ukraine, the North-Western Caucasus, and others), is in isostatic equilibrium. However, it is still debated what the depth of isostatic compensation really is. Non-uniqueness of gravity interpretations is a major obstacle in definition of the depth of isostatic compensation. Attempts to estimate isostatic behavior of the crust directly from seismic data are also known. For example, Warner [1987] noticed that on many conventional reflection profiles the Moho appears at a rather constant two-way travel time (TWTT), despite the highly variable structure of the crust above, and he showed that indeed the Moho would appear to be flat on a time section if the crust was in local isostatic equilibrium and if there was a simple relationship between velocity and density for crustal rocks. He did not conclude that the Moho had to be horizontal in depth to produce that effect, but rather that both Airy and Pratt type isostasies may be applicable to explain the observations.

Low-velocity layers in the consolidated part of continental crust have been revealed in many regions in recent years. In this paper we discuss distribution of high- and low-velocity rocks along a vertical profile through the crust and its possible implications for isostasy. To what extent seismic results presented here characterise isostatic behavior of the crust depends on the correlation between seismic velocity and density. We use direct pressure estimates based on the density data from the Kola Super Deep Borehole (KSDBH) to discuss a relationship between our seismic results and conventional isostasy, but we do not make any attempt in this paper to justify a global character of this relationship.

We present seismic data which were collected in two stable shield areas by different techniques: deep seismic sounding data from the eastern (Russian) part of the Baltic Shield, which were recorded by closely spaced geophones; unique vertical seismic profiling (VSP) and sonic log data from the KSDBH in Northern Russia; refraction/wide-angle seismic data from the Mount Isa Inlier in Northern Australia, and refraction data from the adjacent Mount Isa - Tennant Creek region.

2. DEEP SEISMIC PROFILES IN THE EASTERN PART OF THE BALTIC SHIELD

2.1 Data Coverage and Geometry of Observations

More than 3000 km of reflection and deep seismic sounding (DSS) profiles have been studied in the eastern part of the Baltic Shield during the last 35 years. The eastern (or Russian) part of the Baltic Shield includes the Kola-Karelian region with its western boundary approximately coinciding with the Russian - Norwegian and Russian - Finnish border (Figure 1).

The seismic data from the eastern part of the Baltic Shield are especially significant because about 800 km of DSS and reflection seismic profiles have been studied in the KSDBH area, and this gives a key to calibrate deep seismic boundaries intersected by the borehole and to extrapolate KSDBH results to adjacent regions.

The recording parameters on numerous DSS profiles in the eastern part of the Baltic Shield were similar. Average shot interval was 50-70 km, distance between recording positions was 2-5 km, reversed and overlapping coverage of profiles was a common practice, and the separation of geophone groups within each recording position was 100-200 m. The last feature of the recording geometry is very important, as it shows that the coverage of the DSS profiles in the eastern part of the Baltic Shield is denser than the western part of the Baltic Shield and many other parts of the world. This enabled the development of detailed seismic models of the crust and crust-mantle transition zone in the eastern part of the Baltic Shield [Goncharov et al., 1991; Goncharov, 1993].

2.2 Recorded Wave Field: Evidence for Low-Velocity Layers

The uppermost part of the crust from 0 to 5-10 km depth in the eastern part of the Baltic Shield is very inhomogeneous with velocity variations up to 1 km/s which is almost 20% of average values [Goncharov et al., 1991] (Figures 1 and 6 this paper). The middle crust is more homogeneous, while the lower crust and crust-mantle transition zone both show significant lateral variation in seismic velocity. In this paper we will concentrate mainly on the seismic model of the upper crust in the eastern part of the Baltic Shield. The seismic model of the upper crust is based on the interpretation and modeling of the first arrivals P_0 and subsequent phases P1 and P2. These phases are commonly recorded on the DSS profiles in this area

The absolute times and apparent velocities of the first arrivals recorded in different parts of this region depend on the specific local velocity structure of the crust. Important similarities between the data from different DSS profiles in the eastern part of the Baltic Shield are the discontinuity and time delays of the first arrivals (Figure 2). Both features can be clearly seen, despite the very different travel-time curves of the first arrivals. Time delays of the first arrivals are observed in many cases at equal distances from the shots in reversed and overlapping recording geometry, thus indicating velocity decreases in the medium. The fact that first arrivals remain discontinuous and time-shifted regardless of the specific velocity model indicates that layers with decreased velocity are ubiquitous.

The discontinuity of the first-arrival travel times (Figure 2) results from the discontinuous character of the

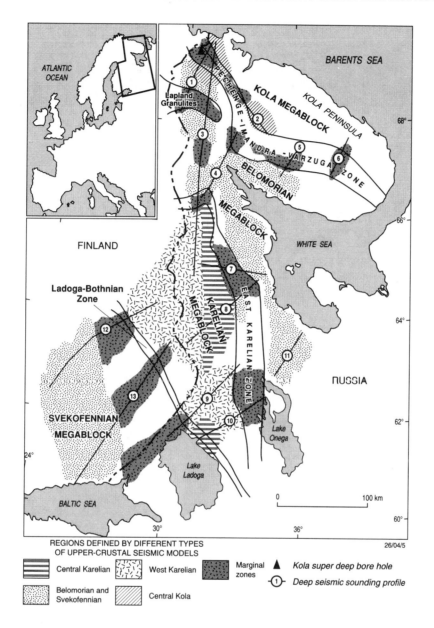

Figure 1. Main geotectonic provinces in the eastern part of the Baltic Shield, locations of deep seismic sounding profiles and regions defined by different types of upper-crustal seismic models. DSS profiles: 1 - Pechenga - Lovno, 2 - Zapolyarny - Umbozero, 3 - Pechenga - Kostomuksha, 4 - Monchegorsky, 5 - Imandra - Varzugsky, 6 - Keivsky, 7 - Kem' - Ukhta, 8 - Kem' - Tulos, 9 - Lahdenpoh'ja - Segozero, 10 - Lake Ladoga - Lake Onega, 11 - Lake Onega - White Sea, 12 - Sveka, 13 - Baltic. Velocity-depth functions typical for different regions shown in Figure 5.

first arrivals recorded on field seismograms (Figure 3). The P_0 wave is the wave refracted in the very top part of the crust. It forms first arrivals from 0 km distance from the source. The outer limit of its recording as a first arrival varies from 90 to 120 km and depends on the specific velocity structures of different profiles. Commonly, the subsequent wavelets (marked "P1" and "P2" on Figure 3) become first arrivals at larger distances because of attenuation of earlier P_0 waves rather than a velocity increase with depth. Separate phases within the P1 and P2

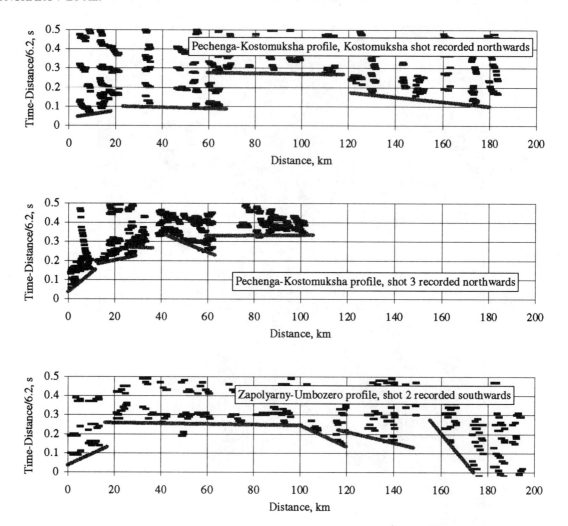

Figure 2. Travel times of seismic waves along the deep seismic sounding profiles in the eastern part of the Baltic Shield. Straight lines show schematic correlation of the first arrivals.

wave groups cannot be traced for more than several kilometers, although the groups in general are normally traced reliably from distances of 20-40 to 120-170 km.

2.3 P1 and P2 Upper-Crustal Phases as Indicators of Velocity-Depth Distribution

The P1 and P2 waves were recorded on numerous DSS profiles, including those in the vicinity of the KSDBH. They originate in the depth intervals of 5-7 and 9-12 km respectively. Both waves are of complicated nature and they are produced by interference of numerous interconversions and peg-legs. The velocity structure of the depth intervals 5-7 and 9-12 km, with several high- and low velocity layers revealed by the VSP and sonic log in the KSDBH (see below), provides favorable conditions for the formation of high-amplitude phases. No lateral variation in seismic velocity was detected in the middle crust ion the eastern part of the Baltic Shield,, and therefore P2 can be taken to mark the bottom of the lateral velocity variation in the upper crust.

Analysis of kinematics of the P_0, P1 and P2 phases and wave-field modeling show that the apparent velocities of all crustal first arrivals depend mainly on the velocity distribution in the uppermost part of the crust, to a depth of 5-7 km. It is necessary to emphasise that a prevailing apparent velocity of 6.2 km/s is characteristic of the first arrivals in the distance interval from 20-70 km up to 170-220 km, where upper mantle waves first arrive. Such a configuration of first arrivals is typical for the eastern part of the Baltic Shield.

On the other hand, the P1 and P2 travel times are concentrated within narrow time-space domains (Figure 4), i.e. they remain very similar in different regions, despite

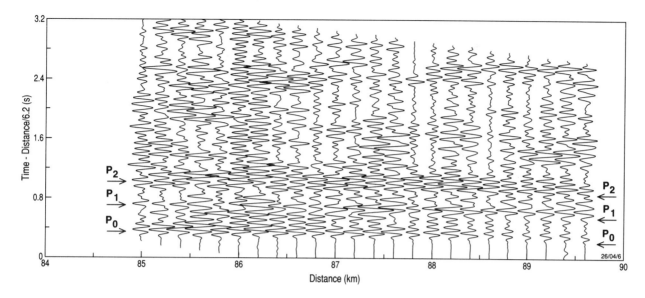

Figure 3. Seismic refraction data from the northern part of the Pechenga-Kostomuksha profile, illustrating typical upper-crustal phases P_0, P1 and P2. Trace spacing is 200 m.

significant lateral velocity variations in the uppermost part of the crust from 0 to 5-7 km depth. The similarity of these travel times, especially of those observed at distances less than 40 km from the shot, indicates that the average velocity through the crust to the depth where P1 and P2 originate remains constant, or varies insignificantly in different blocks of the crust.

2.4 Average Velocity Definition and What is "Seismic Isostasy"?

The term "average velocity" will be widely used in this paper. We define the average velocity to a certain depth in a multi-layered seismic model is the ratio of the total thickness of layers above that depth to the total vertical travel time of seismic waves to that depth. The average velocity-depth functions give an estimate of the degree of high- and low-velocity rocks balance ("seismic isostasy") at any given depth.

Similarly the degree of "seismic" isostatic compensation can be estimated from the two-way-time-depth functions: if two-way-time - depth functions from different regions converge at a certain depth, the velocity distributions in these regions are balanced, and "seismic isostasy" is achieved at the depth of their convergence.

2.5 Seismic Isostasy in the Baltic Shield from Deep Seismic Profiles

Low-velocity layers must be included in the upper-crustal seismic models in the eastern part of the Baltic Shield to explain the similarity of the P1 and P2 travel times and other features mentioned above. The prevailing maximum velocity in the covering medium, which is responsible for the asymptotic apparent velocity values of the P1 and P2 waves, is 6.2 km/s. Several types of seismic models were proposed to explain the kinematical features of the P_0 waves in different regions of the Baltic Shield [Goncharov et al., 1991]. The lowest velocities and the latest arrivals are typical for the Archean rocks in Central Karelia; the highest velocities and the earliest arrivals are typical for the marginal zones of different blocks of the crust, such as the Pechenga-Imandra-Varzuga Zone, East Karelian Zone, Ladoga-Bothnian Zone, and others (Figure 1). There are also several intermediate seismic models (Figure 5).

Variations of average velocity in the upper crustal models of different blocks decrease with depth. At a depth of 12 km, they are already limited to a narrow range of 6.00-6.05 km/s, with the exception of the model for marginal zones (Figure 5b), which indicates that high and low velocities along a vertical profile through the crust are balanced. Comparison of the travel times of the P1 and P2 phases computed for the velocity models from different crustal blocks with the corresponding observed travel times (Figure 6) shows good agreement (again with the exception of marginal zones).

The deviation from this trend by the model of the Pechenga-Imandra-Varzuga Zone and other marginal zones, is due to the lateral extent of these zones, commonly less than 100 km (see Figure 1), and the P1 and P2 phases can never be observed entirely within such zones. Thus the one-dimensional velocity model of marginal zones used to compute the travel times in Figure 6 was not a reasonable

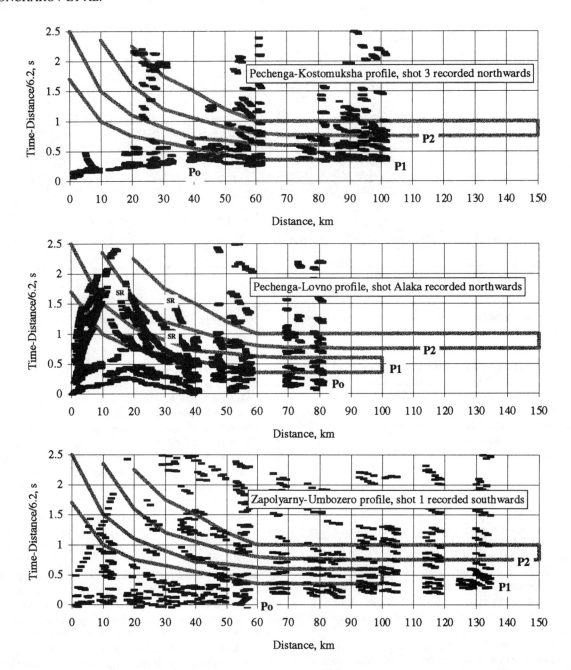

Figure 4. Travel times of main upper-crustal waves P1 and P2 in the eastern part of the Baltic Shield. Solid grey lines represent time-space domains where P1 and P2 arrivals concentrate most commonly. SR - reflections from steep boundaries in the upper crust which sometimes interfere with P1 and P2 phases and complicate their tracking.

approximation of the real geological structure. However, we have reason to believe that the average velocity-depth function, in at least some of the higher-velocity upper-crustal anomalies, differs from those in other blocks of the crust, thus indicating that seismic isostasy is achieved deeper in the crust. Where this happens can be estimated from the KSDBH data.

3. THE KOLA SUPERDEEP BORE HOLE RESULTS

3.1 Geology

The KSDBH is located in the north-western part of the Kola Peninsula within the Pechenga-Imandra-Varzuga Proterozoic Mobile Belt (Figure 1). The detailed geological

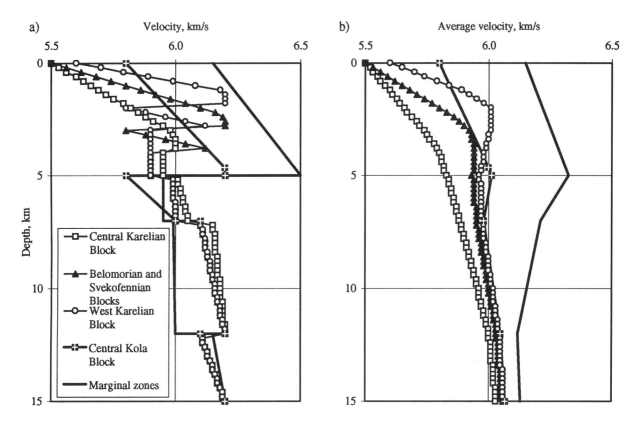

Figure 5. (a) Seismic models of the upper crust in different regions of the Baltic Shield, and (b) average velocity-depth functions calculated for these models. Regions defined by the different types of upper-crustal seismic models are shown in Figure 1.

description of the KSDBH site can be found in Kozlovsky [1984]. The bore hole reached a depth of more than 12 km. There are two main geological complexes revealed by drilling and which are correlated to geological observations at the surface - Archean and Proterozoic (Figure 7a).

The upper part of the cross-section (0-6840 m) corresponds to the Pechenga Proterozoic graben-syncline, and is composed of alternating sequences of meta-sedimentary and meta-magmatic rocks. The lower part of the cross-section corresponds to the Archean basement, and is represented by biotite-plagioclase gneisses with or without high-aluminium minerals and amphibolites. Less abundant are biotite-amphibolite-plagioclase gneisses and schists. The thickness of the Archean complex penetrated by the KSDBH is around 5 km (Figure 7a).

We have already noted in the previous section that there is generally a balancing of high and low velocities along any vertical profile through the crust: often anomalously high-velocity rocks are underlain by anomalously low-velocity rocks and vice-versa. The KSDBH site is typical in this respect. High-velocity Proterozoic greenstones in the upper crust are underlain here by lower-velocity Proterozoic rocks and Archean basement.

3.2 Seismic Studies

Several data sets are available to estimate seismic velocity in the upper crust of the KSDBH site: deep seismic sounding data from three profiles intersecting near this site (Figure 1), vertical seismic profiling data, and sonic log data obtained in the bore hole. VSP and sonic log provide better resolution than the DSS data. VSP data were recorded to a maximum depth of the bore hole (12 km), and sonic-log data were recorded to a depth of 10.62 km.

3.2.1 Velocity-depth functions from sonic log and vertical seismic profiling.

VSP data from the KSDBH have radically altered the conventional idea of a monotonous velocity increase with depth in crystalline crust [Kozlovsky, 1984; Lizinsky and Lanev, 1991]. On the contrary, these data show practically an inverse correlation: the velocity profile has many low-velocity intervals of different scale and complexity (Figure 7c). Low-velocity anomalies are due to variations in both composition and fabric of the rock and are sometimes very large: velocity

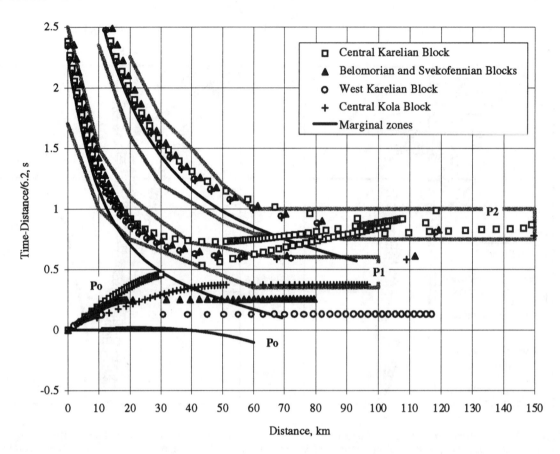

Figure 6. Travel times of upper crustal waves P_0, P1 and P2, computed for the velocity models of Figure 5. Solid grey lines represent time-space domains where experimental P1 and P2 arrivals concentrate most commonly.

decreases reach 1 km/s. The largest velocity decrease occurs at a depth of around 4.5 km. Low-velocity layers were found even at a depth of more than 10 km (Figure 7c).

Sonic log data obtained in the KSDBH were averaged so that they could be compared with the VSP and DSS data. This is important because the travel times in the sonic log were measured by the receiver at an offset of only 1 m from the source. Consequently, the sonic velocity-depth function contains numerous high-frequency oscillations which cannot be resolved in the VSP and DSS data.

Sonic travel times were summed to obtain accumulative vertical travel times. These were further reduced for reduction velocity values in the range 5.6-7.2 km/s, which covers the possible velocity values for this cross-section. Reduced vertical travel-time curves (Figure 7b) were used to detect contacts of layers with different seismic velocity, and to define velocity values in these layers. The interpretation took into consideration only the layers with a thickness comparable to the prevailing wave length expected a priori. For example, in the VSP measurements (prevailing frequency around 40 Hz, background velocity values close to 6.0 km/s), estimated wave length is about 0.15 km. Hence layers less than about 0.15 km thick were ignored in reducing sonic log accuracy to that of the VSP.

The resulting velocity-depth function shows a good correlation with the function derived from the VSP (Figure 7c). Although the sonic log data are less reliable beneath a depth of 6 km (Figure 7b), the velocity values here appear to be consistent with generally low velocity background values close to 6.0 km/s. The top of this macro low-velocity layer, incorporating the lower part of the Proterozoic section and the Archean basement beneath, is defined at a depth of around 4.5 km. According to the VSP data, a velocity of 6.1 km/s prevails in the depth interval 4.5-12.0 km. It is important to note that relatively low velocities are not confined to the Archean basement but also occur in the Proterozoic complex within the depth interval 4.5-6.8 km, where rather high velocities would be expected from the petrology of the rocks and the PT-conditions: the proportion of meta-sedimentary rocks, generally characterised by relatively low velocities, is less in this depth interval than in the upper units of the

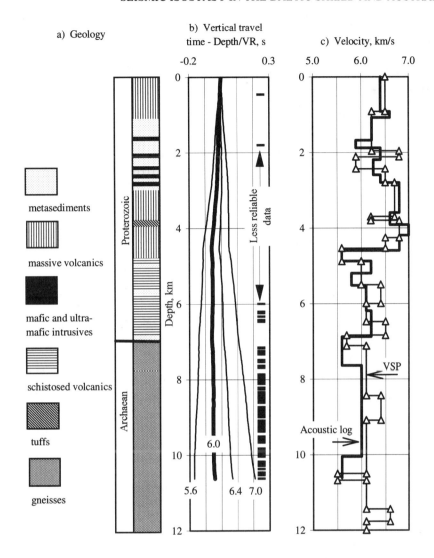

Figure 7. (a) Schematic geological section of the Kola Super Deep Bore Hole, (b) vertical travel times from acoustic log data reduced by different reduction velocity values (VR) plotted near the curves (in km/s), and (c) velocity-depth functions from acoustic log and vertical seismic profiling data.

Proterozoic complex (Figure 7a). The preferred explanation of the velocity decrease in the lower part of the Proterozoic section is the influence of a major fault zone which runs through it [Lizinsky and Lanev, 1991].

3.2.2 Seismic isostasy. The good correlation between VSP and sonic log-derived velocities (Figure 7c) enabled us to use only one data set to estimate the average velocity distribution in the KSDBH region. The VSP velocities were used for this purpose. Remarkably, intervals where the average velocity decreases with depth obviously prevail in the KSDBH (Figure 8a). The most continuous average velocity decrease starts at a depth of 4.5 km and continues to the maximum depth of the bore hole (12 km), where the average velocity is close to 6.2 km/s - which is about 0.1 km/s higher than in the typical model of a marginal zone (Figure 5b).

Nevertheless, the decreasing velocity trend can be extrapolated further down to meet the average velocity-depth curve, typical for the crust in the nearby region, at a depth less than 25.5 km (Figure 8a). This is about 15 km above the Moho in this region [Goncharov, 1993] and means that even objects with anomalous seismic velocity and density like the Pechenga structure may be isostatically compensated well above the Moho.

The estimated range of uncertainty of extrapolations from linear-regression analysis is within the 16-25.5 km depth interval. The latter value corresponds to the most

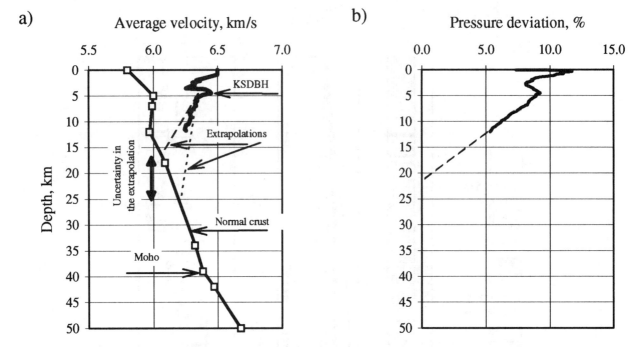

Figure 8. (a) Average velocity, and (b) geostatic pressure, in the Kola Super Deep Bore Hole (KSDBH), compared with the normal crust in the nearby region where there is no upper-crustal high-velocity anomaly. Average velocities in the normal crust and Moho level were derived from deep seismic sounding data; average velocity in the KSDBH was calculated from VSP data. Geostatic pressure is represented by its deviation from a model with homogeneous upper crust of 2.72 t/m^3 density. Dashed lines show extrapolations of the trends marked by the KSDBH data.

"pessimistic" extrapolation, when only data points in the depth range 5.5-7.0 km were used for regression. In this depth interval, the vertical gradient of average velocity decrease with depth is the lowest below 4.5 km depth (Figure 8a).

Low velocity layers also play an important role in the upper-crustal seismic response recorded at large distances (50-150 km). Constructive interference of numerous converted waves and multiples originating in models with low-velocity layers completely changes the style of the wave field in comparison to that derived from primary reflections only. Later phases produced by this interference have very high amplitudes and will produce false boundaries in the cross-section if misinterpreted as primary reflections [Drummond et al., 1995].

3.3 Is Conventional Isostasy Achieved above the Moho in the KSDBH Region?

The KSDBH data also enable the direct estimation of the geostatic pressure in situ. Density measurements were made on the core samples from 0 to 12 km depth. These measurements were averaged in a similar way to the sonic log measurements discussed above. The only difference was that accumulative travel times calculated from the sonic log measurements (Figure 7b) were replaced by accumulative pressure-depth functions (Figure 9b) in density processing.

The general appearance of the curves in the reduced pressure-depth chart (Figure 9b) is remarkable. The continuous downward shift of vertical segments of the reduced pressure-depth curves with reduction density decrease (emphasised by arrows) can be seen clearly. This means that a systematic density decrease with depth is a major trend in the lower 4.5-12.0 km of the upper-crustal section penetrated by the KSDBH. Density increases certainly occur in separate layers (Figure 9c), but they turn out to be only local effects within this general trend.

Geostatic pressure calculated from the density measurements deviates from the "normal upper-crustal values" by up to 12% in the very top part of the Pechenga structure, and decreases to ~6% at a depth of 12 km (Figure 8b). "Normal upper-crustal values" are the pressure values that would be measured in a homogeneous upper crust of density 2.72 t/m^3. This density is characteristic of the prevailing granite-gneiss rock complexes in the eastern part of the Baltic Shield [Sharov, 1993]. It is confirmed by numerous laboratory measurements, and is commonly used in regional-scale gravity modeling of shield regions [Kartvelishvili, 1982].

It is important to note that up to 50% of the excessive pressure produced by high-density Proterozoic rock units in the upper part of the Pechenga structure is isostatically

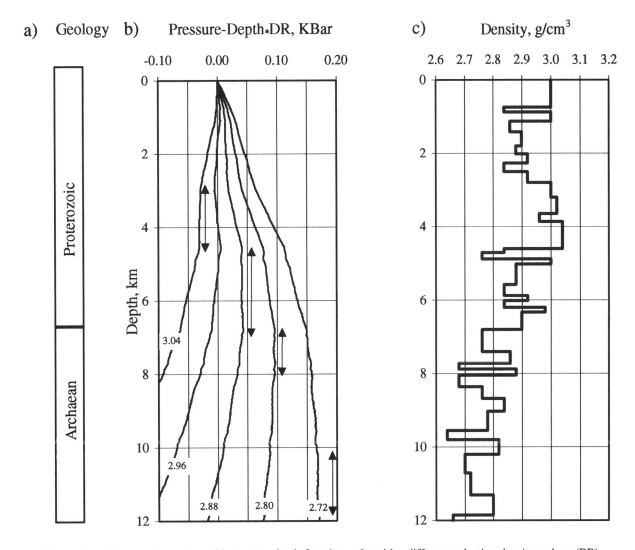

Figure 9. (a) Schematic geology, (b) pressure-depth function reduced by different reduction density values (DR) plotted near the curves (in t/m^3), and (c) density-depth function in the Kola Super Deep Bore Hole (KSDBH) calculated from density measurements on the core samples. For more details on the geology see Figure 7.

compensated within the top 12 km of the crust (12% pressure deviation near the surface compared to ~6% deviation at 12 km depth, Figure 8b). The trend of decreasing pressure deviation with depth can be extrapolated (uncertainty in the extrapolation does not exceed 0.5 km) to 0% deviation from the normal crust of the Baltic Shield at a depth around 22 km (Figure 8b). This is close to the average seismic velocity deviation discussed above (Figure 8a), i.e. about 15 km shallower than the Moho in the nearby region. Therefore we can conclude that direct pressure estimates based on the KSDBH data confirm that objects with anomalous seismic velocity and density in the upper crust may be isostatically compensated well above the Moho.

All the data which have been presented so far refer to the velocity and density anomalies in the Precambrian upper crust of the Baltic Shield. It is important to estimate the isostatic behavior of regions with velocity/density anomalies at deeper levels in the crust. Some Australian Precambrian terranes provide a good opportunity for such estimates.

4. MOUNT ISA REGION, AUSTRALIA

4.1 Geology

The Mount Isa Inlier, northwest Queensland, Australia (Figure 10), comprises Early to Middle Proterozoic sediments, bimodal volcanic rocks, and plutons. The geology of this area has been studied by many authors [Blake and Stewart, 1992; Blake, 1987; Etheridge et al., 1987; and many others] and can be briefly summarised as

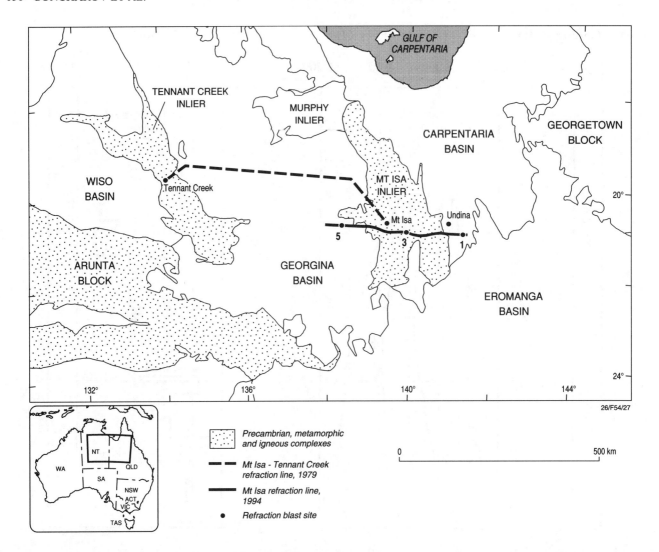

Figure 10. Tectonic provinces and location of refraction lines in the Mount Isa - Tennant Creek region. Mount Isa - Tennant Creek line after Finlayson [1982]; Mount Isa refraction line 1994, studied by the Australian Geodynamic Cooperative Research Centre [Goncharov et al., 1996].

follows. The Mount Isa Inlier has undergone a complex geological evolution, with two major Proterozoic tectonostratigraphic cycles separated by an orogenic event, the Barramundi Orogeny (around 1870 Ma). The first cycle is represented by the basement sequence, and the second cycle by three cover sequences. The second cycle was terminated by regional deformation and metamorphism - the compressional Isan Orogeny (1620 - 1500 Ma). Fluid-rock interaction associated with low-pressure metamorphism prior to the Isan Orogeny (at 1653 Ma, S.-S. Sun, personal communication) produced world-class ore deposits and extensive zones of metasomatism. The Mount Isa Inlier is subdivided by major north-striking faults or fault zones into a series of meridional belts.

4.2 Seismic Studies

Refraction/wide-angle seismic data were recorded along the 450-km line of the Mount Isa transect as a part of a broader geoscience project undertaken by the Australian Geodynamics Cooperative Research Centre in 1994. The geometry of observations included reversed and overlapping recording from five shot locations; two shot locations were quarry blasts at the Mount Isa and Undina mines (Figure 10). A total of 70 recording sites were occupied during the survey, and recording station separation across the Mount Isa Inlier was ~6 km.

The recorded wave field showed significant variation for different sources. An important common feature observed

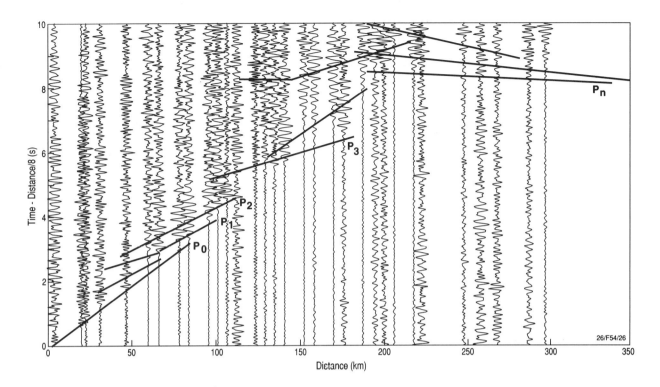

Figure 11. Record section for the Mount Isa shot recorded eastwards on the Mount Isa 1994 refraction line. Band-pass filtered 4-16 Hz, trace normalised. Straight lines show schematic correlation of main arrivals used to construct seismic model. P_0, P1, P2 - upper crustal waves, P3 - mid-crustal refraction penetrating high-velocity body shown in Figure 12a, P_n - crust-mantle transition zone and upper-mantle waves.

from different shots was a clear discontinuous character of the first arrivals. Similar to what was observed in the Baltic Shield data, upper crustal waves P1 and P2 become visible first arrivals because of attenuation of earlier P_0 waves and are time-delayed (Figure 11). This pattern is even more clearly seen where the P_3 phase comes to the first arrivals at larger distances. The P_3 phase, characterised by an apparent velocity of ~7.0 km/s, was modeled as a mid-crustal refraction. This wave was traced from the Mount Isa shot only in a limited range of distance (100-180 km). It is completely attenuated at larger distances, and the P_n waves from the crust-mantle transition zone and the upper mantle become first arrivals at greater distances after significant time delay (Figure 11).

The seismic velocity distribution along the Mount Isa refraction line was derived by iterative ray-trace modeling which minimised the difference between the observed and computed travel times. The interpretation was supported by computing synthetic seismograms by different algorithms [Luetgert, 1988; Ha, 1984]. The final velocity model (Figure 12a) explains not only the travel times but also the amplitudes of the main waves used to construct it (Goncharov et al., in preparation).

The seismic velocity distribution in the crust and crust-mantle transition zone is complicated and varies significantly along the line. Low-velocity layers are common in the crust and in the crust-mantle transition zone. Along the whole transect, there is no sharp velocity boundary between the crust and mantle; instead, there is a thick (up to 15 km) transitional zone above a Moho at 40-55 km depth. A high-velocity (6.9 - 7.3 km/s) body in the middle crust at the center of the transect is an important feature of the model (Figure 12a). The P_3 wave discussed above penetrates this body; it is not observed at distances of more than 180 km, owing to the discontinuity of the high-velocity structure at mid-crustal level.

There are few areas globally which have seismic features similar to those of the Mount Isa area. Deep seismic results from other Australian Precambrian terranes are consistent with the concept of thickened Proterozoic crust compared to Archean crust [Drummond and Collins, 1986]. Where thickening of the crust occurs in Australia, it is due to the thickening of the lower crust with high velocity (more than 7 km/s). Velocity distributions from 0 to around 30 km depth are similar in Australian Precambrian areas. Velocities higher than 7 km/s at depths less than 20 km have not been reported in Australia, making the middle part

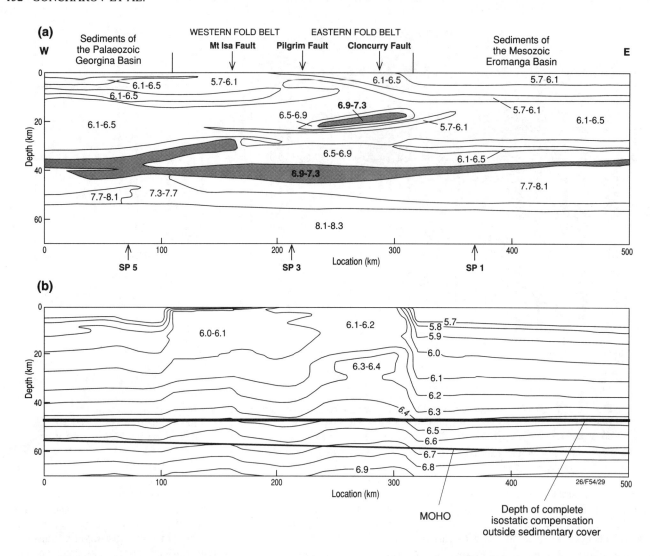

Figure 12. (a) Seismic-velocity distribution along the Mount Isa 1994 refraction line [Goncharov et al., 1996], and (b) average velocity distribution calculated for this line. Velocities in km/s. Velocity structure of thin sediments on the flanks of the line is not represented in (a) owing to scale considerations. Straight line representing the Moho in (b) approximates the top of the layer of 8.1 - 8.3 km/s velocity in (a).

of the Mount Isa Inlier unique, at least on an Australia-wide scale.

It is not only a high-velocity anomaly in the middle crust which differentiates the Mount Isa velocity model from those of other Precambrian regions in Australia. The background velocity values in the depth range 25-45 km are systematically lower than those in other Australian Proterozoic and Archean average models. Similar differences can be seen between Mount Isa seismic velocities and the global average, and between Mount Isa and world average models for shields and platforms [Goncharov et al., 1996].

There are several possible options to explain when and how high-velocity material was emplaced at a mid-crustal level and what it means in terms of rock composition. The emplacement options can be subdivided to two groups - magmatic and/or tectonic (structural). For the Mount Isa transect model, the straightness of the high-velocity layers suggests structural control; however, models of magmatic emplacement are more consistent with an overall seismic velocity distribution in the region [Goncharov et al., 1997].

4.3 Seismic Isostasy

Geological interpretations must be constrained by isostatic considerations. There is a noticeable balancing of high- and low-velocity rocks along any given vertical profile through the crust in the Mount Isa region. For example, the high velocity in the middle crust is underlain

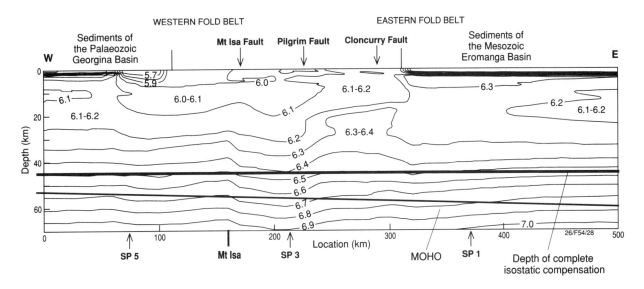

Figure 13. Average velocity distribution along the Mount Isa 1994 refraction line with upper 2 km of the section removed. Velocities in km/s. Straight line representing Moho approximates the top of the layer of 8.1-8.3 km/s velocity in Figure 12(a), adjusted for removal of the top 2 km.

by a zone of decreased velocities (Figure 12a). The crust-mantle transitional layer, characterised by 7.7-8.1 km/s velocities, is also thinner underneath this mid-crustal high-velocity anomaly. The general uplift of high-velocity (6.9-7.3 km/s) rocks in the lower crust in the western part of the Mount Isa transect correlates with an increase in thickness of low-velocity (5.7-6.1 km/s) rocks in the upper crust (Figure 12a). We could possibly continue listing examples of such balancing which is visible in the original seismic velocity model, but the average velocity distribution gives a more accurate estimate of the degree of this balancing.

Average velocity isolines beneath the Mount Isa Inlier deviate strongly from a horizontal position (Figure 12b). Maximum deviation occurs where the mid-crustal high-velocity affects the calculation of the average velocity (between 250 and 320 km along the line, Figure 12b). For example the 6.3 km/s isoline rises to ~20 km depth from ~40 km east of the high-velocity anomaly. But despite significant lateral variation in seismic velocity down to mid-crustal level (Figure 12a), the average velocity isolines (Figure 12b) become almost horizontal at a depth of 47 km except for the regions covered by sediments and a narrow (70 km) block centred around 200 km. The horizontal level taken for the depth of complete isostatic compensation outside the sedimentary cover (Figure 12b) coincides with horizontal segments of the 6.5-km/s isoline, with deviations within 3 km, which is close to the accuracy of the original velocity model, and it occurs at least 8 km above the Moho. In this paper we define the top of the layer characterised by 8.1-8.3 km/s velocity to be the Moho (Figures 12, 13). We do not see any lateral variation in seismic velocity below this boundary in our data.

Average velocity isolines deviate downwards from the horizontal level taken for the depth of complete isostatic compensation underneath regions covered by sediments, thus indicating that these regions may not be isostatically compensated within the crust.

If we calculate the average velocity from 2 km depth downwards, we can exclude the effect of the sediments. Removal of the Paleozoic Georgina Basin and Mesozoic Eromanga Basin sediments on the western and eastern flanks of the line results in flattening of the average velocity isolines along the whole line (Figure 13). Isolines become horizontal (deviations do not exceed 3 km, which is close to the accuracy of the original velocity model) at a depth of 45 (true 47) km, with only one exception of a narrow block centered around 200 km. This confirms that "seismic isostasy" in crystalline crust is achieved above the Moho, and is disrupted where sediments are deposited on the top of crystalline basement.

The 70-km wide block centered around 200 km remains an unusual feature in terms of average velocity distribution. The average velocity to any given depth larger than 20 km within this block remains systematically lower than elsewhere on the line (Figures 12 and 13). The low average velocity may be explained by the lack of three-dimensional information in our interpretation. We suggest that significant lateral variation in seismic velocity distribution is not limited to the east-west direction but also exists in the north-south direction. In this case, low average velocity values between ~160 and 230 km along the line may well result from not recognising some high-velocity rock units to the north or to the south of the refraction line. Moreover, if the strike of the mid-crustal high-velocity anomaly detected in our data is not orthogonal to the seismic line, a continuation of this anomaly in the third dimension could possibly compensate for the (false) decrease in average velocity produced in the vertical section by the two-dimensional calculation of average velocities.

5. OTHER PRECAMBRIAN REGIONS OF AUSTRALIA

Intra-crustal low-velocity layers have been recognised in some other interpretations of seismic data in Australia; for a compilation of these data see Collins [1988]. Preliminary analysis of average velocity-depth functions calculated for these models shows that they generally agree with the concept of high and low velocities balancing down a vertical profile through the crust (Goncharov, in preparation). A spectacular example of such balancing can be taken from the Tennant Creek - Mount Isa region adjacent to the Mount Isa Inlier (Figure 10).

5.1 Geology of the Tennant Creek Inlier

The Tennant Creek Inlier is an outcrop of Precambrian basement of the North Australian Craton similar to the Mount Isa Inlier. Pelitic gneisses were metamorphosed around 1920 Ma, and these possibly form basement to the Warramunga Group sedimentary and volcanic Early Proterozoic rocks [Black, 1977]. The Warramunga Group and associated plutonic rocks underlie much of the Georgina Paleozoic Basin east of Tennant Creek [Tucker et al., 1979].

5.2 Seismic Studies

A refraction survey along the 600-km Tennant Creek - Mount Isa traverse was completed in 1979 [Finlayson, 1982]. Two shots were used to obtain seismic data. Within 100 km of the shots, recorders were placed at 10-km spacing, and between 100 and 300 km, the recorders were about 20 km apart - less dense than that on the Mount Isa 1994 refraction line.

Results of the interpretation were presented as two one-dimensional seismic models, one for Tennant Creek to Mount Isa and the other for the reverse direction [Finlayson, 1982]. Significant differences between the models can be seen down to the lower-crustal level. For example, compressional wave velocities of 6.85 km/s occur at a depth of 26 km near Tennant Creek, whereas such velocities are not evident until depths of about 37 km near Mount Isa.

5.3 Seismic Isostasy

Original seismic models for the Tennant Creek - Mount Isa traverse [Finlayson, 1982] show a trend for mid- to lower-crustal velocities to increase from east to west. At the same time the trend for upper-crustal velocities is reversed: velocities in a depth range 0 - 20 km are noticeably lower underneath Tennant Creek than those underneath Mount Isa. Superimposition of both trends results in the balancing of high and low velocities down vertical profiles through the crust. Average velocity-depth functions calculated from the original seismic models quantify this observation. These functions vary significantly within the top 10-15 km of the crust, but they quickly merge at larger depths and almost coincide at a depth of 35 km (Figure 14). Both average velocity-depth curves become practically indistinguishable from the Mount Isa 1994 representative curves at a depth of 45-48 km. The Mount Isa 1994 refraction line is represented in this comparison (Figure 14) by two average velocity-depth functions derived from the original two-dimensional model (Figure 12). One corresponds to the vertical slice of the section in Figure 12 taken at the 280-km location which runs through the center of the mid-crustal high-velocity anomaly; the second function corresponds to the vertical slice of the same section taken at the 160-km location where the thickness of low-velocity (5.7-6.1 km/s) material in the upper crust is increased as well as the thickness of high-velocity lower crust and crust-mantle transition zone. The shallowest Moho position in the region is about 50 km deep (Figure 14). Despite significant lateral variations in seismic velocity distribution at the upper- to lower-crustal levels, the average velocity-depth functions merge above the Moho, thus indicating that "seismic" isostatic equilibrium is achieved also above the Moho.

6. CONCLUSIONS

We have demonstrated that seismic data obtained by different techniques in various Precambrian provinces in the Baltic Shield and Australia indicate that the average seismic velocity-depth functions merge with depth to form a single trend well above the Moho (see Table 1), even in cases of significant velocity variations within the crust. We interpret this observation as an indication of complete "seismic isostasy" achieved above the Moho. Translation of "seismic isostasy" estimated by us from average velocity-depth functions into conventional isostasy depends on correlation between seismic velocity and density.

Direct pressure estimates based on the density data from the KSDBH confirm that objects with anomalous seismic velocity and density in the upper crust may be isostatically compensated also well above the Moho, but possibility of global translation of seismic isostasy into conventional isostasy is a subject for further studies. Intra-crustal isostasy may eliminate or reduce the need for isostatic equilibrium to be achieved at the Moho or at the lithosphere-asthenosphere boundary in Precambrian crust.

Several mechanisms can be proposed to explain the emplacement of high-velocity bodies in the crust which are

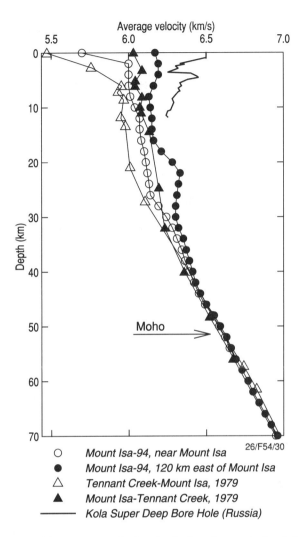

Figure 14. Average velocity-depth functions calculated for the Mount Isa - Tennant Creek refraction line [Finlayson, 1982], compared with those from the Mount Isa 1994 refraction line and the Kola Super Deep Bore Hole (KSDBH). "Mount Isa-94, near Mount Isa" average velocity-depth function corresponds to the vertical slice of the section in Figure 12 taken at the 160 km location, where the relatively low-velocity (5.7-6.1 km/s) material in the upper crust, as well as the high-velocity lower crust and crust-mantle transition zone, are thicker; "Mount Isa-94, 120 km east of Mount Isa" function corresponds to the vertical slice of the same section taken at the 280-km location, which runs through the center of the mid-crustal high-velocity anomaly. The Moho is shown at the shallowest level for the analysed models.

compensated by associated low-velocity rocks immediately underneath (or vice-versa): tectonics, extrusion/intrusion with subsequent burial, fractionation during melting, serpentinisation. Dehydration, phase transition, and partial melt of water-bearing minerals, such as hornblende and biotite, may be responsible for the compressional wave velocity decreases in the crust [Zhao et al., 1996]. Whatever the geological scenario explaining our observations, it must be consistent with a concept of mass transfer in Precambrian crust being a balanced process even in its local aspects.

Our observations of seismic isostasy are more consistent with Pratt-type conventional isostasy (horizontal isobars below a certain depth) than with Airy-type isostasy. A partial revival of the Pratt scheme is in recognition of isostatic compensation being achieved at a depth shallower than the Moho. However, the problem of velocity-density correlation has to be reassessed in the future.

A concept of multi-level seismic isostasy in Precambrian crust emphasises the essence of our observations: wherever a high- or low-velocity anomaly occurs in the crust, it tends to be compensated by its counterpart immediately underneath.

The results presented in this paper may have implications for studies of lithospheric deformation, where various assumptions about the rheology of the crust are often made. These results can also be useful for modeling elastic properties of the crust from an assumed petrological composition or from a xenolith-derived composition. Monotonous velocity increase with depth is often used in such studies as a reasonable approximation of the real

Table 1. Range of average seismic velocity variation as a function of depth in the Precambrian crust (based on the data presented in this paper).

Depth, km	Minimum average velocity, km/s	Maximum* average velocity, km/s
0	5.50	6.50
5	5.91	6.38
10	5.95	6.27
15	5.99	6.24
20	6.00	6.33
25	6.07	6.31
30	6.16	6.30
35	6.27	6.37
40	6.39	6.39
45	6.49	6.49
50	6.57	6.57
55	6.70	6.70
60	6.78	6.78
65	6.87	6.87
70	6.96	6.96

* Note that minimum and maximum average velocity values converge at a depth of 40 km and form a single trend further downwards. Maximum average velocity values in the depth range 0-12 km are taken from the Kola Super Deep Bore Hole data; those in the depth range 20-40 km from the middle part of the Mount Isa 1994 refraction line; average velocity distribution in the depth range 12-20 km was derived by extrapolation and is not well constrained.

situation. We hope to have demonstrated that this is not always true. We also expect that the number of regions where intra-crustal low velocity layers will be interpreted as major elements of achieving isostasy in the (Precambrian?) crust will increase with the increase in density of refraction/wide-angle observations and with the improvement of interpretation techniques.

Acknowledgements. Most of the data presented in this paper were collected by the St.-Petersburg Mining Institute and the Australian Geodynamics Cooperative Research Centre group at the Australian Geological Survey Organisation. Sonic log and density measurements in the Kola Superdeep Bore Hole were carried out by the staff of the Kola Geological Prospecting Expedition for the Superdeep Drilling and provided for our reprocessing by L. I. Faryga, R. V. Medvedev, Yu. P. Smirnov and F. F. Gorbatsevich. Reprocessing of these data was performed by the Department of Geophysics at the St.-Petersburg Mining Institute in 1991-1992 as a part of a research contract 98/91 between the Mining Institute and the Science Industrial Organisation "Nedra" (Yaroslavl, Russia). Dr L. A. Pevzner ("Nedra") strongly stimulated reprocessing and generalisation of the sonic log and density data. The methodology of this work was developed earlier by M. D. Lizinsky. Peter Wellman, Ken Muirhead, Doug Finlayson and Shen-Su Sun provided valuable criticism of an early version of the manuscript. Comments by Rob van der Hilst and two anonymous reviewers helped us to focus the paper mainly on seismic aspect of the problem of intra-crustal isostasy. Alexey Goncharov, Tanya Fomin, Clive Collins, Bruce Goleby and Barry Drummond publish with permission of the Executive Director of the Australian Geological Survey Organisation.

REFERENCES

Andreyev, B. A., and Klushin, I. G., *Geological Interpretation of Gravity Anomalies* (in Russian), 495 pp., Nedra Publishers, Leningrad, 1965.

Black, L. P., A Rb-Sr geochronological study in the Proterozoic Tennant Creek Block, central Australia, *BMR J. Aust. Geol. Geophys.*, 2, 111-122, 1977.

Blake, D. H., Geology of the Mount Isa Inlier and environs, Queensland and Northern Territory, *Aust. Bur. Miner. Resour. Bull.*, 225, 1987.

Blake, D. H., and Stewart, A. J., Stratigraphic and tectonic framework, Mount Isa Inlier, in *Detailed Studies of the Mount Isa Inlier, AGSO Bull.*, 243, edited by A. J. Stewart and D. H. Blake, pp. 1-11, 1992.

Collins, C. D. N., Seismic velocities in the crust and upper mantle of Australia, *Aust. Bur. Miner. Resour. Rep.*, 277, 160, 1988.

Drummond, B. J., and Collins, C. D. N., Seismic evidence for underplating of the lower continental crust of Australia, *Earth Planet. Sci. Lett.*, 79, 361-372, 1986.

Drummond, B. J., Goncharov, A. G., and Collins, C. D. N., Upper crustal heterogeneities in Australian Precambrian Provinces interpreted from deep seismic profiles and the Kola Superdeep Bore Hole, *AGSO J. Aust. Geol. Geophys.*, 15, 519-527, 1995.

Etheridge, M. A., Rutland, R. W. R., and Wyborn, L. A. I., Orogenesis and tectonic process in the Early to Middle Proterozoic of northern Australia, in *AGU Geodynamic Series*, 17, 131-147, 1987.

Finlayson, D. M., Seismic crustal structure of the Proterozoic North Australian Craton between Tennant Creek and Mount Isa, *J. Geophys. Res.*, 87, 10569 - 10578, 1982.

Goncharov, A. G., Seismic models of the crust-mantle transition zone (in Russian), in *Structure of the Lithosphere in the Baltic Shield*, edited by N. V. Sharov, pp. 42-44, VINITI Publishers, Moscow, 1993.

Goncharov, A. G., Kalnin, K. A., Lizinsky, M. D., Lobach-Zhuchenko, S. B., Platonenkova, L. N., and Chekulayev, V. P., Seismogeological characteristics of the Earth's crust in Karelia (in Russian), in *Problems of Integrated Geological/Geophysical Data Interpretation,*, pp. 53-84, Nauka Publishers, Leningrad, 1991.

Goncharov, A. G., Collins, C. D. N., Goleby, B. R., Drummond, B. J., and MacCready, T., The Mount Isa Geodynamic Transect: Implications of the seismic refraction model, *Aust. Geol. Surv. Org. Res. Newsl.*, 24, 9-10, 1996.

Goncharov, A. G., Sun, S-S., and Wyborn, L. A. I., Balanced petrology of the crust in the Mount Isa region, *Aust. Geol. Surv. Org. Res. Newsl.*, 26, 13-16, 1997.

Ha, J., Recurrence relations for computing complete P and SV seismograms, *Geophys. J. Roy. Astron. Soc.*, 79, 863-873, 1984.

Heiskanen, W. A., and F. A. Vening-Meinesz, *The Earth and its Gravity Field.*, 470 pp., McGraw-Hill, New York, 1958.

Kartvelishvili, K. M., *Planetary Density Model and Normal Gravity Field of the Earth* (in Russian), 95 pp., Moscow, Nauka Publishers, 1982.

Kozlovsky, Ye. A. (Ed.), *Exploration of the Deep Continental Crust: The Superdeep Well of the Kola Peninsula,* 558 pp., Springer - Verlag, 1984.

Lizinsky, M. D., and Lanev, V. S., Seismic section of a drilling site of the Kola superdeep bore hole (in Russian), in *Problems of Integrated Geological/Geophysical Data Interpretation*, pp. 131-148, Nauka Publishers, Leningrad, 1991.

Luetgert, J. H., User's Manual for RAY84/R83PLT - Interactive Two-Dimensional Raytracing/Synthetic Seismogram Package, U.S. Geological Survey Open File Report 88-238, 1988.

Sharov, N. V. (Ed.), *Structure of the Lithosphere in the Baltic Shield* (in Russian), 166 pp., VINITI Publishers, Moscow, 1993.

Tucker, D. H., Wyatt, B. W., Druce, E. C., Mathur, S. P., and Harrison, P. L., The upper crustal geology of the Georgina Basin region, *BMR J. Aust. Geol. Geophys.*, 4, 209-226, 1979.

Warner, M. R., Seismic reflections from the Moho - the effect of isostasy, *Geophys. J. Roy. Astron. Soc.*, 88, 425-435, 1987.

Zhao, Zh., Gao, 'S., Luo, T., and Zhang, B., The origin of crustal low-velocity layers: evidence from laboratory measurement of P-wave velocity of rocks at high PT conditions, *30th Int. Geol. Congr. Abstracts, vol.1,* 108, 1996.

___

C. D. N. Collins, B. J. Drummond, T. N. Fomin, B. R. Goleby, and A. G. Goncharov, Australian Geological Survey Organisation, GPO Box 378, Canberra, ACT, 2601, Australia

K. A. Kalnin, M. D. Lizinsky, and L. N. Platonenkova,, Department of Geophysics, St.-Petersburg Mining Institute, 21 line, 2, St.-Petersburg,199026, Russia

Contrasting Styles of Lithospheric Deformation Along the Northern Margin of the Amadeus Basin, Central Australia

Jean Braun

Research School of Earth Sciences, The Australian National University, Canberra, Australia

Russell Shaw

Australian Geological Survey Organisation, Canberra, Australia

We present the results of two numerical experiments in which the continental lithosphere is subjected to compression driven by an imposed basal velocity discontinuity. This discontinuity represents the reactivation of an intracratonic weak zone by in-plane stresses. The lower part of the lithosphere is assumed to be decoupled from the upper part by a weak detachment surface at sub-Moho depths. The two experiments differ by the assumed initial geothermal gradient, which results in the presence or absence of intracrustal décollements. The results of the two numerical experiments are compared to the complex structures observed along the northern margin of the Amadeus Basin. We concentrate on the structures that developed during the late Paleozoic Alice Springs Orogeny. The major differences in structural style, denudation patterns, and the distribution of metamorphic rocks near the surface between the central and eastern parts of the basin margins, are compared to the numerical model predictions. We conclude that those differences can easily be explained in terms of local variations in the initial thermal, and hence mechanical, state of the continental lithosphere.

1. INTRODUCTION

The center of the Australian continent is host to a foreland-like thrust belt which extends east-west along the northern margin of the Amadeus Basin (Figure 1a). Major mylonitic thrust zones on the hinterland side of this thrust belt mark a jump in regional metamorphic grade and a steep Bouguer anomaly gradient (Figure 1b). Dip-slip stretching lineations characterize most of the thrusts within the belt [Forman and Shaw, 1973; Shaw and Black, 1991].

This thrust belt was active during the Alice Springs Orogeny (ASO), a late-Devonian to mid-Carboniferous compressional event that affected most parts of central Australia. Movement was concentrated along the Redbank Thrust Zone (RTZ), a mylonitic shear zone that extends to depths of at least 50 km [Shaw et al., 1992a]. North-over-south thrusting along the RTZ resulted in a major offset of the crust-mantle boundary, deposition of several kilometers of conglomerate along the northern margin of the Amadeus Basin, and exhumation of granulite facies rocks in the southern part of the Arunta Block (Figure 1).

Over most of its length, this structure has a "thick-skinned" tectonic style, in that a single lithospheric-scale structure, the RTZ, dipping 50-60° north, accommodated most of the lithospheric shortening during the ASO [Shaw et al., 1992a].

Along the far northeastern margin of the basin, however, deformation was more complex and distributed among several structures, namely, from south to north, the Ruby Gap Duplex, the Illogwa Shear Zone, and the highly deformed Bruna Gneiss, mantling the upper amphibolite facies rocks of the Entia Dome [Collins and Teyssier, 1989].

Thus marked differences exist, both in structural character and slip on the master thrusts, and also in the

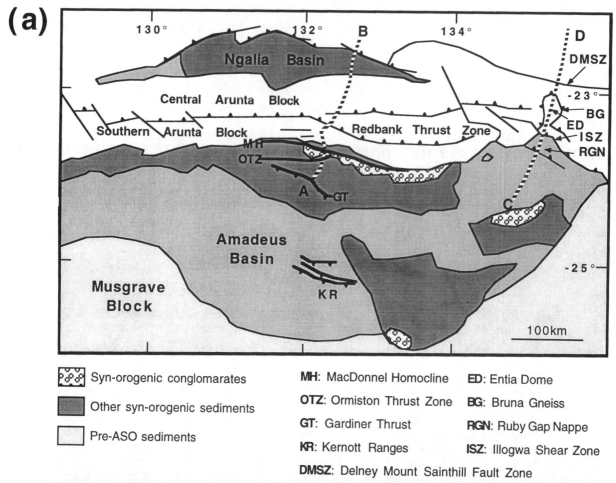
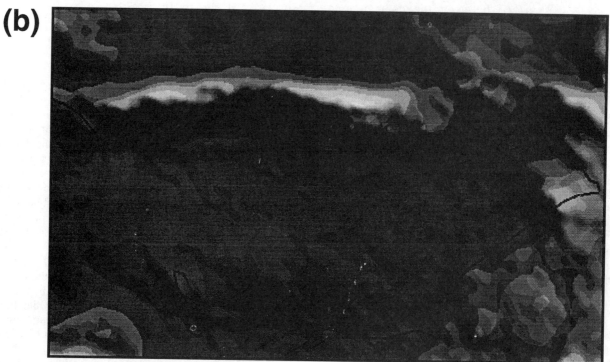

Figure 1. (a) Tectonic and structural elements of the northern margin of the Amadeus Basin and the southern Arunta Block. (b) Artificially illuminated map of the Bouguer gravity anomaly over the same area as (a); illumination is from the northeast.

distribution of synorogenic sediments between the central and eastern parts of the basin: whereas crustal shortening across the RTZ was partly accommodated by progressive denudation of the upthrust wedge [Shaw et al., 1992a], the cover nappes of the northeastern Amadeus Basin represent tectonic denudation of a sedimentary sequence originally overlying the basement thrust structures to the north [Stewart et al., 1991].

In the main "central" part of the thrust belt, the dominant controlling structure, the RTZ (Figure 1a), underwent 12-15 km of uplift [Shaw et al., 1992a] and dips at 50-60° north; from the Moho offset across the RTZ, crustal shortening is estimated at between 10 and 30 km [Lambeck et al., 1988; Shaw et al., 1992a]. In the northeast, several thrust structures were involved (Figure 1a), characterized by dips of 20-35°; the uplift was in excess of 20 km [Shaw et al., 1992a] and accumulated shortening, between the Delney Mount Sainthill Fault Zone and the Ruby Gap Duplex, is thought to be about 60 km [Teyssier, 1985].

Although less pronounced, the large Bouguer gravity anomaly gradient that characterizes the RTZ is also observed along the northeastern margin, suggesting that a deep-seated structure was also active there during the ASO (Figure 1b). The question then arises as to why a single lithospheric-scale structure has such different surface expressions along the central and eastern parts of the northern margin of the Amadeus Basin.

In this paper we suggest a simple explanation for this difference in tectonic style between the Redbank and Ruby Gap areas, based on the results of a numerical model of the thermo-mechanical evolution of the continental lithosphere during continental compression. We suggest that, at the time of the Alice Springs Orogeny, the central part of the basin was characterized by an unusually cold geotherm, whereas the eastern regions were characterized by a more "normal" geotherm. Using the results of two numerical experiments, we demonstrate that variations in tectonic style and resulting surface geology can be attributed to variations in initial geothermal gradient. We compare our results to geological and geophysical evidence from central Australia.

2. GENERALIZATION OF THE MANTLE SUBDUCTION MODEL

In many regions of continental convergence, shortening may be accommodated in different ways in the crust and mantle [Bird et al., 1975]. As suggested by Willett et al. [1993], the mantle part of one of the two colliding continents may detach from the overlying crust and subduct beneath the other (see Figure 2a). Mantle subduction may be facilitated by the presence of a lower crustal low-strength detachment layer [Beaumont et al. 1994]. The lighter crust resists subduction and accommodates the shortening by thickening.

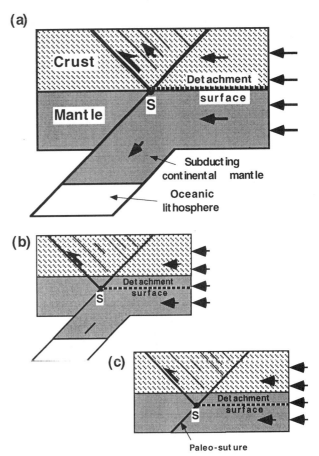

Figure 2. (a) The so-called "mantle subduction model" for continental collision at convergent plate boundaries regarded as the end-product of the closure of an oceanic basin; as the mantle part of the lithosphere subducts, the crust is forced to shorten by thrusting along a crustal-scale shear zone; detachment is likely to occur at the crust-mantle boundary at point "S". (b) The mantle subduction model applied to a "cold" lithosphere; the detachment surface has been pushed down below the crust-mantle boundary. (c) A similar lithospheric-scale response is expected in situations where a mantle heterogeneity ("paleo-suture") is reactivated by in-plane stresses.

From a crustal point of view, mantle subduction may be regarded as a basal velocity discontinuity as shown in Figure 2. Numerical [Willett et al., 1993] and analog [Malavielle, 1984] models of this type of continental collision suggest that the crust overlying the subducting mantle is pushed over the crust overlying the stable mantle along a crustal-scale shear zone (see Figure 2), the so-called *retro-shear* of Willett et al. [1993]. This shear zone is rooted in the detachment layer at the location of the velocity discontinuity. A second shear zone develops on the other (or *pro-*) side of the velocity discontinuity and accommodates more diffuse, lower amplitude deformation [Willett et al., 1993].

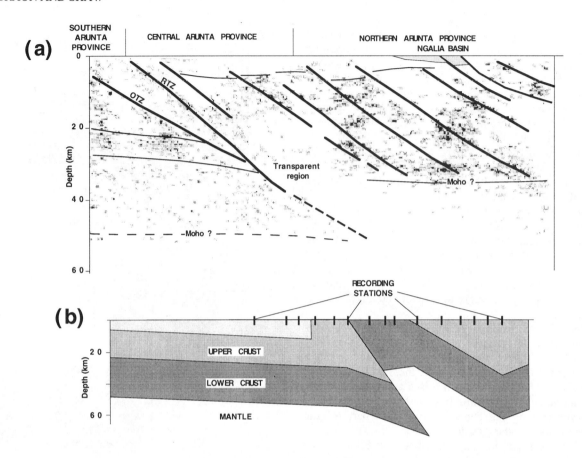

Figure 3. (a) Interpreted deep seismic reflection line across the central Redbank area (modified from Goleby et al. [1989]); RTZ is the Redbank Thrust Zone; OTZ is the Ormiston Thrust Zone. (b) Crustal structure beneath the central Redbank area as determined by seismic travel-time anomalies (modified from Lambeck [1991]).

In central Australia, evidence points to a rather different scenario. The large east-west trending Bouguer gravity anomalies (Figure 1), the results of a recent deep seismic reflection profile (Figure 3a) [Goleby et al., 1989], and seismic travel-time anomalies (Figure 3b) [McQueen and Lambeck, 1996], clearly demonstrate that lithospheric-scale thrusting accommodated north-south continental shortening. It has been postulated that during the Alice Springs Orogeny lithospheric-scale suture zones between continental units were reactivated [Shaw et al., 1992a]. The crust-mantle boundary (or Moho discontinuity) has been offset [Myers et al., 1996], locally by up to 25 km [Lambeck et al., 1988; Goleby et al., 1989; McQueen and Lambeck, 1996].

It is for these reasons that lithospheric deformation in central Australia has often been termed "thick-skinned" [Goleby et al., 1989] in contrast to the more classical style where surface structures do not penetrate beyond the lower crust across the crust-mantle boundary.

We propose to apply the mantle subduction model to thick-skinned tectonics by assuming that the detachment level is located within the mantle part of the continental lithosphere, beneath the crust-mantle boundary. In Figure 4, three stress envelopes corresponding to a uniform strain rate of 10^{-15} s^{-1} are shown, demonstrating that the presence of intra-crustal detachments depends on the assumed geothermal gradient. In relatively old, and therefore cold cratons, the temperature in the crust is such that ductile deformation (by dislocation creep of quartz-rich or feldspar-rich rocks) is not activated. Consequently, the crust-mantle boundary is not a rheological discontinuity and, during continental deformation, detachment along a weak horizontal layer is possible only at the brittle-ductile transition in the olivine-rich mantle.

We present the results of a set of numerical experiments of continental shortening, driven by subduction of the lower part of the lithospheric mantle. We adopt the approach of Willett et al. [1993] in which subduction of the bottom layer is represented by imposing a velocity discontinuity at the base of the top layer; in our study, however, we impose the velocity discontinuity near the brittle-ductile transition in the olivine-rich mantle.

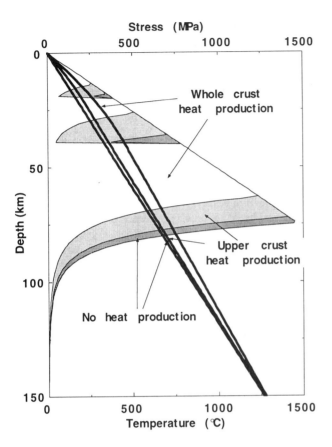

Figure 4. Deviatoric stress profiles for the continental lithosphere based on a layered rheology for three different geotherms assuming a strain rate of 10^{-15} s^{-1}; the three geotherms correspond to cases where there is (1) no heat production, (2) heat production in the upper crust only and (3) heat production in the whole crust; the upper crust rheology is assumed to be quartz-dominated, the lower crust is feldspar-dominated and the mantle is olivine-dominated.

3. THE NUMERICAL MODEL

To study the response of the Earth's lithosphere to tectonic forces, we have developed a finite-element model based on the Dynamic Lagrangian Remeshing (DLR) method [Braun and Sambridge, 1994] to solve the two-dimensional version of the static force balance equations:

$$\sigma_{xx,x} + \sigma_{xz,z} = 0$$
$$\sigma_{xz,z} + \sigma_{zz,z} = -\rho g$$

where the σ's are the components of the stress tensor and ρ is the density.

In our model, the lithosphere is regarded as a complex isotropic elastic rheoid capable of viscous creep at elevated temperature and brittle failure at low pressures. The viscous rheology is ruled by the following non-linear stress-strain rate relationship:

$$\sigma = B \, \varepsilon^{1/n} \exp(Q/nRT)$$

where B, n and Q are rheological parameters derived from laboratory experiments. The model assumes that the lithosphere is compositionally (hence mechanically) layered: in the upper crust, we use a quartz-dominated rheology based on the quartzite rheology of Paterson and Luan [1990]; in the lower crust, we use a feldspar-dominated rheology based on the Adirondak granulite rheology of Wilks and Carter [1990] and, in the mantle, we use an olivine-dominated rheology based on the Aheim dunite rheology of Chopra and Paterson [1981]. Brittle deformation is represented by an associative plastic flow law derived from Griffith's failure criterion [Griffith, 1921] which may be easily expressed in terms of the invariants of the stress tensor [Prager and Hodge, 1951]:

$$J_{2D} + 12 \, T_0 \, p = 0$$

where J_{2D} is the second invariant of the deviatoric part of the stress tensor, p is the pressure, and T_0 is the tensile strength, a rock property assumed constant at 10 MPa in all numerical experiments. In our model, the pressure incorporates the lithostatic pressure (resulting from the weight of the overburden) and the dynamic pressure (resulting from deformation driven by the imposed boundary conditions). Justification for the use of Griffith's failure criterion to model the brittle behavior of crustal rocks is given in Braun [1994] and Braun and Beaumont [1995].

Although based on a continuum description of the material, the model allows for extreme localized deformation. In the DLR method (Braun and Sambridge, 1994), the nodes making the numerical mesh are Lagrangian particles attached to the deforming body; the connections between the nodes on which the finite-element discretization is based are dynamically recomputed at each time step; this leads to an accurate solution of the basic force balance equations even after very large deformation.

The mechanical equations are coupled to the equations of heat transfer by advection and conduction:

$$\partial T/\partial t = \kappa \nabla^2 T + H/c$$

where c is the heat capacity, κ is the thermal diffusivity (assumed constant at 1 mm^{2}s^{-1}) and H is the radiogenic heat production per unit mass (assumed constant in the crust). Advection of heat is incorporated by solving the heat transfer equation on a Lagrangian grid attached to the deforming lithosphere. The temperature is, in turn, used to update the thermally activated viscosity of the rocks. Because we do not attempt to model the dynamics of the mantle, it is a reasonable assumption to neglect the effect of thermal expansion. The effect of lateral temperature changes on crustal rock density is assumed to be small.

Because the nodes on which the spatial discretization is based are attached to the deforming material, it is easy to follow material boundaries such as the crust-mantle boundary or any intra-lithospheric material interface. At the surface of the model, mass is redistributed according to a one-dimensional erosion/deposition algorithm based on the fluvial and diffusion transport equations described in Kooi and Beaumont [1994].

By analogy to the mantle subduction model [Willett et al., 1993], we shall assume that lithospheric shortening is driven by a velocity discontinuity at point "S" at the base of the modeled 50 km-thick lithospheric layer (Figure 5). As discussed in the introduction, this basal velocity boundary condition implies that the lowermost part of the continental lithosphere "subducts", while the uppermost part undergoes thrusting. One could also regard this boundary condition as equivalent to the reactivation of an ancient suture zone by remote in-plane forces originating along one of the margins of the continent. In this case, the imposed discontinuity represents a mantle discontinuity (a Proterozoic suture?) which triggers localisation of the deformation.

The 50 km-thick lithospheric layer rests on a thin elastic plate foundation [Beaumont et al., 1994] characterized by a flexural rigidity of 8.9×10^{21} Pa m^3.

The assumed initial crustal thickness is 40 km. The temperature at the base of the model is held constant and uniform heat production is assumed in the crust. The time step length is 0.02 Myr and the initial spatial discretization is 5 km. The finite element mesh is allowed to evolve in regions of high strain until successive self-refinements lead to a minimum element size of 0.5 km. The imposed shortening velocity is 5 mm yr^{-1}. This velocity is meant to represent a "typical" tectonic velocity rather than the value that may be regarded as the most appropriate for the ASO; the reasons for this are two-fold: firstly, the principal purpose of the numerical simulations is not to present the best possible mechanical model of lithospheric deformation for the ASO, but rather to illustrate in a quantitative manner the differences in tectonic style that may result from different thermal (and thus mechanical) states of the lithosphere; secondly, we only have an approximate idea of how long the active tectonic phase lasted during the ASO [Shaw et al., 1992a], and therefore the rate of convergence is poorly constrained.

4. LOW GEOTHERM EXPERIMENT

In the first of two experiments, the geothermal gradient is chosen to be very low such that no ductile flow is activated in the crust and crustal rheology is dominated by a highly non-linear but temperature-independent deformation mechanism, in our case frictional brittle deformation. This is incorporated in the numerical model by imposing an artificially low temperature at the base of the model (350°C). As shown in Figure 4, this behavior is likely to be encountered in the Earth's lithosphere in situations where the thermal gradient is very low and deformation takes place at a "reasonable" strain rate. It is likely however that intense, localized deformation will lead to grain-size reduction which, in turn, may result in ductile deformation by diffusion creep becoming the dominant deformation mechanism [Drury et al., 1991]. The net effect is a decrease in rock strength and further localization of the deformation along existing brittle "faults" to form highly strained, small-grained mylonitic shear zones.

The results of the computations are shown in Figure 5. In the early stages of the experiment (Figure 5a), a set of narrow conjugate shear zones has developed from the imposed velocity discontinuity at the base of the model, cutting through the crust-mantle boundary and propagating up to the free surface. Both shear zones dip at approximately 45°. The dip of the shear zones is not imposed *a priori,* but is a direct result of the assumed highly non-linear (brittle) rheology for the crust and uppermost mantle. Ductile deformation is activated along the base of the model.

The solution rapidly reaches a quasi-steady state in which rocks travel laterally from regions outside of the orogen, cross the pro-shear zone and are exhumed as a result of thrusting along the retro-shear zone (Figure 5d).

This predicted deformation pattern leads to a major offset in the Moho discontinuity as a narrow "tongue" of uppermost mantle material is driven into the lower crust (Figure 5b). The mantle tongue imposes a load on the lithosphere and causes downward isostatic adjustment and flexure. Two large foreland basins form on either side of the orogen. Because, in this model experiment, the parameters in the erosion/deposition model are such that little surface topography is allowed to develop, it is clear that the foreland basins formed partly in response to intra-lithospheric loading by the mantle tongue; these basins will therefore partly survive post-orogenic erosion of the surface topography.

Following finite deformation, surface uplift occurs which is not totally compensated by erosion, such that a small, yet finite amplitude, topography is allowed to develop (Figure 5c). Subsequently, the dip of the retro-shear zone decreases and the deformation front propagates into the retro-foreland basin, upturning the sedimentary layers deposited in the early stages of the experiment. In contrast, the pro-foreland basin is passively advected with the pro-basement into the active orogen; its contact with the deforming orogen is consequently mostly erosional and shows little evidence of synorogenic deformation.

The total strain accumulated during the experiment is shown in Figure 5d. The retro-shear appears as a zone of relatively focused deformation which has propagated into the retro-foreland basin during the latest stages of

(a) Strain rate at 1 Myr

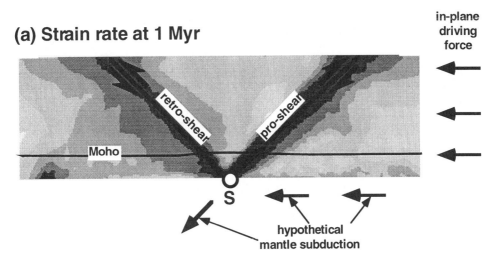

(b) Strain rate at 5 Myr

(c) Strain rate at 10 Myr

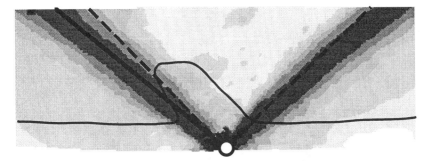

Figure 5. Results of the first numerical experiment based on a uniform rheology. (a)-(c) Contour plots of the second invariant of the deviatoric part of the instantaneous strain rate. Dark shades correspond to high strain rate values, light shades correspond to low strain rates. (d)-(f) Contour plots of the second invariant of the deviatoric part of the total accumulated strain. Superposed are: (e) stratigraphy, or the computed geometry of initially flat-lying passive markers, and (f) an interpretation of the numerical model results compared to structures and other geological features along the Redbank transect.

(d) Total strain at 10 Myr

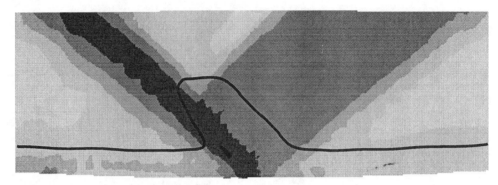

(e) Stratigraphy at 10 Myr

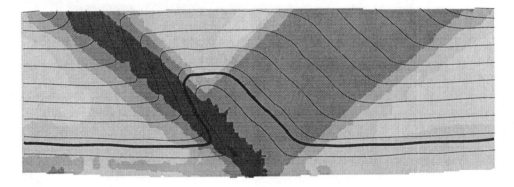

(f) Interpretation of (d) and (e)

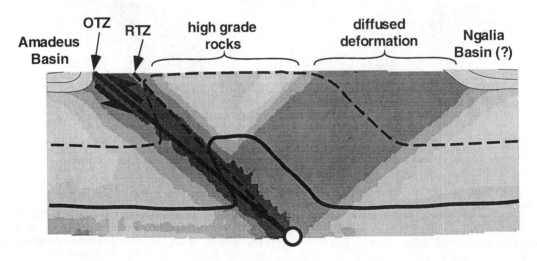

Figure 5. Continued from previous page.

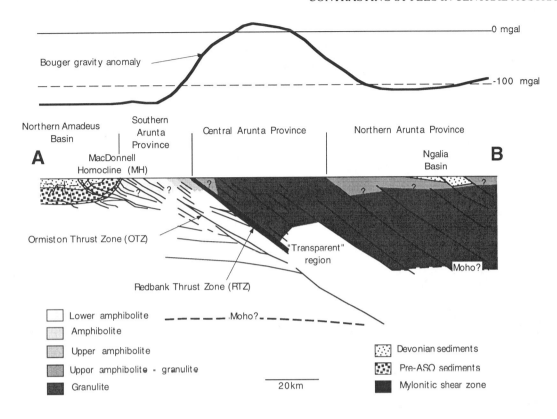

Figure 6. Deep crustal structure beneath the RTZ and distribution of metamorphic grades based on Shaw et al. [1991]'s interpretation of deep reflection seismic line and field observations.

deformation; movement along the pro-shear zone has led to the formation of a broad zone of diffuse deformation which extends from the central regions of the orogen to the most recently active pro-shear. The exact distribution of this diffused deformation is difficult to predict from the numerical model; the results of the second experiment will show that any horizontal or sub-horizontal weakness is likely to be used to accommodate that deformation and form a series of imbricate thrusts parallel to the retro-shear zone as indicated on Figure 2.

In Figure 5e the final geometry of a set of originally flat stratigraphic markers is shown. The results of the experiment clearly show that denudation is greatest in the upthrust wedge on the pro-side of the retro-shear zone and diminishes gradually across a wide zone. The retro-shear zone is therefore the locus of a large gradient in surface-rock metamorphic grade; the highest metamorphic grade rocks are found in a relatively wide zone from the retro-shear zone to the center of the orogen; on the pro-side of the orogen, the transition to preorogenic conditions takes place across a wide zone from the center of the orogen to the pro-shear zone (Figure 5f).

5. THE REDBANK TRAVERSE

Figure 6 presents a previously published model of the crustal structure beneath the RTZ based on extensive geological and geophysical studies, including structural mapping [Shaw, 1991] integrated with an interpretation of a deep seismic profile [Goleby et al., 1990; Shaw, 1991], studies of teleseismic travel-time residuals [Lambeck, 1991], and interpretation of Bouguer gravity anomaly data.

The main feature of this interpretation is the large offset (10-20 km) in the Moho discontinuity, resulting from thrusting along a north-dipping lithospheric-scale discontinuity which links with the RTZ near the surface. This, together with the coincident large gradient in surface-rock metamorphic grade, from lower amphibolite facies near the northern margin of the Amadeus Basin to granulite facies just to the north of the RTZ, point to the RTZ being a crustal-scale retro-shear zone.

Geochronological data [Collins and Shaw, 1995] clearly show that the RTZ was the most active structure during the ASO, and that tectonic uplift was limited to the area between the northern margin of the Amadeus Basin and the

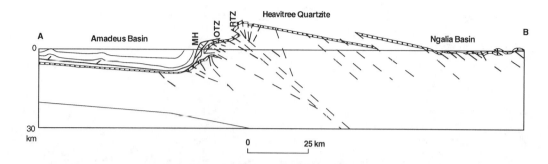

Figure 7. Estimates of total ASO uplift and denudation across the Redbank Thrust Zone along transect A-B (Figure 1) derived from Ar/Ar and Rb/Sr dating [Shaw et al., 1992b].

southern margin of the Ngalia Basin (Figure 7). To the north of the RTZ, differential uplift between the RTZ and the southern margin of the Ngalia Basin was accommodated by a series of north-dipping imbricate thrusts. Because, in our numerical model, the width of the region undergoing uplift is directly related to its initial thickness (or the assumed depth to the detachment point, "S", Figure 2), one may infer from this simple observation that the layer accommodating shortening by thrusting along the RTZ is approximately 50 km thick.

The large gradient in Bouguer gravity anomaly [Forman and Shaw, 1973] coincident with the RTZ results from the major offset along the Moho caused by thick-skinned deformation before and during the ASO. This implies that the detachment point must lie below the Moho which, in turn, provides us with a well-constrained lower limit of 40 km for the detachment depth (or, more importantly, the thickness of the layer involved in thrusting during the ASO).

It is interesting to note that despite the age (>300 Ma) of the last major tectonic movement across the RTZ and the lack of present-day topography in the southern Arunta Block, a substantial late Paleozoic sedimentary section (>3 km) has been preserved along the northern margin of the Amadeus Basin [Shaw et al., 1992a]. Our numerical model suggests that the large mantle wedge which was thrust up along the RTZ is the intra-lithospheric load which keeps the lithosphere in a state of downward flexure. Unlike loads resulting from synorogenic surface topography, the mantle load is a long-lived feature which is not affected by postorogenic erosion. As suggested by Lambeck [1983], Stephenson and Lambeck [1985], and Lambeck et al. [1988], substantial in-plane stresses are required to sustain the large offset in the Moho which causes the large Bouguer gravity anomaly.

During the latest stages of the ASO, shortening was accommodated by movement along a system of shallow-dipping thrusts [Shaw et al., 1992a]. These form the Ormiston Fault System (OTZ, see Figures 1, 6 and 7), which led to major reworking of the pre- and syn-ASO sediments along the northern margin of the Amadeus Basin to form the MacDonnell Homocline (MH, see Figure 6). The results of the numerical model show that this widening of the zone of deformation, into the retro-foreland basin by shallowing of the dip of the active thrust system, is a consequence of the syntectonic thickening of the layer involved in the shortening. This thickening has also led to a deepening of the "S" point (attached to the base of the modeled layer) from 50 to 62 km.

The absence of igneous activity at the time of the ASO in the Redbank area, combined with the limited synorogenic metamorphic retrogression, leads to the conclusion that the lithosphere, or at least the part of the lithosphere that underwent thrusting along the RTZ, was characterized by a relatively low geothermal gradient, in agreement with ^{40}Ar-^{39}Ar and conodont maturation data from the basin cover interface [Shaw et al., 1992b]

All these observations point to a relatively simple, yet unusual, behavior for the continental lithosphere under compressive tectonic stresses during the ASO, which may be reproduced numerically by assuming a low initial geothermal gradient. We conclude that the RTZ may, therefore, be a lithospheric-scale retro-shear zone that roots into a detachment point at depths of 50 to 60 km.

6. NORMAL GEOTHERM EXPERIMENT

In the second experiment, the temperature at the base of the model is fixed at 500°C such that, unlike in the first experiment, dislocation creep is activated at the base of the upper and lower crustal layers.

The results of this numerical experiment are shown in Figure 8. As in the uniform rheology experiment, two lithospheric-scale shear zones have developed, dipping in opposite directions and rooting into the velocity discontinuity (Figure 8a). Most of the shortening is accommodated by thrusting of the pro-side of the orogen along the retro-shear zone. The pro-shear zone is a transient feature that does not accumulate finite strain (Figure 8d).

Unlike the uniform rheology experiment, however, thrusting along the two conjugate shear zones is accompanied by deformation along two originally flat

CONTRASTING STYLES IN CENTRAL AUSTRALIA 149

(a) Strain rate at 1 Myr

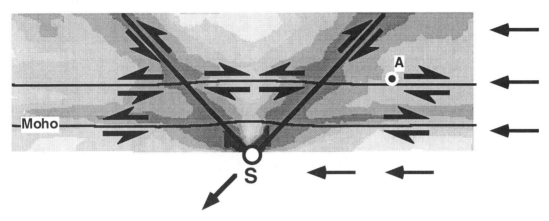

(b) Strain rate at 5 Myr

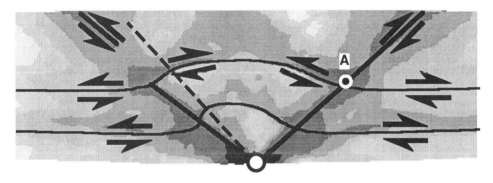

(c) Strain rate at 10 Myr

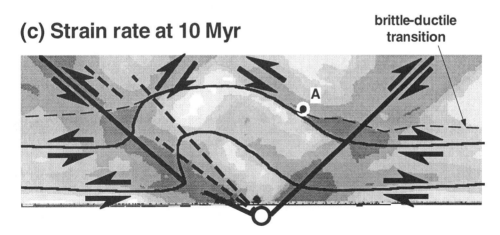

Figure 8. Results of the second numerical experiment based on a layered rheology. (a)-(c) Contour plots of the second invariant of the deviatoric part of the instantaneous strain rate. Dark shades correspond to high strain rate values, light shades correspond to low strain rates. Arrows indicate sense of shear. The thin dashed line in (c) indicates the approximate location of the brittle-ductile transition in the quartz-rich upper crustal layer; the thick dashed line indicates the position of the retro-shear zone in the early stages of deformation. (d)-(f) Contour plots of the second invariant of the deviatoric part of the total accumulated strain. Superposed are: (e) stratigraphy, or the computed geometry of initially flat-lying passive markers and (f) an interpretation of the numerical model results compared to structures and other geological features along the Redbank transect.

(d) Total strain at 15 Myr

(e) Stratigraphy at 15 Myr

(f) Interpretation of (d) and (e)

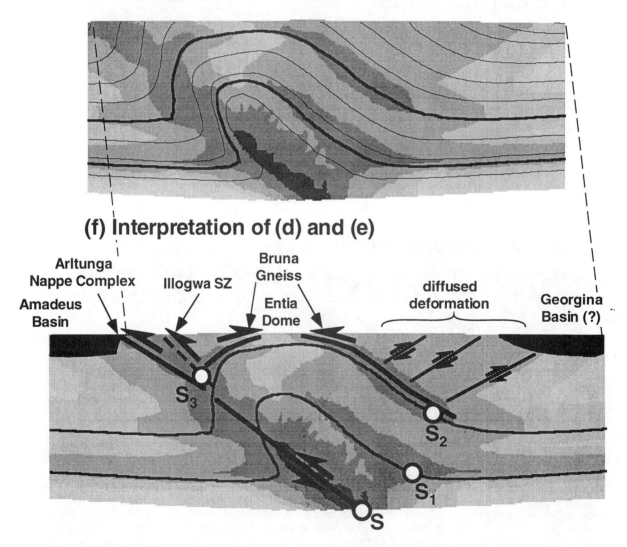

Figure 8. Continued from previous page.

intracrustal detachments: one at the base of the quartz-rich upper-crust and one at the base of the feldspar-rich lower crust. The sense of shear along these detachments is variable and is indicated on Figure 8a. Outside of the orogenic wedge, the sense of shear is similar to the imposed basal shearing: sinistral on the retro-side of the orogen and dextral on the pro-side of the orogen. Inside the wedge, the sense of shear is in the opposite direction: dextral on the retro-side of the orogen and sinistral on the pro-side. Rocks that originate at the base of the quartz-rich upper crustal layer or at the base of the crust therefore experience a complex deformation history as they traverse the orogen (Figure 8a-c). Particle A, for instance, experiences horizontal dextral shearing as it approaches the pro-shear zone (Figure 8a), undergoes a rapid phase of left-over-right thrusting as it travels through the pro-shear zone (Figure 8b) and, from then on, experiences right-over-left thrusting (hence sinistral shearing) as it is uplifted towards the surface (Figure 8c).

As deformation accumulates inside the orogen, the originally flat-lying intracrustal detachments are rotated and progressively brought closer to the surface (Figure 8b). In the early stages of deformation, shear in the intracrustal detachments is dominated by thermally activated ductile flow; later in the deformation history, parts of the detachments experience cooling, and the dominant deformation mechanism progressively switches to frictional brittle behavior (Figure 8c). When the intracrustal detachments are finally exhumed at the surface of the orogen (Figure 8c), two fabrics are thus recorded in the rocks: early ductile deformation overprinted by late brittle deformation.

The depth at which the transition from ductile to brittle behavior takes place is determined not only by the assumed geothermal gradient but also by the assumed convergence velocity. Indeed, because we assume that surface erosion prevents the build up of a large topography, the rate at which rocks are exhumed in the center of the orogen is directly proportional to the convergence velocity. As rocks are exhumed, they transport heat by advection more efficiently than they lose it by conduction [Batt and Braun, 1997]. Rock exhumation therefore leads to a local increase of the geothermal gradient near the surface of the orogen and, consequently, a shallowing of the brittle-ductile transition (Figure 8c) [Koons, 1987].

As shown in Figure 8c, the dip of the main retro-shear zone decreases with the accumulating deformation. As in the uniform rheology experiment, this results from the thickening of the deforming layer and the subsequent change in stress conditions along the shear zone. The kinematic consequence of this shallowing is a migration of the orogen to the left into the developing retro-foreland basin.

The second experiment was conducted for a longer time (15 Myr) than the first experiment (10 Myr) in order to expose the mid-crustal detachment. The total shortenings imposed by the boundary conditions are therefore 50 km in the first experiment and 75 km in the second.

At the end of the experiment (Figure 8d), the predicted deformation field is complex. Going from the retro- to the pro-side of the orogen, one observes:
• a foreland basin, the margin of which has been strongly reworked by thrusting with the deformation "younging" from right (pro) to left (retro);
• a major ductile shear zone overprinted by brittle deformation marks the boundary between the sedimentary basin and pre-orogenic basement rocks;
• an unusual dome-like structure comprising a high-grade core bounded by two oppositely dipping ductile shear zones overprinted by more recent brittle deformation; the sense of shear along the shear zone suggests thrusting of the lower-grade rocks over the higher-grade central core;
• on the pro-flank of the orogen, a zone of diffused deformation which extends to the margin of the pro-foreland basin.

As in the uniform rheology experiment, but to a lesser degree, the crust-mantle boundary (or Moho) is offset by movement along the retro-shear zone. A mantle tongue penetrates into the crust and, together with the resulting surface topography, creates a large isostatic load. This load results in turn, in a long wavelength deflection of the lithosphere and is responsible for the formation (and long-term survival) of the two "foreland" basins on either side of the orogen.

7. THE RUBY GAP TRAVERSE

Unlike the central Amadeus/Redbank area, there is no direct information on lithospheric-scale structures beneath the northeastern margin of the basin along the Ruby Gap traverse. Attempts at defining the tectonic history of the area have therefore been based on inferences from surface geology, the distribution of metamorphic facies [Forman and Shaw, 1973] structural analysis and geochronology [Collins and Teyssier, 1989; Dunlap and Teyssier, 1995].

Going from south to north from the northern margin of the Amadeus Basin into the Arunta Block along the Ruby Gap traverse (Figure 1), one observes the following structures (Figure 9):
• The Arltunga Nappe Complex formed by shortening of pre- and syn-ASO sedimentary units in the Amadeus Basin; based on an extensive geochronological study of the area, Dunlap and Teyssier [1995] report a younging of the deformation across the Arltunga Nappe Complex towards the south, implying that intra-basinal deformation is a consequence of the progressive southward migration of the deformation, or shallowing of the main retro-shear zone, as predicted by the numerical model. The presence of a relatively weak décollement surface near the base of the Amadeus Basin sedimentary sequence (the evaporite horizons in the Neoproterozoic Bitter Springs Formation) [Wells et al., 1970] is likely to be responsible for the

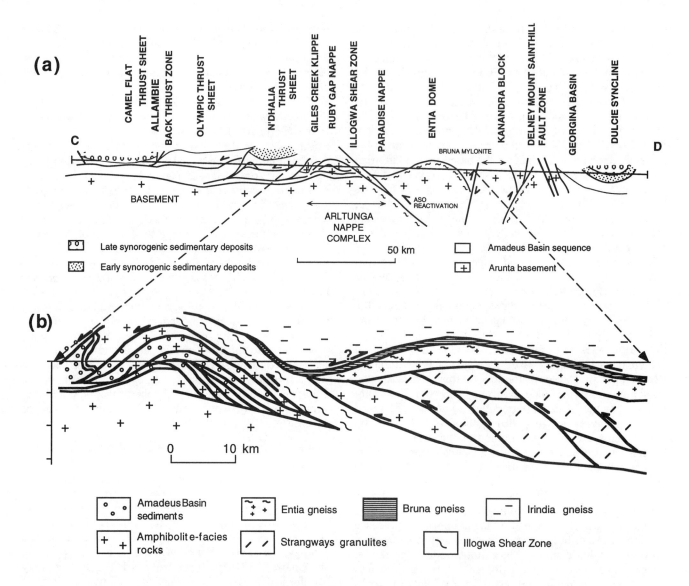

Figure 9. (a) Main structures along transect C-D (see Figure 1). (b) Details of the structures across the northern margin of the Amadeus Basin and the southeastern Arunta Block (modified from Collins and Teyssier [1989]).

extensive propagation of the deformation front into the central parts of the Amadeus Basin that caused southward movement of the N'Dhalia, Olympic and Camel Flat thrust sheets. Such intra-basinal deformation, accommodated by sheet sliding along a low-strength décollement, is also observed in the central parts of the Amadeus Basin, possibly as far south as the Kernot Range Thrust System (Figure 1a).

• The Illogwa Shear Zone is the locus of intense upper-crustal deformation where ductile fabrics are overprinted by brittle deformation [Collins and Teyssier, 1989]. We regard the Illogwa Shear Zone as the main retro-shear zone along the Ruby Gap traverse.

• The Entia Dome is composed of mildly deformed high-grade (upper amphibolite-facies) rocks, which we interpret as the crustal-scale anticlinal structure which, in the numerical model, formed by shearing along a mid-crustal low-strength detachment surface, and was exhumed by movement along the main retro-shear zone.

• The Bruna Gneiss mylonitic shear zone surrounds the Entia Dome; the mylonitic fabric formed at high temperature, and was overprinted by cooler brittle deformation [Collins and Teyssier, 1989; James et al., 1989]. Structures cropping out along the Bruna Gneiss on opposite sides of the Entia Dome dip in opposite directions away from the center of the dome [Collins and Teyssier,

1989]. In the light of our second numerical experiment, we therefore regard the Bruna Gneiss as a mid-crustal detachment surface that accommodated some of the imposed crustal shortening during the early stages of the ASO, and that was later exhumed by movement along the main retro-shear zone. The sense of shear along structures in the northern part of the Bruna Gneiss is relatively well constrained by field evidence and indicates north-over-south thrusting [Collins and Teyssier, 1989; Foden et al., 1995]. Further investigation is required to determine if the sense of shear in the mylonitic fabric in the Bruna Gneiss south of the Entia Dome is south-over-north as predicted by our numerical experiments.

The surface distribution of metamorphic rocks along the Ruby Gap traverse (Figure 9) is different from the one observed in the Redbank area (Figure 6) with the highest grade rocks exhumed being located near the center of the Entia Dome [Collins and Teyssier, 1989] some 40 km on the pro-side of the retro-shear zone (the Illogwa Shear Zone). Along the Redbank traverse, the highest grade rocks are found in the near vicinity (<10 km) of the retro-shear zone (in this case, the RTZ). Results from the second numerical experiment predict a pattern of exhumation that is very similar to the one observed along the Ruby Gap traverse, suggesting that our choice of 50 km for the initial detachment level is appropriate.

By comparing the numerical model results with the available structural, metamorphic and surface geology data, we conclude that the differences in the distribution of uplift and total denudation observed between the Redbank and the Ruby Gap areas are a direct consequence of the difference in mechanical behavior between the two regions at the time of the ASO. These differences should not be interpreted as indicative of variations in the thickness of the layer involved in the thrusting. Indeed, in both our numerical experiments, the initial depth of the detachment point is identical and set at 50 km. Caution should therefore be used in associating the differences between the two areas with variations in tectonic style, i.e. "thick-skinned" versus crustal tectonics.

8. CONCLUSIONS AND DISCUSSION

8.1. Evidence for Difference in Thermal Regime

Our modeling suggests that the thickness of the layer involved in the shortening during the ASO is similar in both areas investigated. The main reason for the differences observed between the Redbank and Ruby Gap areas is, according to our modeling, the difference in the lithospheric thermal state prior to deformation.

ASO synorogenic metamorphism is relatively widespread along the Ruby Gap traverse [Arnold et al., 1995; Foden et al., 1995], whereas it is confined to the main shear zone (RTZ) along the Redbank traverse [Shaw, 1991], suggesting significantly warmer conditions inside the orogenic belt in the east compared to the center.

Metamorphic assemblages that could be associated with late Paleozoic crustal reworking during the ASO give peak temperature estimates of 600-700°C for the Entia Dome amphibolites [Arnold et al., 1995]. In contrast, Rb-Sr whole-rock isochrons from the RTZ mylonites yield consistent mid-Proterozoic ages (1500-1400 Ma) [Shaw and Black, 1991], whereas biotite Rb-Sr reset ages yield ASO ages (350-300 Ma), suggesting a peak temperature of only 300°C. This difference in maximum temperature experienced during the ASO, by rocks now located at the surface of the crust, may be explained in part by the apparently smaller amount of ASO shortening (hence exhumation) in the Redbank area, but is also consistent with the hypothesis that the Redbank area was characterized by a lower geothermal gradient during the ASO.

Finally, the only evidence of igneous activity at the time of the ASO is a series of thick pegmatite dykes in the southern part of the Entia Dome [Joklik, 1955; Shaw et al., 1984], which also suggests that, during the ASO, the eastern part of the southern margin of the Arunta Block was characterized by a higher geothermal gradient than the central Redbank area.

We can only speculate why there may have existed a difference in the lithospheric thermal regime along the northern margin of the Amadeus Basin at the time of the ASO. A possibility is the presence of a thick (5 to 6 km) proto-Paleozoic sedimentary cover in the east, as suggested by Stewart et al. [1991].

8.2. Total Amount of Shortening

An interesting by-product of our numerical modeling is the predicted efficiency of the retro-shear zone at accommodating the imposed shortening. This efficiency may be obtained by comparing the amount of Moho offset predicted by the numerical model with the total amount of shortening imposed by the boundary conditions. In the first "cold-lithosphere" experiment (Figure 5e), total imposed shortening is 50 km, and predicted movement along the main retro-shear zone (measured as the predicted offset in the originally flat Moho discontinuity) is only 35 km, implying that only 70% of the imposed shortening is taken up by thrusting along the retro-shear, and that the remaining 30% is accommodated by pure-shear thickening. In the "normal lithosphere" experiment (Figure 8e), total imposed shortening is 75 km, of which 60% only is taken up by movement along the retro-shear zone. Ductile lithosphere accommodates more shortening by pure shear than by simple shear.

That deformation is more diffuse in the normal geotherm experiment may also explain why the gravity anomaly beneath the Ruby Gap traverse is less pronounced than beneath the central Redbank area (Figure 1b).

Finally, that even in a cold, mostly brittle, lithosphere deformation tends to partition itself between thrusting along a discrete lithospheric-scale shear zone, and distributed shear inside the orogen, demonstrates that great care must be taken in interpreting offset in the Moho (or any other clear stratigraphic marker) as an accurate measure of total tectonic shortening.

8.3. Timing of deformation in Ruby Gap Transect

While the thorough geochronological work of Dunlap and Teyssier [1995] clearly demonstrated that much deformation in the Ruby Gap Duplex and within the Illogwa Shear Zone took place in the late Paleozoic ASO, much debate has taken place over recent years on the age of the deformation in and around the Entia Dome. James et al. [1989] define the Harts Range Detachment Zone (HRDZ) as "the zone between the Entia Gneiss and the Irindina Gneiss containing the intrusive Bruna Gneiss, ... a zone of intense and long-lived, but mostly Proterozoic, shearing". Collins and Teyssier [1989] proposed that the Paleozoic deformation that caused the mylonitic fabric observed in the Illogwa Shear Zone was also responsible for the intense shearing in the Irindina and Bruna Gneiss. Recent Rb-Sr, Ar-Ar and Sm-Nd isotopic analyses by Foden et al. [1995] convincingly support the view that the HRDZ experienced rapid cooling and deformation during the ASO.

The results of the numerical model we present in this paper cannot be used directly to discriminate between the two possible ages for the deformation in the HRDZ. One may, however, deduce from the numerical model that deformation must have been coeval along the main retro-shear zone (the Illogwa Shear Zone along the Ruby Gap transect) and the originally flat-lying intracrustal shear zone (the Bruna Gneiss), to form the structures exposed today. Indeed, if the intracrustal shear zone had been actively deforming before movement along the retro-shear zone and had been later "passively" exhumed by movement along the retro-shear zone, the resulting pattern of exhumation (and hence the distribution of metamorphic rocks at the surface) would have been similar to the one observed along the Redbank transect (Figures 5 and 6). That the highest metamorphic grade rocks are found inside the Entia Dome (Figure 9) therefore suggests that movement along the Bruna Gneiss was contemporaneous with thrusting along the Illogwa Shear Zone.

8.4. How Applicable is the Mantle Subduction Model to Intracratonic Deformation in Central Australia

In the mantle subduction model of Willett et al. [1993] it is assumed that the heavier mantle part of the lithosphere undergoes subduction and therefore drives shortening of the lighter overlying crust. In this paper, we have investigated the consequences of imposing subduction along a surface located below the crust-mantle boundary.

One may question however, the applicability of the mantle subduction model to an intracontinental setting; indeed, there is convincing evidence that cratonization between northern and central Australia dates back to at least 1700-1600 Ma [Myers et al., 1996]. There is no evidence of the ASO being the end product of subduction of Paleozoic oceanic crust; hence, there is no evidence for syn- or pre-ASO mantle subduction. As pointed out by Braun et al. [1991], the most likely source of compressional stresses in central Australia at the time of the ASO is along one of the margins of the continent. The ASO is, therefore, a prime example of reactivation tectonics driven by in-plane "remote" forces rather than an example of crustal shortening at an active plate margin driven by "local" mantle subduction.

Under these conditions, the imposed velocity discontinuity (or "S" point) used in the numerical model may be regarded as the location of a mechanical heterogeneity in the lithosphere which causes in-plane stresses to concentrate and leads to localized deformation along a 45° dipping shear zone (Figure 2). An alternative approach to modeling lithospheric deformation at the time of the ASO in central Australia would be to include, *a priori*, a 45° dipping lithospheric-scale weak zone and reactivate it by in-plane stresses. We suspect that both approaches would yield very similar results.

More important than the exact tectonic significance of the "S" point (Figure 2) is the clear dependence of the model results on the depth of the detachment layer or, more precisely, the thickness of the lithospheric layer involved in the thrusting. We have clearly shown that locating the detachment beneath the crust-mantle boundary leads to an offset in the Moho which has implications for the isostatic balance of the orogen. More importantly, the thickness of the deforming layer is the primary factor which determines the width of the region of syn-orogenic exhumation. All geological, petrological and geochronological indicators point to a maximum value of 100 km for the width of the zone of significant exhumation along the southern margin of the Arunta Block [Shaw et al., 1992a]. This, in turn, implies a maximum value of 50 km for the original thickness of the layer involved in the ASO compressional event. Following the imposed shortening, the layer thickness (and hence the depth to the "S" point, attached to the based of the deforming layer) increases to 62 km (Figure 5f).

This estimate is not consistent with the conclusions of several seismic travel-time anomaly studies [Lambeck, 1991; McQueen et al., 1996], which demonstrate that the suture between the northern and southern parts of the Arunta Block (the continuation at depth of the north-dipping RTZ) has a clear seismic expression to depths of at least 70 km. Further work is needed to reconcile these apparently conflicting observations.

8.5. Variations in Tectonic Styles Explained by Initial Conditions, not Variations in Tectonic Forcing

Our numerical modeling demonstrates that major variations in tectonic style within a single paleo-orogen may be explained in terms of variations in the initial thermal, and thus mechanical, state of the continental lithosphere. This is a particularly attractive proposition for an intracratonic orogen that was likely to be driven by in-plane stresses originating along one of the margins of the continent. It would be futile to call upon lateral variations in the tectonic driving mechanism (active oceanic subduction or island-arc accretion) to explain variations in the style of deformation along the strike of an orogen located more than 1000 km away from the active plate boundary.

Acknowledgments. We wish to thank Greg Houseman, Geoffrey Nicholls, George Gibson, Alastair Stewart, Herb McQueen, Kurt Lambeck, Christopher Beaumont and an anonymous reviewer for their constructive comments on the work presented in this manuscript. Clementine Kraishek's help in drafting some of the figures is much appreciated. Russell Shaw publishes with the permission of the Executive Director, Australian Geological Survey Organisation.

REFERENCES

Arnold, J., M. Sandiford, and S. Wetherley, Metamorphic events in the eastern Arunta Inlier, Part 1, Metamorphic petrology, *Precambrian. Res.*, *71*, 183-205, 1995.

Batt G. E., and J. Braun, On the thermo-mechanical evolution of compressional orogens, *Geophys. J. Int.*, *128*, 364-382, 1997.

Beaumont, C., P. Fullsack, and J. Hamilton, Styles of crustal deformation caused by subduction of the underlying mantle, *Tectonophysics*, *232*, 119-132, 1994.

Bird, P., N. Toksoz, and N. Sleep, Thermal and mechanical models of continent-continent convergence zones, *J. Geophys. Res.*, *80*, 4405-4416, 1975.

Braun, J., Three-dimensional numerical simulations of crustal-scale wrenching using a non-linear failure criterion, *J. Struct. Geol.*, *16*, 1173-1186, 1994.

Braun, J., and C. Beaumont, Three-dimensional numerical experiments of strain partitioning at oblique plate boundaries: implications for contrasting tectonic styles in the southern Coast Ranges, California, and central South Island, New Zealand, *J. Geophys. Res.*, *100*, 18,059-18,074, 1995.

Braun, J., and M. Sambridge, Dynamical Lagrangian Remeshing (DLR): a new algorithm for solving large strain deformation problems and its application to fault-propagation folding, *Earth Planet. Sci. Lett.*, *124*, 211-220, 1994.

Braun, J., H. McQueen, and M. A. Etheridge, A fresh look at the Late Palaeozoic tectonic history of west-central Australia. *Explor. Geophys.*, *22*, 49-54, 1991.

Chopra, P. N., and M. S. Paterson, The experimental deformation of dunite, *Tectonophysics*, *78*, 453-473, 1981.

Collins, W. J., and R. D. Shaw, Geochronological constraints on orogenic events in the Arunta Inlier: a review, *Precambrian Res.*, *71*, 315-346, 1995.

Collins, W. J., and C. Teyssier, Crustal scale ductile fault systems in the Arunta inlier, central Australia. *Tectonophysics*, *158*, 49-66, 1989.

Drury, M. R., R. L. M. Vissers, D. Van der Wal, and E. H. Hoogerduijn Strating, Shear localisation in upper mantle peridotites, *Pageoph*, *137*, 439-460, 1991.

Dunlap, W. J., and C. Teyssier, Paleozoic deformation and isotopic disturbance in the southeastern Arunta Block, central Australia, *Precambrian Res.*, *71*, 229-250, 1995.

Foden, J., J. Mawby, S. Kelley, S. Turner, and D. Bruce, Metamorphic events in the eastern Arunta Inlier, Part 2. Nd-Sr-Ar isotopic constraints, *Precambrian Res.*, *71*, 207-227, 1995.

Forman, D. J., and R. D. Shaw, Deformation of the crust and mantle in central Australia, *Bur. Miner. Resour. Aust. Bull.*, *144*, 1973.

Goleby, B. R., R. D. Shaw, C. Wright, B. L. N. Kennett, and K. Lambeck, Geophysical evidence for 'thick-skinned' crustal deformation in central Australia, *Nature*, *337*, 325-330, 1989.

Goleby, B. R., B. L. N. Kennett, C. Wright, R. D. Shaw, and K. Lambeck, Seismic reflection profiling in the Proterozoic Arunta Block, central Australia: processing for testing models of tectonic evolution, *Tectonophysics*, *173*, 257-268, 1990.

Griffith, A. A., 1921. The phenomena of rupture and flow in solids, *Roy. Soc. London Philos. Trans. ser. A*, *221*, 163-198.

James, P. R., P. MacDonald, and M. Parker, Strain and displacement in the Harts Range Detachment Zone: a structural study of the Bruna Gneiss from the western margin of the Entia Dome, central Australia, *Tectonophysics*, *158*, 23-48, 1989.

Joklik, G. F., The geology and mica-fields of the Harts Range, central Australia, *Bur. Miner. Resour. Aust. Bull.*, *26*, 1955.

Kooi, H., and C. Beaumont, Escarpment evolution on high elevated rifted margins: insights derived from a surface process model that combines diffusion, advection and reaction, *J. Geophys. Res.*, *99*, 12,191-12,209, 1994.

Koons, P., Some thermal and mechanical consequences of rapid uplift: an example from the Southern Alps, New Zealand, *Earth Planet. Sci. Lett.*, *86*, 307-319, 1987.

Lambeck, K., Structure and evolution of the intracratonic basins of central Australia, *Geophys. J. Roy. Astron. Soc.*, *74*, 843-886, 1983.

Lambeck, K., Teleseismic travel-time anomalies and deep crustal structure of northern and southern margins of the Amadeus Basin, *Bur. Miner. Resour. Aust. Bull.*, *236*, 409-427, 1991.

Lambeck, K., G. Burgess, and R. D. Shaw, Teleseismic travel-time anomalies and deep crustal structure in central Australia, *Geophys. J. Int.*, *94*, 105-124, 1988.

Malavielle, J., Modélisation expérimentale des chevauchements imbriqués: application aux chaînes de montagnes, *Soc. Géol. France Bull.*, *26*, 129-138, 1984.

McQueen, H. W. S., and K. Lambeck, Determination of crustal structure in central Australia by inversion of travel-time residuals, *Geophys. J. Int.*, *126*, 645-662, 1996.

Myers, J. S., R. D. Shaw., and I. M. Tyler, Tectonic evolution of Proterozoic Australia, *Tectonics*, *15*, 1431-1446, 1996.

Paterson, M. S., and F. C. Luan, Quartzite rheology under geological conditions, in *Deformation mechanisms, rheology and tectonics*, edited by R. J. Knipe and E. H. Rutter, Geol. Soc. Spec. Pub., 42, London, 299-307, 1990.

Prager, W., and Hodge, P.G., *Theory of perfectly plastic solids,* 264pp., John Wiley & Sons, New York,.1951.

Shaw, R. D., The tectonic development of the Amadeus Basin, central Australia., in *Geological and Geophysical Studies in the Amadeus Basin, Central Australia*, edited by R. J. Korsch and J. M. Kennard, Bur. Miner. Resour. Aust. Bull., *236*, pp. 429-461, 1991.

Shaw, R. D., and L. P. Black, The history and tectonic implications of the Redbank Thrust Zone, central Australia, based on structural, metamorphic and Rb-Sr isotopic evidence, *Aust. J. Earth Sci.*, *38*, 307-332, 1991.

Shaw, R. D., A. J. Stewart, and L. P. Black, The Arunta Inlier: a complex ensialic mobile belt in central Australia, part 2, tectonic history, *Aust. J. Earth Sci.*, *31*, 457-484, 1984.

Shaw, R. D., B. R. Goleby, R. J. Korsch, and C. Wright, Basement and cover thrust tectonics in central Australia based on the Arunta-Amadeus seismic reflection profile, in *Basement Tectonics*, edited by M. J. Rickard, H. J. Harrington, and P. R. Williams, pp. 55-84, Kluwer Academic Publishers, Dordrecht, 1992a.

Shaw, R. D., P. K. Zietler, I. McDougall, and P. Tingate, The Palaeozoic history of an unusual intracratonic thrust belt in central Australia based on ^{40}Ar-^{39}Ar, K-Ar and fission track dating, *J. Geol. Soc. London*, *149*, 937-954, 1992b.

Shaw, R. D., P. Wellman, P. Gunn, A. J. Whitaker, C. Tarlowski, and M. Morse, Users Guide to the Australian Crustal Elements map, *Aust. Geol. Surv. Org. Rec.*, 1996/30, 93pp, 1996.

Stephenson, R., and K. Lambeck, Isostatic response of the lithosphere with in-plane stress: application to central Australia, *J. Geophys. Res.*, *90*, 8581-8588, 1985.

Stewart, A. J., R. Q. Oaks, J. A. Deckelman, and R. D. Shaw, "Mesothrust" versus "megathrust" interpretations of the structure of the northeastern Amadeus Basin, central Australia, in *Geological and Geophysical Studies in the Amadeus Basin, Central Australia*, edited by R. J. Korsch and J. M. Kennard, Bur. Miner. Resour. Aust. Bull., *236*, 361-383, 1991.

Teyssier, C., A crustal thrust system in an intracratonic environment, *J. Struct. Geol.*, *7*, 689-700, 1985.

Wells, A. T., D. J. Forman, L. C. Ranford, and P. J. Cook, Geology of the Amadeus Basin, central Australia, *Bur. Miner. Resour. Aust. Rec.*, *1970/15*, 1970.

Wilks, K., and N. L. Carter, Rheology of some continental lower crustal rocks, *Tectonophysics*, *182*, 57-77, 1990.

Willett, S., C. Beaumont, and P. Fullsack, Mechanical model for the tectonics of doubly vergent compressional orogens, *Geology*, *21*, 371-374, 1993.

J. Braun, Research School of Earth Sciences, The Australian National University, Canberra, ACT 0200, AUSTRALIA.

R. D. Shaw, Australian Geological Survey Organisation, PO Box 378, Canberra, ACT 2601, AUSTRALIA.

Extension in the Fitzroy Trough, Western Australia: an Example of Reactivation Tectonics

Jean Braun

Research School of Earth Sciences, The Australian National University, Canberra, Australia

Russell Shaw

Australian Geological Survey Organisation, Canberra, Australia

We propose a mechanical model to explain the lithospheric-scale deformation style observed beneath the Fitzroy Trough of northern Western Australia. It is proposed that a Proterozoic compressional structure seated deep in the mantle lithosphere has been reactivated by in-plane forces during the late Paleozoic Pillara extension episode; the continental lithosphere is assumed to respond to this reactivation by detachment along an upper mantle low-viscosity shear zone. Using a finite-element model in which this tectonic scenario is reproduced by applying a velocity discontinuity at the base of a 65 km-thick lithospheric layer, we predict the formation of an asymmetric rift basin. This basin is bounded on one side (called here the "retro"-side of the rift) by a crustal-scale listric fault active during all stages of the rifting episode, and on the other side (the "pro"-side of the rift) by an array of listric normal faults active at various stages of the extension episode with the faults "ageing" away from the center of the rift. All faults except the most recently active pro-fault sole into the 65 km-deep low-viscosity detachment surface; the most recent pro-fault appears as the surface continuation of the reactivated deep-mantle Proterozoic shear zone. We compare the predicted distribution of crustal- and lithospheric-scale faults, basin stratigraphy and Moho geometry with those observed in the Fitzroy Trough to conclude that our proposed mechanism can explain an apparently complex episode of continental extension. We also describe the circumstances under which this style of continental deformation is likely to be activated.

1. INTRODUCTION

The continental lithosphere is characterized by a complex layered rheology and, therefore, accommodates deformation in a wide range of styles. When subjected to horizontal extension, continents may respond by shearing along a shallow-dipping lithospheric-scale structure (Figure 1) [Wernicke, 1985]. This model, called the "simple shear model", and its close relative the "detachment model" [Lister et al., 1991], assumes that continental extension at the lithospheric scale is inherently an asymmetric process, in which the locus of extension in the mantle part of the lithosphere is laterally offset from the locus of extension in the crust. The two centers of extension are linked by a through-going lithospheric-scale shear zone (in the simple-shear model) or a low-viscosity horizontal detachment surface at the base of the crust (in the detachment model).

It is commonly accepted [Bally and Snelson, 1980; van den Beukel, 1992; Davies and von Blanckenburg, 1995] that oceanic subduction may continue past the closure of an oceanic basin and lead to continental lithosphere (and thus

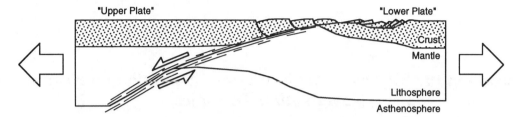

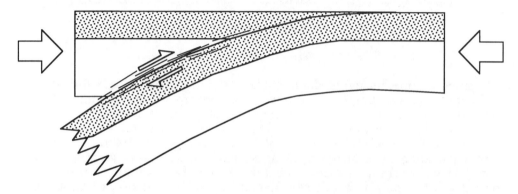

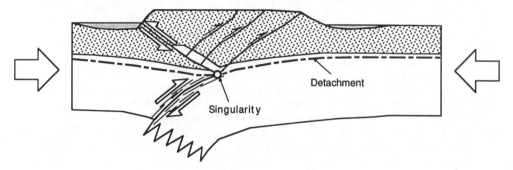

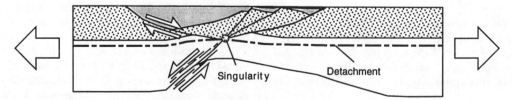

Figure 1. Conceptual models of continental lithosphere deformation : (a) extension by simple shear along a shallow-dipping through-going shear zone; (b) continental crust "subduction", the equivalent of the simple shear model in compression; (c) continental convergence accommodated by subduction in the mantle and thrusting and mountain building in the crust; (d) the equivalent in extension by reactivation of a pre-existing mantle shear zone. The singularity in strain rate (or discontinuity in velocity) is felt at the base of the upper layer as a result of subduction of the lower layer.

crustal) subduction (Figure 1b). To first order, continental subduction at convergent plate margins may be regarded as the compressional equivalent of the simple shear model. In this situation, however, the large buoyancy of the continental crust acts to limit this subduction to relatively shallow depths; yet, the process is thought to be sufficient to form very high pressure rocks (> 2 GPa) found in many orogenic belts [Davies and von Blanckenburg, 1995].

A more viable mode of continental convergence calls upon subduction of the denser mantle while the lighter continental crust accommodates convergence by thrusting and thickening [Willett et al., 1993]. Detachment between crust and mantle is facilitated by the likely presence of low-viscosity channels at or near the base of the crust (Figure 1c). Numerical and analog models have shown that, when mantle subduction occurs, crustal shortening is accommodated by thrusting along a crustal-scale shear zone dipping approximately at 45°, conjugate to, and originating at the location of the assumed mantle subduction, the so-called "strain singularity" or "velocity discontinuity" of Willett et al. [1993] (Figure 1c).

Our limited knowledge of the rheology of the continental lithosphere points to the base of the crust as the most likely candidate for the location of a low-viscosity ductile layer along which detachment may take place. One cannot preclude the possibility, however, that such detachments may be activated at mid-crustal levels or, in the case of a very "cold" continental block, within the upper mantle at the depth of the olivine brittle-ductile transition [Braun and Shaw, this volume].

We remarked earlier that, to a first order, the simple shear model (Figure 1a) may be regarded as the equivalent in extension of the continental subduction model (Figure 1b). In this paper, we postulate and demonstrate that lithospheric extension may be accommodated by "inverted subduction" of the lower part of the continental lithosphere (Figure 1d). In other words, we suggest that the mantle subduction model of Willett et al. [1993] may also be applicable to continental lithospheric extension. We also suggest that this particular mode of deformation is most likely to be activated in regions of past continental shortening which later are subjected to regional extension.

In this paper, we present evidence in support of this hypothesis from a well documented example of continental extension resulting from reactivation of a pre-existing convergent plate boundary: the late Paleozoic Fitzroy Trough of northwestern Australia. We also present the results of a complex, finite deformation numerical model of continental extension, driven by reactivation of an ancient lithospheric-scale suture zone, and accommodated by shear along a sub-crustal low-viscosity layer. By comparing results from the numerical model with field (seismic) observations from the Fitzroy Trough, we are able to draw conclusions about the tectonic and structural implications of reactivation of a lithospheric structure during continental extension.

2. THE FITZROY TROUGH

2.1. Tectonic Setting

The Fitzroy Trough, together with its southeastern continuation known as the Gregory Sub-Basin, is flanked to the northeast and southwest by a series of terraces, which pass, in turn, into two platform regions (Figure 2a): the Lennard Shelf in the northeast, and the Broome and Crossland platforms in the southwest [Shaw et al., 1994]. Both platforms are comprised of shallow basement, capped by a thin succession of Ordovician, Devonian and Permian rocks. The Lennard Shelf incorporates several small basement inliers and borders the King Leopold province, a Paleoproterozoic basement inlier comprised of metamorphosed turbidites intruded by granites. The Broome and Crossland platforms form an arch separating the predominantly Devonian Fitzroy Trough from a southern trough region (the Willara and Kidson Sub-Basins) containing 4-5 km of Ordovician and Silurian rocks, capped by less than one kilometer of Devonian and younger sedimentary rocks.

In the images of the deep seismic reflection data (Figure 3), a discontinuous set of quasi-planar reflectors appears to track a major southwest-dipping fault zone that extends deep into the mantle [Shaw et al., 1994]. Its surface projection corresponds to the Cambrian Spielers Shear Zone and western segment of the Beagle Bay - Pinnacles Fault Complex (Figure 2b), the major fault system at the northern margin of the Fitzroy Trough, active in the Ordovician and the mid-Devonian. It seems likely that this composite feature marks a complex, composite ancient Proterozoic province boundary that has undergone several phases of reactivation [Shaw et al., 1996; Myers et al., 1996]. The current view is that this region was the locus of convergence between the North Australian craton and a somewhat older and complex orogenic belt, which forms part of the central Australian terranes of Myers et al. [1996].

2.2. Gravity Expression of the Region

Long-wavelength (>20 km) gravity features (Figure 4) outline a series of southeast to east-trending geophysical domains that suggest continuity of crustal structure between central Australia and northwestern Australia [Shaw et al., 1996]. This suggests that the tectonic framework in the west may well be thick-skinned in style, as has been documented for central Australia [Shaw et al., 1992].

A definite swing in the trend of magnetic and gravity anomalies and a drop-off in their magnitude in the central-west, corresponds to the Lasseter shear zone [Braun et al., 1991], but the en-echelon, discontinuous and distributed nature of lineaments in this zone suggests a multiphase, intracratonic feature [Shaw et al., 1996] (Figure 4b).

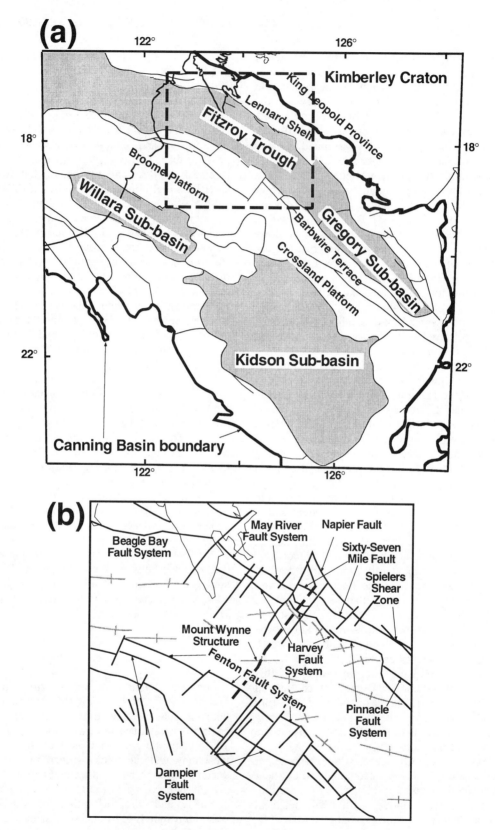

Figure 2. (a) Tectonic elements and main depocentres (in grey shade) during Palaeozoic extension in the Canning Basin; (b) main faults (thick black lines) and anticlines (thick grey lines) in the Fitzroy Trough; the thick dashed line shows the location of the deep reflection seismic line shown in Figure 3.

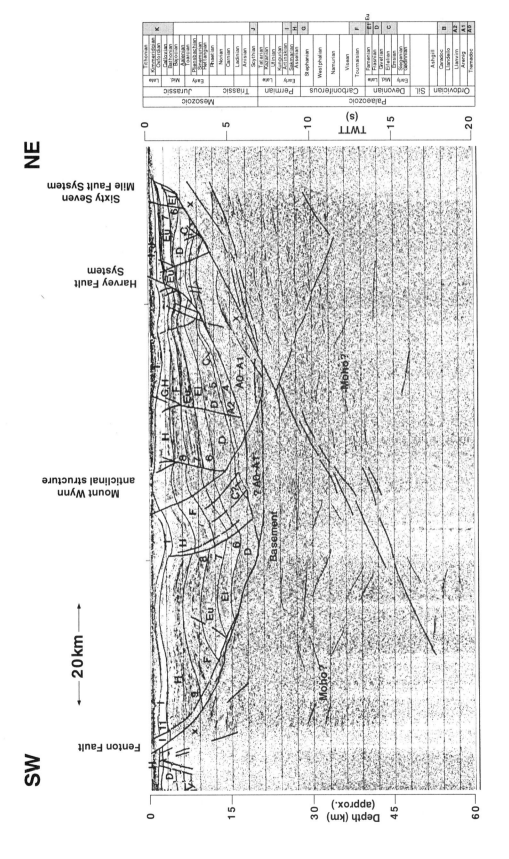

Figure 3. Interpreted unmigrated deep reflection seismic transect across the Fitzroy Trough (from Shaw et al, 1994) and stratigraphic relationship of the various units marked on the seismic line. Stratigraphic units defined by Kennard et al. [*1994a*]. TWTT: two way travel time; the depth axis is only approximate.

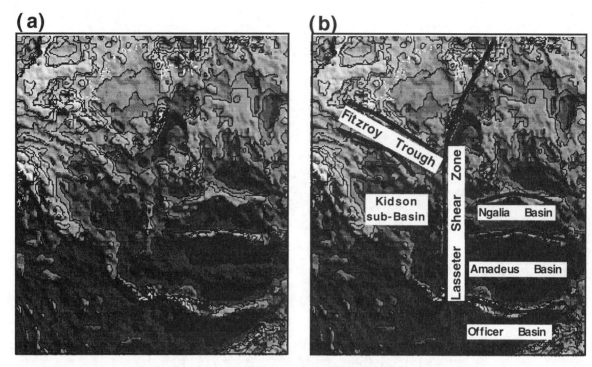

Figure 4. Artificially illuminated contour map of the Bouguer gravity anomaly over central and northwestern Australia; the illumination is from the northeast; light shaded areas correspond to positive anomalies, dark shaded areas correspond to negative anomalies; (b) interpreted version of (a) showing the major tectonic elements in the area. Bouguer anomaly data from Australian Geological Survey Organisation database.

2.3. Structure of the Fitzroy Trough

The structure of the 500-km long by 100-km wide, main onshore section of the Fitzroy Trough is summarized in our cross-section (Figure 3) along BMR deep seismic reflection line 3 (dashed line in Figure 2b, BMR03). Our portrayal builds on that initially reported by Drummond et al. [1991] and interpreted in detail by Shaw et al. [1994]. Unfortunately, no migrated version of this deep seismic line has been published. The depth axis on the left-hand side of the figure is therefore very approximate, as the thick sedimentary layer is likely to be characterized by much lower seismic velocities than the adjacent basement. The true dip of seismic reflectors is also likely to be greater than the value suggested.

The trough is asymmetrical: the thickest section of Devonian-Carboniferous in the Canning Basin is preserved against the Fenton Fault System on the southwestern margin of the Fitzroy Trough, whereas the same units appear to drape across the northeastern margin of the trough. The gravity expression of the region suggests that the southern margin is a well-defined, single structure, the Fenton Fault System, whereas the northern margin is complex, comprising many subparallel structures. The absence of a more general 'stag-horn' rift geometry, with a rift phase followed by a thermally driven sag phase, suggests that extension was relatively slow and did not lead to any major thermal perturbation, or that the sag-phase sedimentary section has been later removed by erosion [Kennard et al., 1994a].

The Fenton Fault System (Figure 3) forms a listric fault, which soles into a horizontal detachment at or near the basement contact, at depths of 7 s two-way travel time (TWTT; ≈15 km). The consistent thickening of the Devonian-Carboniferous section against the Fenton Fault System suggests that it was active from the mid-Devonian to the mid-Carboniferous, during an event known as the Pillara Extension. The Dampier Fault System comprises a discontinuous set of faults, which appears to represent local backstepping of the Fenton Fault, and outlines the southern terraces (Figure 2b). These faults were also active during the Pillara Extension Phase, but to a much lesser degree than the Fenton Fault System. Within this system, individual fault compartments, showing differing degrees and styles of extensional faulting, are separated by offsets interpreted as accommodation zones.

The northern margin of the Fitzroy Trough is not a simple draped hinge, as it appears at first sight (Figure 3). It is a complex fault zone, including the Beagle Bay, May River, Harvey and Pinnacle Fault Systems [Jackson et al., 1993] separated and linked by accommodation zones (Figure 2). At its eastern and western ends these faults converge, then diverge as a set of splay faults. In the central part of the graben, near the BMR03 deep seismic line (Figure 3), the Harvey Fault System marks the southern limit of the northern terraces and is paralleled by the Sixty-Seven Mile Fault System to the north (Figure 2b). The most northerly faults, such as the Sixty-Seven Mile Fault System, date from the initiation of extension in the Pillara Phase (the Gogo movement). However, the Harvey Fault System (HFS) was initiated later during the Van Emmerick movement, and reactivated during the Red Bluffs movement. Horst structures formed behind the Harvey Fault System and, at the same time, the trough deepened in front of the HFS. The Sixty-Seven Mile and Harvey Fault Systems appear to join near the base of the basin with a band of southwest-dipping reflectors (Figure 3) which extend deep into the mantle.

The accommodation zones appear to be individually linked to the bounding fault systems, as they are orthogonal to them and do not extend across the Fitzroy Trough (Figure 2b).

The Mount Wynne Structure (Figure 3) aligns with an important geophysical domain boundary [Shaw et al., 1996]. It shows evidence of growth during the Carboniferous and appears to act as a thrust during the Fitzroy Transpressional Movement in the latest Triassic.

2.4. Evolutionary Phases

2.4.1. Neoproterozoic intracratonic orogeny. Intracratonic reactivation of a structurally weak zone (King Leopold Province; Figure 2a), lying between the province boundary marked by the Spielers Shear Zone and the Kimberley Craton, resulted in several episodes of subsidence, the most important of which ended as a result of north-directed thrusting at about 1200-1000 Ma (Yampi Orogeny). Further subsidence followed, to be again terminated by south-directed thrusting at 560-530 Ma (King Leopold Orogeny) [Myers et al., 1996].

2.4.2. Cambrian activity on the Spielers Shear Zone. At about 500 Ma, north-directed thrusting is recognized on the Spielers Shear Zone (Table 1), along what appears to be a reactivated Proterozoic structure at the deformed margin of the Kimberley craton (now corresponding to the northern margin of the Fitzroy Trough). The lack of preserved syntectonic deposits dating from this compressional episode suggests that displacement was minor, did not result in important uplift or subsidence, and there was probably no major perturbation of the Moho.

2.4.3. Ordovician extension and thermal relaxation. Initiation of the Canning Basin during the Samphire Marsh Extension (490-470 Ma, Table 1) [Kennard et al., 1994b; Romine et al., 1994] was followed by the prolonged Carribuddy Sag Phase (470-434 Ma; Table 1), characterized by widespread evaporite accumulation (Mallowa Salt). The resulting Ordovician section is roughly uniform across the Canning Basin. Normal faulting was concentrated either at the northern basin margin or south of the Broome Platform. This period of thermal relaxation (or sag phase) probably continued until the end of the Silurian and resulted in increasing lithospheric strength [Braun, 1992].

2.4.4. Silurian and Early Devonian intraplate adjustments. The Silurian sequence (Table 1) displays a thinning across the basin to the north [Kennard et al., 1994b; Romine et al., 1994]: a relationship implying a slight and progressive regional tilting, possibly generated by renewed transpressional movement on the Spielers Shear Zone (Figure 2) and related structures (e.g. Beagle Bay and Pinnacle faults).

This tilting was a precursor to the Prices Creek Compressional Movement (410-390 Ma, Table 1), a phase of minor folding, basin-wide uplift and erosion, centered on the Barbwire Terrace and the outer margin of the Lennard Shelf. No syntectonic sedimentary rock appears to be associated with this event.

2.4.5. Tandalgoo Sag Phase. Laterally extensive aeolian and terrestrial deposits, grouped in the Tandalgoo Sag Phase (390-375 Ma, Table 1) thicken towards the eastern margin of the basin. This phase represents a rapid change in shape of the Canning Basin, and suggests that tectonic activity has shifted from faults at the northern margin of the Fitzroy Trough to faults related to the Lasseter Shear Zone Complex (Figure 4). This new successor basin may be the expression of "pre-rift" tectonism, seated deep in the crust or lithosphere.

2.4.6. Pillara Extension. The Pillara Extension (375-326 Ma, Table 2) was the major period of extension and rapid subsidence that formed the Fitzroy Trough. The main driving structure, the Fenton Fault System, was active from the mid-Devonian to the mid-Carboniferous. Several pulses make up this event, the most obvious of which are the Gogo, Van Emmerick and Red Bluffs movements. Each pulse was followed by a relaxation or mini-sag phase, which may be one reason for the lack of a "classical" coupled thermal-sag cycle accompanying the Pillara Extension as a whole.

The Gogo movement (≈ 375 Ma, Table 2) refers to the event that initiated subsidence in the Fitzroy Trough. In addition to movement on the Fenton Fault System, it involved activity on segments of the Beagle Bay-Pinnacle Fault Complex, including the Sixty-Seven Mile Fault System (Figure 2). Deposition spilled into the Kidson Sub-Basin, implying an additional, more widespread subsidence mechanism.

The Van Emmerick movement (366-363 Ma, latest Frasnian earliest Famennian, Table 2) involved syn-extensional fault-block tilting, and was marked by a

Table 1. The Neoproterozoic to Mid-Devonian Tectonic Evolution of the Structural Elements of the Fitzroy Trough; Based on Myers et al. [1996].

Event	Age (Ma)	Broome & Crosslands platforms	Southern terraces	Central Fitzroy Trough	Northern terraces	Lennard Shelf
Point Moody Extensional Movement [H]	298-285	Fluvial and lacustrine deposition	Fluvial and lacustrine deposition	Thickening towards Fenton fault system, then sag phase	Fluvial and lacustrine deposition	Section thins rapidly to north
Drosera Erosion	310-298	Non-deposition	Non-deposition	Non-deposition	Non-deposition	Non-deposition
Subsidence [G]	320-310	Section missing	Section missing	Syn-tectonic fluvial deposits	Syn-tectonic fluvial deposits	Syn-tectonic fluvial deposits
Meda Transpressional Movement	326-320	Exhumation?	Uplift, non-deposition	Thickening of section along Harvey fault system	Positive & negative 'flower structures' along north margin	Syn-tectonic fluvial deposits
Subsidence (two phases) [F]	354-326	Exhumation	Uplift, non-deposition	Apparent growth of section towards Fenton & Stansmore faults (eg., Laurel Formation [354-350 Ma] & Anderson Formation [350-326 Ma]	Carbonate ramp succession, followed by deltaic sand deposition	Carbonate ramp succession, followed by emergence
PILLARA EXTENSION: Red Bluffs extensional movement	357-354	Exhumation	Uplift, non-deposition	Fenton Fault active; slope fan deposits on northeast margin	Syn-faulting carbonate debris; tilting up to north; conglomerate blanket	Exhumation in southeast (northeast edge of Margaret Embayment)
Subsidence [El, Eu]	363-357	Section missing	Nullara cycle of reef development	Basinal clastic deposition (low-stand wedges)	Mixed clastics, forming slope fans and prograding calcareous siltstone complexes	Nullara reef (high-stand system tracts)
PILLARA EXTENSION: Van Emmerick extensional movement	366-363	Section missing	Uplift, non-deposition	Activity shifts to Fenton Fault; thick wedges of slope fan deposits on northeast margin	Localised conglomerate deposition (Napier Embayment)	Exhumation (Titting, uplift & erosion) of Blackstone, May River & Sixty-Seven Mile highs
Subsidence [D]	~375-366	Section missing	Pillara cycle of reef development	Basinal clastic deposition (low-stand wedges)	Pillara (high-stand) carbonate sedimentation	Transgressive (high-stand) Pillara reef
PILLARA EXTENSION: Gogo extensional movement	~375	Continued, but limited subsidence	Continued, but limited subsidence	Thickening towards active segments of Beagle Bay - Pinnacles fault complex	Beagle Bay - Pinnacles fault complex active	Emergent

Table 2. The Mid-Devonian to Permian Tectonic Evolution of the Structural Elements of the Fitzroy Trough; Based on Romine et al. [1994] and Kennard et al. [1994b].

Event	Age (Ma)	Broome & Crosslands platforms	Southern terraces	Central Fitzroy Trough	Northern terraces	Lennard Shelf
Tandalgoo Sag Phase [C]	390-375	Section thickens to east	Section thickens to east	Thin section	Thin section	Section missing
Prices Creek Compressional Movement	410-390	Regional tilting, up to north	Regional tilting, up to north	Regional tilting, up to north	Regional tilting, up to north	Regional tilting, up to north
Unnamed regional tilting (up to north)	434-410	Reduced Silurian section thins onto arch	Section subsequently eroded	Section subsequently eroded	Section missing	Section missing
Carrabuddy Sag Phase	470-434	Salt accumulation	Salt accumulation	Salt accumulation?	Section missing	Section missing
Samphire Marsh Extension Movement	490-470	Subtidal to peritidal deposition	Subtidal to peritidal deposition	Blackstone half-graben	Section missing	Section missing
Exhumation	500-490	Regional hiatus	Regional hiatus	Regional hiatus	Regional hiatus	Regional hiatus
Spielers compressional movement	500-510	Regional hiatus	Regional hiatus	Regional hiatus	Regional hiatus	Regional hiatus, North-directed thrusting on Spielers Shear Zone
Antrim Plateau basaltic volcanism	540-510?	Extruded in easternmost part of basin	Extruded in easternmost part of basin	Extruded in easternmost part of basin	Extruded in easternmost part of basin	Extruded in easternmost part of basin
King Leopold Orogeny, Precipice event	560-530	Uplift & erosion	Uplift & erosion	Uplift & erosion	Uplift & erosion	Uplift & erosion
Lousia subsidence		No record	No record	No record	No record	Subsidence in northeasternmost part of basin

spectacular influx of conglomerate on the Lennard Shelf and the buildup of thick wedges of slope fan deposits along the northern slopes of the Fitzroy Trough. The clast content and stratal geometry of these deposits point to uplift of the trough flanks [Jackson et al., 1992; Shaw et al., 1994]. The Fenton Fault System continued to be the main driving structure on the southwestern side of the Fitzroy Trough. On the northeastern side of the trough, activity switched from the Sixty-Seven Mile Fault System to the Harvey Fault System, as shown by an unconformity and thinning of section in the hanging wall north of the Harvey Fault System.

The Red Bluffs movement (357-354 Ma, Table 2) was a similar event, involving syn-extensional fault-block tilting and renewed uplift of the trough flanks. It also resulted in a mantle of conglomerate and overflow of detritus onto the trough slopes to form wedges of fan deposits [Jackson et al., 1992]. Renewed movement took place on both the Harvey and Fenton Fault Systems.

2.4.7. Meda Transpressional Movement. The Meda Transpressional Movement (326-320 Ma, Table 2) resulted in the inversion of normal faults, inducing uplift and erosion at the northeastern margin of the Fitzroy Trough. It led to a marked hiatus throughout the Canning Basin, particularly on the Lennard Shelf (base-Grant Group unconformity). Permian strata appear to onlap the Mount Wynne Structure, implying growth of this structure during the Meda Transpressional Movement. The event may correspond to the peak of the Alice Springs Orogeny in central Australia.

2.4.8. Point Moody Extension. The Point Moody Extension (298-285 Ma) marks renewed extension and subsidence of the Fitzroy Trough at the time of the onset of glacial conditions in the Permian.

2.4.9. Fitzroy Transpressional Movement. In the late Triassic, the Fitzroy Transpressional Movement (?200-180 Ma) resulted in dextral wrench movement, fault reactivation and anticlinal uplifts. It was the last major event to affect the Canning Basin region.

3. THE NUMERICAL MODEL

To study the response of the Earth's lithosphere to tectonic forces, we have developed a finite-element model based on the Dynamic Lagrangian Remeshing (DLR) method [Braun and Sambridge, 1994]. The model regards the lithosphere as a complex visco-elasto-plastic pressure-dependent and thermally-activated material and is described in Braun and Shaw [this volume].

As discussed earlier, there is strong evidence to suggest that late Palaeozoic extension in the Fitzroy Trough was accommodated by reactivation, along its northern margin, of a Neoproterozoic suture, the Spielers Shear Zone.

Reflection seismic data (Figure 3) support the hypothesis that the Spielers Shear Zone extends to mantle depths. In our numerical simulations, we shall therefore assume that a pre-extension lithospheric-scale structure is reactivated by in-plane forces (Figure 5). The purpose of the numerical modeling exercise is therefore to investigate how extension imposed along a mantle discontinuity is accommodated in the overlying crust/upper mantle.

In our models, we will adopt the method proposed by Willett et al. [1993] and many others since [Beaumont et al., 1994; Braun and Beaumont, 1995], in which deformation is driven by a velocity discontinuity imposed at the base of the deforming layer (Figure 5). In our experiment, this layer includes the crust and the uppermost mantle. In practice, the base of the model on the left-hand side of the discontinuity is held fixed while the right-hand side is forced to move at a constant velocity, v_0 (Figure 5), which is also imposed along the far-field right-hand side vertical boundary of the model. It is worth noting that a similar boundary condition was used by Beslier and Brun [1991] in a series of analog experiments.

The initial crustal thickness is 40 km, the velocity discontinuity is placed at a depth of 65 km (i.e. in the upper mantle). The crustal rheology is assumed to be dominated by feldspar-rich rocks and is based on the Adirondak Granulite rheology of Wilks and Carter [1990], whereas the olivine-rich mantle rheology is based on the Aheim Dunite rheology of Chopra and Paterson [1981]. The geotherm used is relatively low (10^{-5} °C yr^{-1}), such that little ductile deformation takes place at the base of the crust. The deep mantle velocity discontinuity is placed near the mantle brittle-ductile transition resulting from the assumed olivine-based rheology and initial geothermal gradient. The imposed extension velocity, v_0, is arbitrarily fixed at 2.5 mm yr^{-1} to produce 50 km of extension in 20 Myr. The time step length is constant and set at 0.04 Myr.

The 65 km-thick lithospheric layer rests on a thin elastic plate foundation [Beaumont et al., 1994] characterized by a flexural rigidity of 8.9×10^{21} Pa m^3.

At the surface of the model, mass is redistributed according to a one-dimensional erosion-deposition algorithm [Kooi and Beaumont, 1994]. The values of erosion parameters are such that no substantial topography is allowed to develop during the modeled extensional episode.

It is clear that the postulated "ancient" tectonic episode, responsible for the presence of the deep mantle structure, would have deformed the crust-mantle boundary. It would be, however, very speculative to predict the exact post-orogenic Moho topography that should be used at the onset of the extensional episode. We therefore opted to start the modeling of the rift with a flat Moho geometry (Figure 5).

The temperature is dynamically updated by solving the equation of transient heat transfer by conduction/advection at each time step. The temperature is used to update the assumed thermally activated rheology; thermal expansion

(a) Tectonic Setting

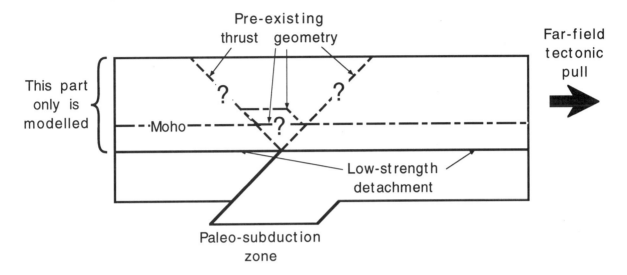

(b) Numerical model boundary conditions

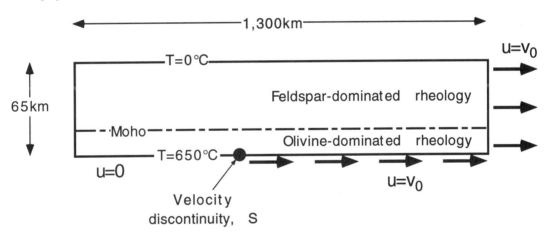

Figure 5. (a) Tectonic setting and (b) its numerical implementation; note that the low-strength detachment is assumed to be located in the upper mantle, below the Moho discontinuity; because it is difficult to make prediction on the pre-extension Moho geometry, it is assumed in the calculations that the Moho is originally flat; the assumed pre-existing thrust has not been included in the numerical model.

effects are neglected, which implies that post-extensional lithospheric contraction and associated surface subsidence are not modeled; to model thermal subsidence properly would require the entire lithosphere to be modeled (not just the upper 65 km).

Sediments are assumed to have thermal and mechanical properties similar to those of the underlying crust.

4. MODEL PREDICTIONS

4.1. Predicted Structures

The results of the model are shown in Figure 6 as contour plots of the total accumulated strain at times 2, 10 and 20 Myr since the onset of extension.

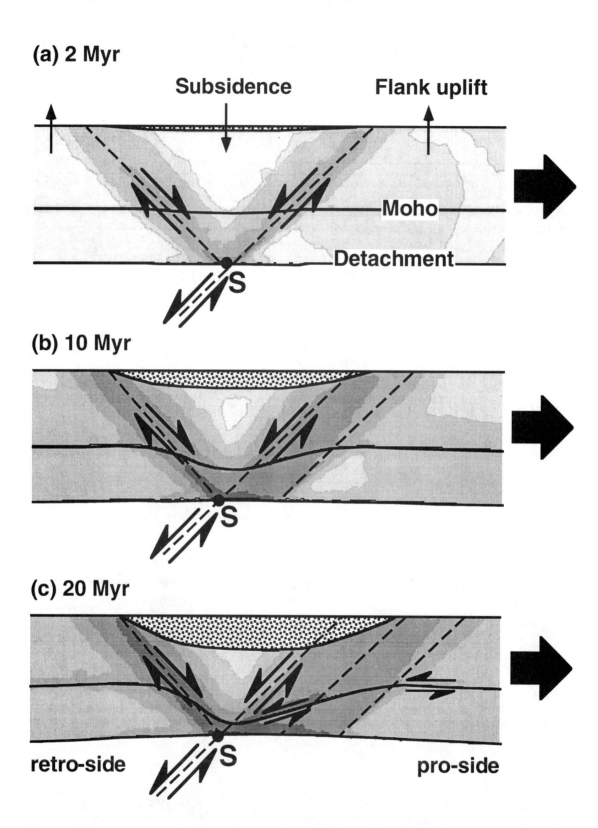

Figure 6. Results of the numerical model as contour plots of computed total shear strain (second invariant of the total deviatoric strain tensor), after (a) 2, (b) 10 and (c) 20 Myr since onset of extension; dark grey shades correspond to high strain values; light grey shades correspond to low strain values; the stippled area corresponds to the predicted rift basin; the dashed lines are inferred faults/shear zones; the arrows indicate the dominant mode of deformation.

A set of oppositely dipping normal faults (or shear zones) develops, originating at the imposed velocity discontinuity at the base of the model (Figure 6a). The shape and dip of these shear zones is not set a priori. They develop as a direct consequence of our assumption of (a) subduction of the lower part of the continental lithosphere along a low-viscosity detachment surface (the base of our model), and (b) the highly non-linear plastic rheology characterizing most of the upper lithosphere.

These conjugate shear zones form a symmetrical pair during the early stages of the experiment. Normal movement along these shear zones results in local offsets of the Moho, crustal thinning, surface subsidence at the center of the rift, and concomitant rift flank uplift. The formation of the rift basin and the uplift of the rift flanks are a direct consequence of the finite flexural strength of the continental lithosphere [Braun and Beaumont, 1995]. Because the shear zones dip at approximately 45°, the predicted width of the rift basin (130 km) is approximately twice the thickness of the deforming layer (65 km). Because the main basin-bounding faults of the Fitzroy Trough appear to dip at approximately 45° (Figure 3), the width of the graben (120-130 km) provides us with a first-order estimate of the depth of detachment (60-65 km).

As deformation progresses, asymmetry develops (Figure 6b). This arises from our assumption that the velocity singularity is attached to one of the two sides of the rift zone, which implies that the reactivated mantle structure dips beneath the side of the model to which the singularity is attached. Following the convention of Willett et al. [1993], we shall term the fixed side of the rift (the left-hand side in our model) the retro-side and the mobile side of the rift (the right-hand side in our model), the pro-side.

The shear zone on the retro-side of the discontinuity accumulates a large amount of deformation which results in a major offset of the Moho. The pro-shear zone accumulates relatively modest deformation before it is rapidly translated away from its root by movement along the retro-shear zone. The pro-fault (the surface expression of the pro-shear zone) is therefore a transient feature and the pro-margin of the basin follows a quasi-cyclic behavior: (1) a pro-fault propagates up from the basal velocity discontinuity, (2) while it accumulates deformation, the pro-fault is translated laterally by movement along the retro-shear zone, and (3) the pro-fault is progressively abandoned as a new pro-fault develops from the velocity discontinuity.

This behavior of the model results in a rather asymmetric sedimentary basin, with the retro-margin characterized by a single structure accumulating extension at all stages of the tectonic episode, while the pro-margin is made of an array of normal faults active at different times during the extension event, with the oldest movement on the fault furthest away from the center of the rift basin.

A similar structural evolution is observed in the Fitzroy Trough, with the Fenton Fault, the retro-fault, accommodating most of the extension at all stages of the Pillara extension episode while, during the same period, fault activity along the pro-side of the basin progressively migrates from the Sixty-Seven Mile Fault during the early Gogo movement, to the Harvey Fault during the Van Emmerick and Red Bluffs movements. It is also possible that extension along the northeastern margin (on the pro-side of the rift) initiated on a structure now to the northeast of the Sixty-Seven Mile Fault, possibly the Napier Fault (Figure 2). This is suggested by a series of strong dipping reflectors in the basement beneath the northeastern margin of the rift basin (see Figure 3), which also suggests that the main lithospheric-scale pro-shear may have been connected to another pro-fault in the early stages of extension.

The Bouguer gravity anomaly over the margins of the Fitzroy Trough support this asymmetry, with the southwestern margin clearly defined by a single lineation (corresponding to the Fenton Fault), whereas the northeastern margin along the Lennard Shelf consists of a complex system of smaller lineaments (Figure 7).

4.2. Rift-flank Uplift and Development of an Unconformity

As pro-faults are translated into the pro-side of the basin, they become affected by the rapid flank uplift. This results in major unconformities developing along the pro-margin of the basin. This is reflected in the predicted rift-basin stratigraphy shown in Figure 8. Growth along the retro-fault results in the accumulation of a thick sedimentary sequence along the retro-side of the basin; the pro-side of the basin is characterized by syn-extensional subsidence along the active pro-fault, and concomitant uplift and erosion along the basin margin as it is translated towards the pro-side of the rift; this results in a series of erosional unconformities on the pro-side of the active pro-fault.

The northeastern margin of the Fitzroy Trough is marked by well defined unconformities to the north of the Harvey Fault during the Gogo (≈ 375 Ma), Van Emmerick (366-363 Ma) and Red Bluffs (357-354 Ma) extension pulses of the Pillara Extension (Figure 3 and Table 2). We associate the uplift and erosion along the Lennard Shelf associated with this unconformity to syn-extensional translation of the basin margin into the region of active rift-flank uplift.

4.3. Moho Geometry

The evolution of the crust-mantle boundary (an originally flat marker located above the detachment) is pictured in Figure 9. Initial strain along the conjugate shear zones results in an apparently symmetrical subsidence of the crustal block between the two shear zones (Figure 9a). Finite deformation along the retro-shear zone causes a marked offset in the Moho, whereas the offset across the pro-shear zone is more diffuse (Figure 9b). Neglecting isostatic compensation, we may predict the geometry of the Moho across the rift system: on either side of the basin, the Moho remains flat at its original depth, while near the

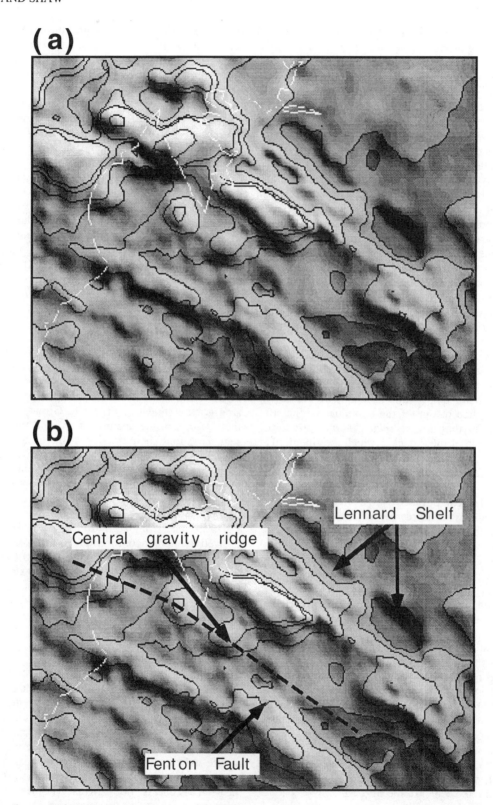

Figure 7. (a) Artificially illuminated map of the Bouguer gravity anomaly in the vicinity of the Fitzroy Trough and (b) interpreted version of (a); light shaded areas correspond to positive anomalies, dark shaded areas correspond to negative anomalies.

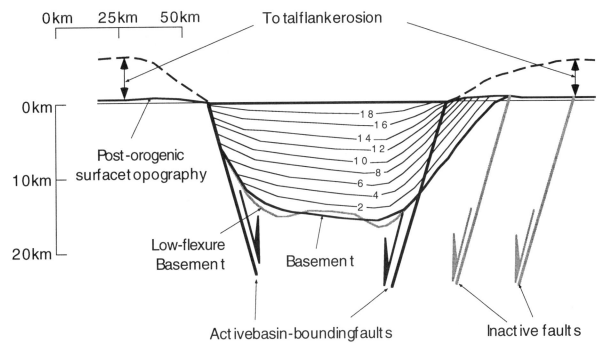

Figure 8. Predicted stratigraphy in the rift basin; the dashed line correspond to the total amount of rift flank erosion; the numbers correspond to the time (since onset of extension) at which the corresponding stratigraphic horizon was formed. Note that very large vertical exaggeration which makes the basin-bounding fault appear as extremely steep faults whereas, in reality, they dip at approximately 45°. The thick grey line correspond to the basement geometry computed from a low-flexural strength experiment (see text for further explanation).

center of the rift, the Moho subsides by an amount Δu, equal to the imposed far-field horizontal displacement; this is because, in our model, the major basin-bounding faults dip at ±45°. The crust is therefore thinned most along the margins of the basin, which results in a complex negative isostatic load characterized by two maxima at the rift margins (Figure 9d). The final rift geometry (Figure 9c and 6c) is the result of convolving the flexural response to the isostatic load (Figure 9d) with the uncompensated rift geometry (Figure 9b). On the retro-side of the rift, the Moho rises gradually from its "normal" pre-rift geometry (40 km) to shallow depths (33 km) beneath the region of rift flank uplift. It is strongly offset by the retro-shear zone. Beneath the central part of the rift, the crust is only moderately thinned (35 km). Beneath the pro-margin of the rift, the Moho gradually ramps upward to reach a minimum depth (32 km) beneath the rift pro-flank. The transition to "normal" Moho depth on the pro-side of the rift is similar to its gradual rise on the retro-side of the rift.

A direct comparison of the predicted Moho geometry and the one observed along the deep reflection seismic traverse (Figure 3) is not possible owing to the lack of clear continuous reflections from the crust-mantle boundary. Furthermore, care should be taken in interpreting the unmigrated seismic line of Figure 3, as the relatively low-velocity basin fill causes the rift basin, and the Moho reflections beneath it, to appear much deeper than they actually are. The seismic profile suggests, however, that the Moho discontinuity rises to relatively shallow depths on the southwestern (retro-)side of the rift A band of southwesterly dipping strong reflectors [Drummond et al., 1991; Shaw et al., 1994] cuts through the Moho suggesting that a narrow wedge of crustal material dips to the southwest beneath the northeastern (pro-)side of the rift. This Moho geometry, which is rather uncommon for an extensional basin, is in very good agreement with the geometry predicted by the numerical simulations suggesting, that the assumed driving mechanism (reactivation of a Proterozoic thrust structure by intra-lithospheric detachment) is a plausible one.

Crustal thinning is maximum along the margins of the basin, not at its center. Local isostatic compensation of the rift would result in maximum subsidence along the basin margins; the finite flexural strength of the lithosphere acts, however, as a "low-pass filter" that forces the center of the basin to subside and the margins to uplift. Lowering the assumed flexural strength of the lithosphere (which is done in our numerical experiment by reducing the effective elastic thickness in the thin elastic beam foundation) results in a reduced subsidence of the central part of the basin, the formation of a mid-basinal arch (thick shaded curve on Figure 8) and reduced uplift and erosion of the rift flanks.

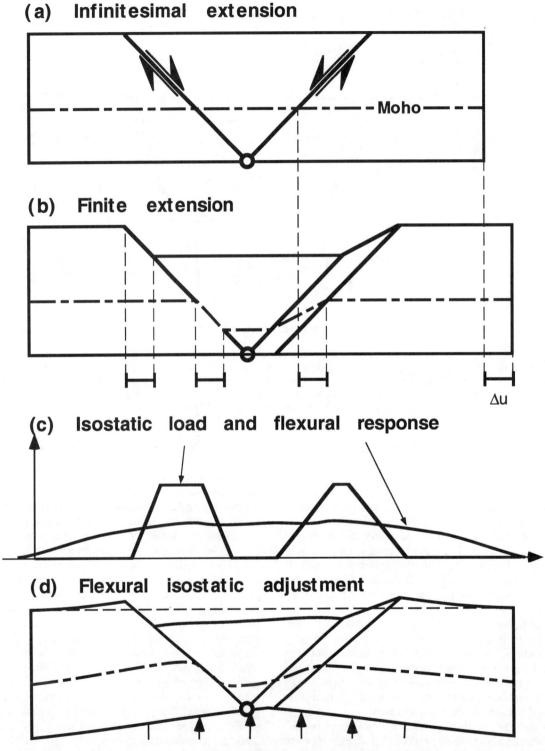

Figure 9. Step-by-step description of the effect of continental lithosphere deformation on the geometry of the crust-mantle boundary as suggested by the results of the numerical model including (a) movement along a set of conjugate thrust faults, (b) finite deformation along the retro-fault and horizontal translation of the pro-fault, (c) local isostatic load and flexural response resulting from Moho offset and (d) final geometry resulting from isostatic flexural adjustment; vertical displacement and load not to scale.

Such a structure is unlikely to be clearly imaged on the unmigrated seismic line (Figure 3), but could be responsible for the relative gravity high observed along the axis of the Fitzroy Trough (Figure 7). Alternatively, the observed central gravity ridge may be the result of the major restructuring which affected the Fitzroy Trough during the late Triassic Fitzroy Transpressional Movement.

5. DISCUSSION

5.1. Lithospheric-Scale Asymmetry

We have presented the results of a numerical model which illustrates the behavior of the continental lithosphere under extension. We have assumed that lithospheric extension is accommodated by reactivation of a pre-existing lithospheric-scale structure, and that the lithosphere is prone to delaminate along a low-viscosity detachment surface. The results of our computations show great similarity with the large-scale structure of the Fitzroy Trough of northwestern Australia, suggesting that this mode of deformation is likely to be activated in the Earth's lithosphere. We therefore propose that this mode of lithospheric deformation is a viable alternative to the simple shear model proposed by Wernicke [1985] or its variants as described by Lister et al. [1991].

Possibly the most remarkable feature of our model is the formation of a long-lived basin-bounding fault on the retro-side of the rift, which dips in the opposing direction from the underlying pre-existing mantle structure on which the rift develops. Such a geometry, which is clearly imaged by a deep seismic reflection transect beneath the Fitzroy Trough, cannot be accounted for by the simple shear model.

5.2. Reactivation Leads to Detachment

Our modeling has demonstrated that the Fitzroy Trough of northwestern Australia is a likely candidate for a case of continental extension by reactivation of mantle subduction. Supporting evidence includes (a) the asymmetric nature of its bounding margins, and (b) the presence of a lithospheric-scale structure dipping in a direction opposite to the dip of the structure most active during the late Palaeozoic extension.

The location of the Fitzroy Trough provides clues on the tectonic setting required to activate this mode of lithospheric deformation: the reactivation of a pre-existing suture. Indeed, the Fitzroy Trough is the northwestern extension of the northern margin of the Amadeus Basin (Figure 4), which was also reactivated in the late Devonian to mid-Carboniferous during the compressional Alice Springs Orogeny. The nature of the mechanism that was responsible for contemporaneous extension in the Fitzroy Trough and compression in central Australia remains conjectural [Braun et al., 1991]; it is not even clear whether the two events took place sequentially rather than simultaneously [Forman and Shaw, 1973]. The important point for the present discussion is that both tectonic events took place along what is regarded as the same suture between two Proterozoic cratonic blocks: north Australia to the north and central Australia to the south [Shaw et al., 1996]. Regardless of the source of the tectonic stresses responsible for late Palaeozoic deformation, we postulate that both regions accommodated shortening/extension by reactivation of a pre-existing suture and detachment along an upper-mantle low-viscosity layer.

Acknowledgments. We wish to thank Peter van der Beek, Gianna Bassi, Greg Houseman, Geoffrey Nicholls, John Kennard, Deborah Scott, and an anonymous reviewer for their constructive comments on the work presented in this manuscript. Clementine Kraishek's help in drafting some of the figures is much appreciated. Russell Shaw publishes with the permission of the Director, Australian Geological Survey Organisation.

REFERENCES

Bally, A. W., and S. Snelson, Facts and principles of world petroleum occurrence: realms of subsidence, in *Facts and Principles of World Petroleum Occurrence*, edited by A. D. Miall, Can. Soc. Petrol. Geol. Memoir, 6, 9-94., 1980.

Batt, G. E., and J. Braun, On the thermo-mechanical evolution of compressional orogens, Geophys. J. Int., *128*, 364-382, 1997.

Beaumont, C., P. Fullsack, and J. Hamilton, Styles of crustal deformation caused by subduction of the underlying mantle, *Tectonophysics*, 232, 119-132, 1994.

Beslier, M.-O., and J.-P. Brun, Boudinage de la lithosphère et formation des marges passives, *Comptes-rend. Acad. Sci. Paris*, 313-2, 951-958, 1991.

Braun, J., Post-extensional mantle healing and episodic extension in the Canning Basin, *J. Geophys. Res.*, 87, 8927-8936, 1992.

Braun, J., and C. Beaumont, A physical explanation of the relationship between flank uplift and the breakup unconformity at passive continental margins, *Geology*, 17, 760-764, 1989.

Braun, J., and C. Beaumont, Three-dimensional numerical experiments of strain partitioning at oblique plate boundaries: implications for contrasting tectonic styles in the Southern Coast Ranges, California, and central South Island, New Zealand, *J. Geophys. Res.*, 100, 18,059-18,074, 1995.

Braun, J., and M. Sambridge, Dynamical Lagrangian Remeshing (DLR): a new algorithm for solving large strain deformation problems and its application to fault-propagation folding, *Earth Planet. Sci. Lett.*, 124, 211-220, 1994.

Braun, J., H. McQueen, and M. A. Etheridge, A fresh look

at the Late Palaeozoic tectonic history of west-central Australia, *Explor. Geophys.*, 22, 49-54, 1991.

Chopra, P. N., and M. S. Paterson, The experimental deformation of dunite, *Tectonophysics*, 78, 453-473, 1981.

Davies, J. H., and F. von Blanckenburg, Slab breakoff: a model of lithosphere detachment and its test in the magmatism and deformation of collisional orogens, *Earth Planet. Sci. Lett.*, 129, 85-102, 1995.

Drummond, B. J., M. J. Sexton, T. J. Barton, and R. D. Shaw, The nature of faulting along the margins of the Fitzroy Trough, Canning Basin, and implications for the tectonic development of the trough, *Explor. Geophys.*, 22, 111-116, 1991.

Forman, D. J.,,,, and R. D. Shaw, Deformation of the crust and mantle in central Australia, *Bur. Min. Res. Aust., Bull.*, 144, 1973.

Jackson, M. J., L. J. Diekman, J. M. Kennard, P. N. Southgate, P. E. O'Brien, and M. J. Sexton, Sequence stratigraphy, basin-floor fans and petroleum plays in the Devonian-Carboniferous of the northern Canning Basin, *Aust. Petrol. Explor. Ass. J.*, 32(1), 214-230, 1992.

Jackson, M. J., J. M. Kennard, M. Moffat, P. E. O'Brien, M. J. Sexton, P. N. Southgate, and I. Zeilinger, Canning Basin Project Stage I, Lennard Shef: Explanatory notes from seismic and map folios. *Aust. Geol. Surv. Org. Rec., 1993/1*, 24pp., 1993.

Kennard, J. M., M. J. Jackson, K. K. Romine, and P. N. Southgate, Canning Basin Project Stage II, geohistory modelling. *Aust. Geol. Surv. Org. Rec., 1994/67*, 243pp., 1994a.

Kennard, J. M., M. J. Jackson, K. K. Romine, R. D. Shaw, and P. N. Southgate, Depositional sequences and associated petroleum systems of the Canning Basin, WA, in *The Sedimentary Basins of Western Australia*, edited by P. G. Purcell and R. R. Purcell, Proc. Petrol. Explor. Soc. Aust. Symp., Perth, 657-676, 1994b.

Kooi, H., and C. Beaumont, Escarpment evolution on high elevated rifted margins: insights derived from a surface process model that combines diffusion, advection and reaction, *J. Geophys. Res.*, 99, 12,191-12,209, 1994.

Lister, G. S., M. A. Etheridge, and P. A. Symonds, 1991. Detachment models for the formation of passive continental margins, *Tectonics*, 10, 1038-1064.

Myers J. S., R. D. Shaw, and I. M. Tyler, Tectonic evolution of Proterozoic Australia, Tectonics, 15, 1431-1446, 1996.

Romine, K. K., P. N. Southgate, J. K. Kennard, and M. J. Jackson, The Ordovician to Silurian phase of the Canning Basin, WA: Structure and sequence evolution, in *The Sedimentary Basins of Western Australia*, edited by P. G. Purcell and R. R. Purcell, Proc. Petrol. Explor. Soc. Aust. Symp., Perth, 677-696, 1994.

Shaw, R. D., B. R. Goleby, R. J. Korsch, and C. Wright, Basement and cover thrust tectonics in central Australia based on the Arunta-Amadeus seismic reflection profile, in *Basement Tectonics* edited by M. J. Rickard, H. J. Harrington and P. R. Williams, Kluwer Academic Publishers, Dordrecht, 55-84, 1992.

Shaw, R. D., M. J. Sexton, and I. Zeilinger, The tectonic framework of the Canning Basin, W.A., including the 1:2 million structural element map of the Canning Basin, *Aust. Geol. Surv. Org. Rec., 1994/48*, 1994.

Shaw, R. D., P. Wellman, P. Gunn, A. J. Whitaker, C. Tarlowski, and M. Morse, Users Guide to the Australian Crustal Elements map, *Aust. Geol. Surv. Org. Rec., 1996/30*, 93 pp, 1996.

van den Beukel, J., . Some thermomechanical aspects of the subduction of continental lithosphere, *Tectonics*, 11, 316-329, 1992.

Wernicke, B., Uniform-sense normal simple shear of the continental lithosphere, *Can. J. Earth Sci.*, 22, 108-125, 1985.

Wilks, K., and N. L. Carter, Rheology of some continental lower crustal rocks, *Tectonophysics*, 182, 57-77, 1990.

Willett, S., C. Beaumont, and P. Fullsack, Mechanical model for the tectonics of doubly vergent compressional orogens, *Geology,* 21, 371-374, 1993.

J. Braun, Research School of Earth Sciences, The Australian National University, Canberra, ACT 0200, AUSTRALIA.

R. D. Shaw, Australian Geological Survey Organisation, PO Box 378, Canberra, ACT 2601, AUSTRALIA.

Granite-Greenstone Zircon U-Pb Chronology of the Gum Creek Greenstone Belt, Southern Cross Province, Yilgarn Craton: Tectonic Implications

Q. Wang
Research School of Earth Sciences, Australia National University, Canberra, ACT 0200, Australia

J. Beeson
Goldfields Exploration Pty. Ltd, Post Box 862, Kalgoorlie, WA 6430, Australia

I.H. Campbell
Research School of Earth Sciences, Australia National University, Canberra, ACT 0200, Australia

SHRIMP U-Pb zircon dating was carried out on the supracrustal rocks and granitoids in the Gum Creek greenstone belt, Southern Cross Province of the Yilgarn Block, for which no previous geochronology is available. An age of ca. 2.7 Ga was obtained for the dominantly felsic part of the greenstone sequence. It clearly indicates that at least the upper units of the sequence in this greenstone belt do not correlate with the ca. 2.9 to ca. 3.0 Ga lower greenstone sequences recognised in both the adjacent Murchison Province and the southern part of the Southern Cross Province. However, the lower tholeiitic basalt and BIF dominant units, which are probably unconformably overlain by the ca. 2.7 Ga component, may have an age of ca. 2.9 to ca. 3.0 Ga. Two pre- to syn-kinematic granitoids were dated at 2722±7 Ma and 2699±7 Ma. The ages obtained for two post-kinematic granitic intrusions are 2682±8 Ma and 2638±10 Ma. The observation of xenocrystic zircons from granites as old as ca. 3.6 Ga indicates that the magmas derived from melting of a pre-existing sialic crust.

INTRODUCTION

The Gum Creek greenstone belt is located at the northern end of the Southern Cross Province, Yilgarn Block, and separated from other greenstone belts in the adjacent Murchison and Eastern Goldfields Provinces by extensive granitoid intrusions (Figure 1). The Southern Cross Province was interpreted to be a distinct granite-greenstone province [Gee et al., 1981], yet the boundaries between it and the other provinces are still imprecisely defined. Although some geochronology has been carried out in the central part of this province, none is available for the Gum Creek greenstone belt in the north of the province. Thus the relationship between it and other greenstone belts, in the Southern Cross, Murchison and Eastern Goldfields Provinces, is unclear.

This paper presents the results of the first geochronological study of the supracrustal rocks and granitoids in the Gum Creek greenstone belt using the SHRIMP - I ion probe at the Research School of Earth Sciences, Australian National University. These data, combined with previous geochronology data, allow a comparison between this and some other greenstone belts in the Yilgarn Block, to be discussed.

GENERAL GEOLOGY

The elongate to arcuate greenstone belts in the Southern Cross Province have a dominant north-northwest trend parallel to the regional fault system, and are surrounded and intruded by extensive granitoids. Metamorphism within the province varies from prehnite-pumpellyite to granulite

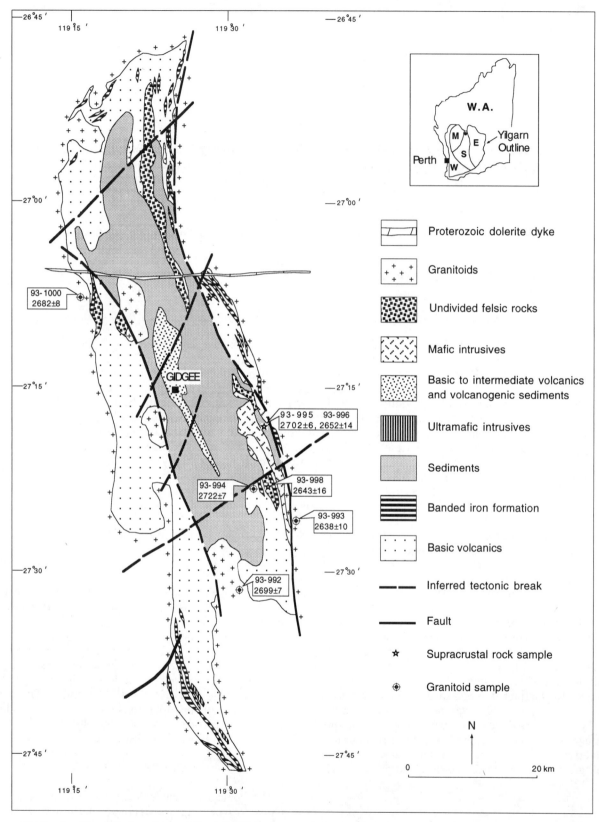

Figure 1. Simplified geology of Gum Creek greenstone belt, showing the approximate localities and ages of the studied samples. The inlet map shows the Yilgarn outline, its subdivision and the approximate locality of the Gum Creek greenstone belt (W - Western Gneiss Terrain, M - Murchison Province, S - Southern Cross Province, E - Eastern Goldfields Province). Modified from Otterman et al. (1990).

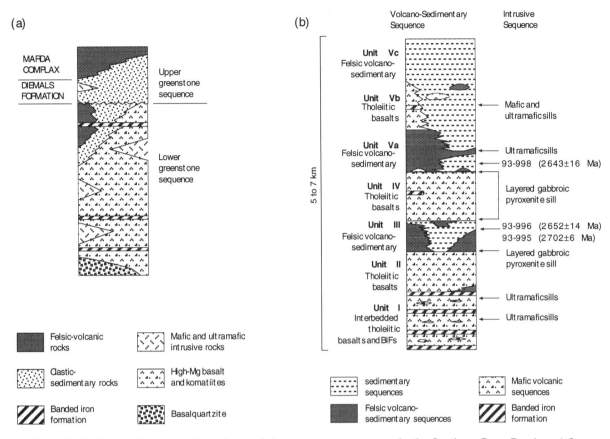

Figure 2. (a) Proposed stratigraphic column of the greenstone sequences in the Southern Cross Province (after Griffin, 1990). (b) Stratigraphic column of the Gum Creek greenstone belt summarised by Beeson et al. (1993).

facies with the higher grades predominant in the south [Gee et al., 1981; Solomon et al., 1994]. Gee et al. [1981] proposed a generalised greenstone sequence for the Southern Cross Province, consisting of a basal quartzitic unit overlain by a lower mafic unit, which in turn is unconformably overlain by a volcanogenic sedimentary unit. Griffin [1990] reviewed the regional geology of the Southern Cross Province and divided the stratigraphy into two greenstone sequences (Figure 2a): a lower mafic and ultramafic sequence dominated by tholeiitic basalts and an upper sequence composed of calc-alkaline volcanic and clastic sedimentary rocks, which is best developed in the Diemals - Marda area in the central part of the province [Hallberg et al., 1976; Chapman et al., 1981; Bickle et al., 1983].

The geology of the Gum Creek greenstone belt has been described by Tingey [1985] and Otterman [1990], and is summarised by Beeson et al. [1993]. It has a NNW trending synclinal structure [Elias et al., 1982]. Some faults and joint sets, which are subparallel to the synclinal axis, were inferred from the aeromagnetic and satellite images [Otterman, 1990; Figure 1]. The greenstone belt is subdivided by Beeson et al. [1993] into five volcano-sedimentary stratigraphic units of felsic and mafic volcanic rocks (Figure 2b). This subdivision is based on the supracrustal rocks of the eastern flank of the greenstone belt, where the best outcrops occur and where all five units are presented. The greenstones have undergone dominantly greenschist facies metamorphism. The felsic volcanic and sedimentary rocks, which host the Gidgee gold deposit, are generally poorly exposed and deeply weathered [Tingey, 1985; Otterman 1990].

ANALYTICAL METHOD

Following crushing and screening of ~1.5 kg rock, zircons were extracted using conventional heavy liquid and magnetic techniques. After hand picking, the zircons were mounted in epoxy together with the Australian National University standard SL13 (572 Ma; $^{206}Pb/^{238}U=0.0928$, 238 ppm U). The mounts were sectioned and polished to expose the interiors of grains. U-Th-Pb measurements were made using the SHRIMP - I ion microprobe. Negative oxygen primary beam was used to sputter material from the surface of analysed single zircon grain. Secondary ions of interest, including Zr_2O^+, $^{204}Pb^+$ background, $^{206}Pb^+$, $^{207}Pb^+$, $^{208}Pb^+$, $^{238}U^+$, ThO^+ and UO^+, were measured by a single collector by cyclically stepping the magnet through the masses. For individual analysed sites, five scans were performed.

The isotopic ratios of the unknowns and their uncertainties were calculated based on the analyses of the standard undertaken at regular intervals. $^{207}Pb/^{206}Pb$ calculation and common Pb correction procedures are those presented by Compston et al. (1984). The $^{206}Pb/^{238}U$ ratio and absolute U content of the unknowns were referenced to the values of standard SL13 respectively. $^{207}Pb/^{235}U$ ratio was calculated based on the determination of $^{206}Pb/^{238}U$, $^{207}Pb/^{206}Pb$ and present day $^{238}U/^{235}U$ ratio valued at 137.88. Once the U content of the unknown is determined, the Th and Pb absolute abundances are calculated from the measured Th/U and Pb/U ratios. The uncertainties of the isotopic ratios and elemental abundances are highly dependent on the accuracy achieved for the measurement of the reference standard. Further details of the SHRIMP analytical procedures and assessment of data are given by Compston et al. [1984], Roddick and Van Breemen [1994] and Claoué-Long et al. [1995].

Unless stated, the quoted ages in the text are weighted means of $^{207}Pb/^{206}Pb$ ages (at ±2σ or 95% confidence) derived from the least disturbed sites in grains of the same zircon population. The ages of individual zircon populations in complex data sets, were determined using a combination of cumulative probability plots of $^{207}Pb/^{206}Pb$ ages and the Mix program of Sambridge and Compston [1995]. Only near concordant analyses with indistinguishable $^{207}Pb/^{206}Pb$ and low common ^{206}Pb values were included in the calculation of the mean ages of populations. Xenocrysts with ages greatly older than the crystallisation age, are easily recognised by this method. Zircons, with ages younger than the crystallisation age, could represent part of the main population of zircons that underwent early Pb loss, or new grains or parts of grains that crystallised at younger ages. Those which are greater than 100 Ma younger than the crystallisation age, are obvious outliers. When the age difference is small, the approach followed by Schiøtte and Campbell [1996] was used to identify the outliers.

The common Pb ratios used in the age calculation are those of Cumming and Richards [1975] at the estimated ages of the rocks. The decay constants of U and Th and present day $^{238}U/^{235}U$ value are those given by Steiger and Jäger [1977].

RESULTS

The localities of the analysed samples are shown in Figure 1. The SHRIMP analytical results are presented in Table 1 and in Figure 3 as conventional concordia plots and cumulative probability plots of $^{207}Pb/^{206}Pb$ age.

Supracrustal Rocks

Sample *93-995* (27°18´S, 119°33´E) is a schistose metarhyolite with quartz, plagioclase and amphibole as phenocrysts in a fine-grained quartz-feldspar groundmass. The foliation is defined by amphibole. This rock was collected from the base of the Unit III, the lowest of the felsic volcano-sedimentary units. The zircons are prismatic, pale pink in colour with euhedral zoning. Forty analyses were carried out on thirty-seven grains. Most lead loss (Figure 3a) is recent but some grains, with high U content and lower $^{207}Pb/^{206}Pb$ ages, show Proterozoic Pb loss. This could be related to the intrusion of the Proterozoic dykes into the greenstone belt. No old zircon cores were observed in this sample. The plot of the cumulative $^{207}Pb/^{206}Pb$ ages shows a strong peak at ca. 2.7 Ga and a few small peaks at younger ages (Figure 3b). An age of 2702±6 Ma was calculated from the sixteen least disturbed analyses (Table 1). This is interpreted as the best estimate of the age for the sample.

93-996 (27°18´S, 119°33´E) is a porphyritic metarhyolite collected from Unit III, ~ 300 m northwest of 93-995 (Figure 1). It contains K- feldspar and quartz phenocrysts in a microcrystalline quartz + feldspar groundmass. Zircons extracted from this sample are equant and prismatic, pale pink in colour with well developed euhedral zoning. Data were obtained for 22 sites from 20 grains (Table 1). Only very recent lead loss is observed for the zircons (Figure 3c). The age of 2652±18 Ma is determined from the seven concordant analyses from the main group of zircons (18.1, 20.1, 6.1, 12.1, 16.1, 3.1 and 11.1). Those data combined with ten other discordant ones, give an age of 2652±14 Ma. This is interpreted to be the crystallization age of the rock, and the 2722±14 Ma age, obtained from four analyses (4.1, 3.2, 9.1 and 7.1) is tentatively interpreted as dating the xenocrystic grains. If all analyses are included, regardless of the criteria mentioned above, an age of 2671±20 Ma is obtained. The cumulative $^{207}Pb/^{206}Pb$ age plot (Figure 3d) shows a left skewed peak. Analysis 13.1 which has very high U, Th contents and large Pb/U uncertainties (Table 1), is not shown in the concordia plot of Figure 3c.

93-998 (27°23´S, 119°34´E) is a foliated volcaniclastic rock from the base of Unit V, the uppermost felsic volcano-sedimentary unit. Most of the zircons are stubby prismatic, pale pink with fine scale zoning and somewhat rounded edges. Cathodoluminescence imaging indicates a complicated growth history for these grains. A total of 22 analyses was performed on 12 grains (Table 1). A number of groups of ages can be distinguished on the concordia plot and the cumulative $^{207}Pb/^{206}Pb$ age plot (Figures 3e and 3f). These are 3134±20 (2.2, 11.1 and 2.1), 2965±21 (9.1, 6.2, 8.1 and 7.2), 2852±21 (7.1, 10.1 and 5.4), 2728±48 (2.3 and 4.4), 2643±16 (5.1, 5.3, 4.2, 5.2, 4.3, and 4.1) Ma. The oldest age of 3243 Ma was obtained for a zircon core (6.1). The youngest age of 2643±16 Ma, which is derived from six clear sites in two grains with no obvious evidence of prehistory, is believed to be the maximum age for the deposition of this rock, and the others the ages of older sedimentary zircons. Three other grains (3.1, 12.1,

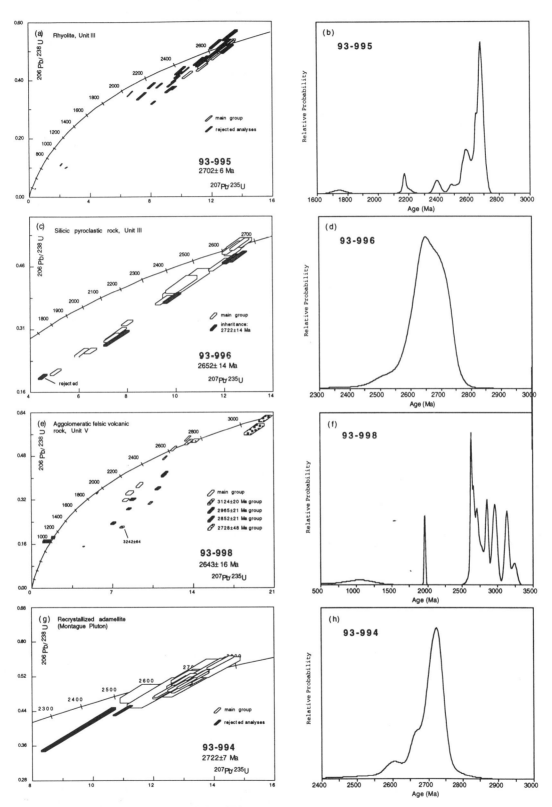

Figure 3. Concordia and cumulative ^{207}Pb/^{206}Pb age plots of the analysed samples. Individual analyses are shown with 1σ error bars.

Table 1. SHRIMP U-Pb zircon data[#]

Site[◊]	U (ppm)	Th (ppm)	^{204}Pb (ppb)	f_{206} (%)[†]	^{208}Pb/^{206}Pb	^{206}Pb/^{238}U	^{207}Pb/^{235}U	^{207}Pb/^{206}Pb	207/206 age (Ma)±1σ	Disc (%)[+]
93-995										
22.1*	225	123	12	0.21	0.1435±24	0.5040±154	13.15±0.43	0.1893±15	2736±13	-4
34.1	247	88	53	1.00	0.0802±31	0.4419±139	11.46±0.39	0.1881±17	2726±15	-14
33.1	219	181	33	0.64	0.1880±27	0.4830±160	12.50±0.43	0.1877±13	2722±12	-7
1.1*	227	124	127	3.04	0.1481±32	0.3719±68	9.60±0.20	0.1872±14	2718±13	-25
31.1	379	297	32	0.35	0.2145±16	0.5063±149	13.04±0.39	0.1868±8	2715±7	-3
35.1	271	149	28	0.41	0.1304±21	0.5153±154	13.23±0.42	0.1863±13	2710±12	-1
30.1	278	157	23	0.33	0.1499±18	0.5169±165	13.24±0.44	0.1858±11	2705±8	-1
11.2	249	92	14	0.24	0.0988±16	0.4891±151	12.52±0.41	0.1857±14	2704±12	-5
16.1	213	216	26	0.52	0.1025±23	0.4747±153	12.15±0.41	0.1856±12	2704±10	-7
23.1	242	223	102	1.90	0.1648±36	0.4519±135	11.54±0.37	0.1852±17	2700±15	-11
4.1*	216	252	37	1.08	0.1482±24	0.3229±59	8.23±0.17	0.1849±14	2698±13	-33
11.1	334	129	14	0.17	0.0972±12	0.5053±103	12.88±0.27	0.1849±7	2696±6	-2
21.1	346	197	41	0.48	0.1483±31	0.5031±188	12.81±0.51	0.1847±18	2695±16	-2
21.2	412	244	103	1.01	0.1661±24	0.5145±153	13.10±0.41	0.1847±12	2695±11	-1
15.1	273	169	35	0.50	0.1563±25	0.5220±166	13.28±0.44	0.1845±13	2694±11	1
9.1	127	76	10	0.32	0.1581±19	0.5097±118	12.95±0.32	0.1843±13	2692±12	-1
27.1*	353	902	191	2.97	0.1600±40	0.3669±112	9.32±0.31	0.1842±18	2691±16	-25
9.2	129	78	18	0.55	0.1568±28	0.5155±170	13.07±0.46	0.1838±15	2688±14	0
32.1	144	97	21	0.59	0.1812±32	0.5215±162	13.19±0.43	0.1834±15	2684±13	1
29.1	213	92	81	1.87	0.1196±38	0.4169±134	10.48±0.36	0.1824±18	2675±18	-16
3.1*	559	164	26	0.20	0.0735±8	0.4846±83	12.14±0.22	0.1817±6	2668±5	-5
17.1*	328	216	120	1.52	0.1356±31	0.4897±153	12.26±0.41	0.1816±16	2667±15	-4
5.1*	348	284	38	0.50	0.1833±30	0.4595±154	11.46±0.42	0.1809±16	2661±16	-8
37.1*	150	137	26	0.68	0.2350±40	0.5196±178	12.91±0.48	0.1802±18	2654±16	2
7.1*	439	299	13	0.13	0.1840±22	0.4714±95	11.69±0.29	0.1798±22	2651±20	-6
8.1*	392	178	88	1.20	0.1139±50	0.3869±119	9.46±0.31	0.1773±16	2628±15	-20
2.1*	938	827	494	9.82	0.2916±54	0.1007±31	2.45±0.09	0.1765±36	2620±34	-76
19.1*	450	285	220	2.15	0.1229±61	0.4624±135	11.25±0.35	0.1764±14	2620±13	-6
24.1*	209	139	100	1.77	0.1498±46	0.5513±190	13.34±0.50	0.1755±21	2611±20	8
36.1*	460	385	36	0.40	0.1850±19	0.4024±121	9.731±0.30	0.1754±10	2610±9	-17
10.1*	377	135	56	0.71	0.0542±13	0.4312±114	10.35±0.28	0.1740±8	2597±8	-11
26.1*	482	369	27	0.28	0.2019±16	0.4168±122	9.905±0.30	0.1724±9	2581±9	-13
20.1*	268	188	159	2.31	0.0945±36	0.5204±163	12.36±0.42	0.1722±17	2579±16	5
12.1*	638	279	41	0.31	0.1177±15	0.4314±147	10.02±0.37	0.1685±18	2543±18	-9
14.1*	630	288	219	1.86	0.1181±26	0.3811±96	8.637±0.24	0.1644±16	2501±16	-17
25.1*	857	828	90	0.63	0.2151±41	0.3452140	7.445±0.33	0.1565±19	2418±21	-21
6.1*	795	494	58	0.40	0.1640±26	0.3807±115	8.125±0.27	0.1548±15	2399±17	-13
18.1*	1418	1513	510	6.32	0.3143±52	0.1111±37	2.123±0.08	0.1386±21	2210±26	-69
28.1*	896	1043	61	0.40	0.3323±21	0.3535±102	6.693±0.20	0.1373±7	2194±9	-11
13.1*	3164	5477	315	6.86	0.4924±58	0.0282±134	0.414±0.02	0.1066±23	1742±40	-90
93-996										
4.1*	273	1200	106	2.72	0.3125±91	0.2889±193	7.56±0.53	0.1898±28	2740±25	-40
3.2 (c)*	177	191	1	0.02	0.1762±36	0.4843±134	12.58±0.38	0.1885±18	2729±16	-7
9.1*	188	224	70	1.59	0.1162±48	0.4782±121	12.30±0.36	0.1865±22	2712±19	-7
7.1*	192	712	36	1.00	0.1581±55	0.3845±132	9.86±0.38	0.1860±25	2708±22	-22
18.1	184	262	19	0.47	0.1098±38	0.4647±154	11.80±0.43	0.1841±22	2690±20	-8
1.1	305	2518	128	2.91	0.2235±66	0.2900±111	7.35±0.32	0.1838±30	2688±28	-39
20.1	264	1182	92	2.38	0.2009±93	0.2990±165	7.54±0.45	0.1828±32	2678±29	-37
12.2	249	824	91	1.97	0.1285±53	0.3791±115	9.52±0.34	0.1822±27	2673±25	-22
2.1	426	4476	199	3.62	0.1629±91	0.2591±74	6.48±0.26	0.1814±44	2666±40	-44
17.1	141	157	24	0.71	0.1467±44	0.4950±99	12.38±0.35	0.1814±32	2665±29	-3
14.1	151	327	42	1.36	0.1055±61	0.4198±349	10.44±0.98	0.1804±61	2657±57	-15
6.1	146	234	61	1.67	0.2621±83	0.5137±144	12.69±0.48	0.1792±40	2646±38	1
12.1	173	105	17	0.39	0.1167±43	0.5129±148	12.66±0.40	0.1790±19	2644±18	1
16.1	104	132	20	0.77	0.1278±53	0.5026±118	12.40±0.46	0.1789±46	2643±43	-1
19.1	269	1152	122	2.82	0.1261±70	0.3253±72	8.01±0.24	0.1786±34	2640±32	-31
8.1	192	211	29	0.75	0.1800±49	0.4169±216	10.26±0.65	0.1784±55	2638±52	-15
3.1 (r)	106	77.9	11	0.44	0.1548±41	0.5047±158	12.38±0.44	0.1779±23	2633±21	0
11.1	156	191	44	1.12	0.1501±49	0.5139±144	12.56±0.44	0.1772±32	2627±30	2
5.1	309	1904	155	4.78	0.3893±84	0.2078±76	5.05±0.22	0.1762±35	2618±33	-53
10.1	513	1956	132	2.05	0.2559±75	0.2546±101	6.11±0.28	0.1740±32	2596±31	-44
13.1*	1049	2596	468	2.36	0.0764±375	0.3852±368	8.96±1.28	0.1678±160	2544±169	-17
15.1*	469	2075	258	5.48	0.1538±95	0.1972±64	4.52±0.19	0.1661±38	2519±39	-54

Table 1. (Continued)

Site◊	U (ppm)	Th (ppm)	^{204}Pb (ppb)	f_{206} (%)†	^{208}Pb/^{206}Pb	^{206}Pb/^{238}U	^{207}Pb/^{235}U	^{207}Pb/^{206}Pb	207/206 age (Ma)±1σ	Disc (%)+
93-998										
6.1 (c)*	245	768	222	7.89	0.6666±116	0.2198±35	7.86±0.22	0.2594±53	3243±33	-60
2.2 (c)*	249	258	293	4.09	0.6452±232	0.5753±213	19.36±0.88	0.2441±55	3147±36	-7
11.1(c)*	201	477	150	4.81	0.4802±72	0.3077±51	10.33±0.22	0.2436±28	3144±18	-45
2.1 (c)*	149	186	96	2.14	0.6065±102	0.6150±123	20.29±0.53	0.2393±34	3115±23	-1
9.1 (c)*	133	343	87	3.61	0.5285±79	0.3620±58	11.00±0.25	0.2205±31	2984±23	-33
6.2 (r)*	324	976	252	6.45	0.6230±138	0.2349±50	7.11±0.24	0.2195±50	2977±37	-54
8.1 (c)*	463	930	313	8.49	0.5466±83	0.1516±19	4.53±0.09	0.2165±32	2955±24	-69
7.2 (c)*	343	952	299	5.97	0.4224±102	0.2867±47	8.46±0.23	0.2140±42	2936±32	-45
7.1 (m)*	143	254	53	1.86	0.3089±48	0.4085±64	11.51±0.22	0.2043±20	2861±16	-23
10.1*	142	213	98	3.33	0.3002±66	0.4147±93	11.59±0.31	0.2027±25	2848±20	-21
5.4*	215	498	244	6.82	0.3691±114	0.3221±55	8.81±0.27	0.1984±48	2813±40	-36
2.3 (r)*	955	396	237	0.96	0.2598±100	0.5352±78	14.23±0.36	0.1928±36	2766±31	0
4.4	277	389	27	0.38	0.3656±75	0.5305±106	13.67±0.33	0.1869±21	2715±18	1
5.1	143	497	149	6.36	0.3104±124	0.3195±65	8.19±0.30	0.1859±52	2706±47	-34
5.3	200	607	137	3.69	0.2784±106	0.3739±94	9.41±0.32	0.1826±37	2676±34	-23
4.2	216	276	9	0.15	0.3225±40	0.5432±103	13.58±0.29	0.1813±14	2665±13	5
5.2	125	258	108	4.95	0.2625±129	0.3456±74	8.55±0.33	0.1794±54	2647±51	-28
4.3	172	223	13	0.32	0.3123±24	0.4736±73	11.62±0.20	0.1780±10	2634±10	-5
4.1	211	277	21	0.40	0.3098±65	0.5065±82	12.40±0.23	0.1776±13	2630±13	0
3.1*	505	378	46	0.55	0.2489±37	0.3452±45	5.73±0.08	0.1204±6	1962±9	-3
12.1*	76	83	13	1.88	0.3427±167	0.1826±53	1.88±0.17	0.0746±61	1058±173	2
1.1 *	33	15	21	7.12	0.0563±35	0.1699±67	1.36±0.35	0.0578±144	1012±379¶	93
93-994										
14.1	195	231	1	0.02	0.3279±34	0.4932±125	12.96±0.35	0.1906±13	2747±11	-6
11.1	214	298	18	0.32	0.3761±117	0.5404±241	14.15±0.76	0.1899±47	2742±41	2
21.1	178	190	20	0.44	0.2867±33	0.5155±144	13.46±0.40	0.1894±14	2737±12	-2
9.1	379	696	22	0.24	0.5107±24	0.5045±120	13.14±0.33	0.1890±10	2733±9	-4
20.1	135	144	17	0.49	0.3001±48	0.5166±153	13.44±0.44	0.1887±20	2731±18	-2
10.1	215	213	21	0.38	0.2722±25	0.5211±133	13.51±0.37	0.1881±12	2726±11	-1
5.1	107	106	12	0.42	0.2716±41	0.5511±158	14.25±0.44	0.1875±17	2720±15	4
22.1	161	169	14	0.37	0.2961±28	0.4972±125	12.84±0.35	0.1873±13	2719±11	-4
15.1	137	112	8	0.23	0.2218±28	0.5042±136	12.98±0.38	0.1867±15	2714±13	-3
23.1	151	155	10	0.30	0.2854±41	0.4839±163	12.45±0.45	0.1866±20	2713±17	-6
6.1	119	122	17	0.59	0.2737±156	0.5145±424	13.23±1.28	0.1865±77	2711±70	-1
8.1	136	171	29	0.84	0.3553±60	0.5318±295	13.66±0.80	0.1862±23	2709±20	2
17.1	238	107	23	0.38	0.1236±112	0.5284±308	13.55±0.99	0.1859±70	2706±64	1
19.1	124	131	9	0.32	0.2878±35	0.4850±135	12.41±0.37	0.1856±14	2704±13	-6
13.1	146	138	25	0.69	0.2541±50	0.5107±143	13.04±0.41	0.1852±22	2700±19	-2
3.1	37	151	27	0.79	0.2966±64	0.5065±149	12.92±0.45	0.1850±29	2699±26	-2
12.1	94	89	19	0.79	0.2519±91	0.5192±163	13.22±0.54	0.1847±40	2696±37	0
2.1	205	207	38	0.80	0.2846±115	0.4816±252	12.17±1.01	0.1832±58	2682±54	-6
7.1	72	71	34	1.80	0.2467±71	0.5189±149	13.07±0.46	0.1827±32	2678±29	1
18.1*	272	160	8	0.14	0.1955±24	0.4412±123	11.01±0.32	0.1810±13	2662±12	11
1.1	133	123	52	1.70	0.2340±217	0.4799±214	11.80±0.89	0.1784±99	2638±95	-4
4.1*	847	352	456	2.70	0.4776±183	0.3971±508	9.54±1.24	0.1743±21	2600±20	17
93-992										
6.2*	267	136	154	3.52	0.1929±63	0.3295±68	8.79±0.24	0.1934±30	2772±25	34
13.1*	365	242	208	2.92	0.2713±44	0.3947±83	10.48±0.26	0.1926±20	2764±17	-22
4.1	360	330	82	0.90	0.2734±82	0.5213±143	13.72±0.48	0.1909±34	2750±30	-2
16.2*	375	115	127	1.50	0.1111±37	0.4631±112	12.17±0.32	0.1906±14	2747±12	-11
11.1*	371	317	148	1.62	0.2840±33	0.5048±122	13.23±0.35	0.1901±15	2743±13	-4
10.1	268	120	55	0.84	0.1433±31	0.5064±109	13.12±0.32	0.1879±17	2724±15	-3
16.1*	318	100	191	2.46	0.1057±41	0.4969±127	12.79±0.36	0.1866±18	2712±16	-4
4.2	521	520	19	0.15	0.2819±20	0.5035±100	12.91±0.27	0.1859±10	2707±9	-3
15.1	225	171	4	0.07	0.2076±25	0.5483±134	14.03±0.37	0.1856±12	2704±11	4
6.1	220	115	5	0.08	0.1317±36	0.5482±129	14.03±0.38	0.1856±19	2704±17	4
1.1	197	105	26	0.52	0.1466±33	0.5374±127	13.72±0.36	0.1852±16	2700±15	3
5.2	238	104	24	0.42	0.1300±26	0.5046±124	12.89±0.34	0.1852±14	2700±12	-2
18.2	248	124	17	0.27	0.1382±20	0.5123±101	13.07±0.28	0.1850±11	2698±10	-1
15.2	205	152	6	0.11	0.2108±27	0.5280±143	13.45±0.39	0.1847±13	2696±11	1
17.1*	156	122	51	1.49	0.1646±115	0.4545±451	11.56±1.28	0.1844±69	2693±63	-10
2.2	255	224	20	0.33	0.2368±34	0.5074±106	12.87±0.32	0.1839±19	2689±17	-2

Table 1. (Continued)

Site◊	U (ppm)	Th (ppm)	^{204}Pb (ppb)	f_{206} (%)†	^{208}Pb/^{206}Pb	^{206}Pb/^{238}U	^{207}Pb/^{235}U	^{207}Pb/^{206}Pb	207/206 age (Ma)±1σ	Disc (%)+
93-992										
5.1	259	106	26	0.40	0.1104±28	0.5244±121	13.24±0.34	0.1832±15	2679±13	1
2.1	157	143	17	0.44	0.2409±41	0.5168±134	13.04±0.38	0.1829±20	2676±12	0
18.1	260	129	20	0.31	0.1343±27	0.5115±128	12.90±0.35	0.1829±14	2662±21	-1
8.1*	464	435	117	1.35	0.2682±30	0.3863±116	9.72±0.31	0.1825±14	2636±9	-21
5.3*	252	106	75	1.36	0.1348±51	0.4506±114	11.24±0.33	0.1810±23	2629±10	-10
7.1*	522	491	97	0.76	0.2440±28	0.5021±104	12.33±0.27	0.1781±10	2612±28	-1
14.1*	476	349	124	1.18	0.2115±24	0.4546±99	11.12±0.26	0.1775±11	2579±10	-8
3.1*	397	274	52	0.65	0.1891±31	0.4219±247	10.21±0.64	0.1756±29	2550±10	-13
12.1*	456	324	101	0.95	0.2849±25	0.4789±116	11.37±0.29	0.1722±11	2531±14	-2
9.1*	753	630	72	0.45	0.2286±19	0.4362±95	10.18±0.24	0.1692±10	2682±14	-8
4.3*	510	439	73	0.82	0.2345±27	0.3589±71	8.28±0.19	0.1673±14	2680±18	-22
93-993										
17.1(c)*	247	162	954	9.66	0.3001±139	0.7523±190	35.46±1.18	0.3419±64	3671±29	-2
6.1 (c)*	244	210	1044	13.6	0.3726±72	0.5671±106	19.97±0.54	0.2554±45	3219±28	-10
4.1 (c)*	52	45	13	0.84	0.2152±115	0.5906±153	19.39±0.86	0.2381±78	3107±53	-4
5.1 (c)*	123	137	80	2.43	0.3341±68	0.5407±135	17.41±0.51	0.2335±30	3076±20	-9
16.1(m)*	238	349	1836	20.3	0.4772±434	0.6307±201	18.28±1.53	0.2103±155	2908±124	8
12.1(m)*	478	172	5837	41.8	0.8097±151	0.3536±158	9.48±0.52	0.1945±53	2780±45	-30
18.1*	725	623	3836	39.8	1.0892±736	0.1669±93	4.47±0.62	0.1943±236	2779±214	-64
21.1	147	136	429	11.0	0.2477±166	0.4918±99	12.69±0.50	0.1871±59	2717±53	-5
4.2 (r)*	296	209	5484	45.3	1.0143±933	0.4652±310	12.00±2.18	0.1871±300	2717±292	-9
9.1*	124	150	151	8.85	0.4914±144	0.2607±61	6.71±0.27	0.1868±59	2714±52	-45
11.1	253	195	65	1.10	0.2181±32	0.4866±91	12.03±0.25	0.1793±14	2646±13	-3
13.1	333	232	5688	42.7	0.9062±186	0.4772±88	11.78±0.50	0.1790±65	2644±62	-5
19.1	329	136	3011	32.7	0.5532±128	0.3924±103	9.68±0.60	0.1789±94	2643±90	-19
8.1	178	160	298	7.18	0.3306±82	0.4511±90	11.11±0.32	0.1787±33	2641±32	-9
2.2	54	59	10	0.78	0.2933±91	0.5114±120	12.59±0.42	0.1786±38	2640±36	1
7.1	176	214	18	0.42	0.3352±35	0.5115±102	12.56±0.28	0.1781±14	2635±14	1
2.1	68	59	53	3.02	0.2663±141	0.5219±110	12.81±0.53	0.1781±59	2635±56	3
14.1	266	394	1207	18.2	0.6469±245	0.4250±96	10.37±0.61	0.1769±91	2624±88	-13
11.2	171	113	143	4.28	0.2569±61	0.3900±81	9.50±0.26	0.1767±26	2622±24	-19
7.2	199	245	55	1.21	0.3452±74	0.4691±112	11.41±0.38	0.1764±35	2620±34	-5
10.1	733	2693	18005	53.4	1.4195±438	0.4461±276	10.73±1.12	0.1745±134	2602±134	-9
15.1	216	434	7	0.12	0.5399±288	0.5275±339	12.36±1.05	0.1700±81	2557±82	7
20.1*	301	162	311	4.85	0.1944±43	0.4210±79	9.79±0.22	0.1687±18	2545±18	-11
3.1*	201	31	1590	49.2	0.9578±941	0.1697±134	3.46±0.82	0.1481±317	2324±422	-57
1.1*	690	66	1865	21.5	0.2604±824	0.2059±120	4.17±0.96	0.1469±315	2310±424	-48
93-1000										
5.2*	293	424	627	12.6	0.7483±120	0.3082±50	10.22±0.32	0.2405±62	3124±41	-45
5.1*	271	465	343	5.63	0.6032±86	0.4412±57	12.79±0.28	0.2103±34	2908±26	-19
20.1*	103	125	30	1.34	0.3799±74	0.4399±219	11.70±0.64	0.1929±34	2767±29	-15
18.2	268	196	9	0.14	0.2042±20	0.4811±110	12.32±0.30	0.1857±13	2705±12	-6
15.1	108	144	9	0.34	0.3471±51	0.4912±125	12.56±0.37	0.1855±22	2702±20	-5
4.1	143	141	3	0.08	0.2728±55	0.5611±103	14.34±0.33	0.1854±21	2702±19	6
1.1	324	269	21	0.30	0.2416±27	0.4576±52	11.62±0.16	0.1842±12	2691±10	-10
9.1	234	400	3	0.05	0.4557±45	0.5225±79	13.24±0.23	0.1838±12	2687±11	1
3.1	189	301	8	0.18	0.4561±47	0.5083±94	12.87±0.28	0.1837±16	2687±15	-1
13.1	168	198	7	0.19	0.3237±50	0.4758±122	12.05±0.34	0.1837±18	2686±16	-7
10.1	225	194	3	0.05	0.2271±27	0.5238±115	13.24±0.32	0.1834±15	2684±14	1
7.2	234	646	17	0.28	0.7298±86	0.5349±85	13.51±0.32	0.1832±20	2682±18	3
8.1*	215	278	236	8.03	0.5581±120	0.2622±38	6.61±0.22	0.1827±50	2678±46	-44
6.1*	332	418	100	1.65	0.3819±54	0.3713±55	9.35±0.19	0.1826±23	2677±21	-24
18.1	124	206	2	0.07	0.4464±80	0.4837±186	12.17±0.52	0.1825±26	2675±23	-5
20.2	125	158	5	0.18	0.3465±68	0.4569±144	11.49±0.42	0.1823±28	2674±26	-9
17.1*	145	278	19	0.61	0.5640±62	0.4372±84	10.97±0.26	0.1820±22	2671±20	-13
11.1	209	171	33	0.66	0.2332±41	0.4866±94	12.19±0.28	0.1817±19	2668±17	-4
14.1	244	210	20	0.34	0.2304±32	0.4936±107	12.34±0.30	0.1814±15	2666±14	-3
16.1*	202	202	56	1.95	0.3114±84	0.2875±167	7.18±0.46	0.1813±39	2665±32	-39
7.1*	211	663	16	0.28	0.8229±121	0.5482±84	13.63±0.24	0.1803±12	2655±11	6
21.1	201	223	10	0.21	0.2969±39	0.4897±89	12.15±0.20	0.1800±18	2653±17	-3
19.1	102	124	13	0.59	0.3141±50	0.4645±104	11.47±0.32	0.1790±24	2644±23	-7
12.1*	214	163	146	3.24	0.2232±77	0.4241±90	10.44±0.31	0.1786±34	2640±31	-14

Table 1. (Continued)

Site◊	U (ppm)	Th (ppm)	^{204}Pb (ppb)	f_{206} (%)†	^{208}Pb/^{206}Pb	^{206}Pb/^{238}U	^{207}Pb/^{235}U	^{207}Pb/^{206}Pb	207/206 age (Ma)±1σ	Disc (%)+
93-1000										
2.1	296	845	23	0.34	0.7751±43	0.4771±82	11.64±0.22	0.1769±11	2625±11	-4
15.2*	421	542	245	4.18	0.4283±180	0.2779±133	5.94±0.40	0.1551±66	2403±74	-34
11.2*	634	630	57	0.67	0.2574±267	0.2774±51	4.77±0.13	0.1246±23	2023±34	22

Analyses are listed in order of decreasing ^{207}Pb/^{206}Pb. All ratios involving Pb refer to radiogenic Pb. Quoted uncertainties (1σ) are the standard error of the mean and refer to the last digits. ◊ Sites with asterisks were rejected from the mean. For those sites in the grains with old cores, c, r and m denote core, rim and mixture respectively. † f_{206} (%) denotes percent of common ^{206}Pb in the total measured ^{206}Pb. + Disc (%) is discordance $(^{206}Pb/^{238}U\ age/^{207}Pb/^{235}U\ age-1)$. ¶ 206Pb/238U age.

1.1), with Proterozoic ages, were probably reset at that time. These grains are clear and structureless, being distinct from the main zircon population.

Granitoids

Sample *93-994* (27°23´S, 119°32´E) is a medium-sized recrystallized adamellite (Figure 1). This pluton (Montague Pluton) does not crop out, but coincides with an elliptical magnetic anomaly, which is clearly visible on airborne magnetic images. It is believed that this intrusion was strongly affected by regional deformation because of its elliptical shape and its sheared contacts with the adjacent greenstones [Beeson et al., 1993], the field evidence indicating that this pluton is pre- or syn-kinematic. The rock sample, collected from drill core, is moderately altered, with biotite altered to chlorite, and feldspar to clay minerals and sericite. The majority of the zircons in this sample are equant and stubby prismatic, pink in colour, euhedral and show oscillatory zoning. The age of 2722±7 Ma obtained from 20 concordant sites out from 23 analyses, is interpreted as dating the crystallization age of the rock. Discordant analyses 7.1 and 4.1 were rejected (Table 1 and Figure 3g). A strong peak at ca. 2720 Ma can be seen in the cumulative ^{207}Pb/^{206}Pb age plot, showing a basically single population of zircon in this sample (Figure 3h).

Sample *93-992* (27°32´S, 119°31´E) is a medium to coarse-grained adamellite that intrudes the greenstone sequence at the southern end of the greenstone belt (Figure 1). It is close to the regional fold axial trace, has an elliptical shape and is foliated near its intrusive contacts with neighbouring greenstones. Beeson et al. [1993] suggested that it was emplaced either prior to or synchronously with regional deformation. The analysed sample is strongly recrystallized and slightly altered. Zircons in the rock are prismatic, pale brown to pink in colour and strongly zoned. Twenty-seven analyses were carried out on 18 grains. An age of 2699±7 Ma was obtained from the 13 least disturbed sites (Table 1), and this is interpreted as the best estimate of the crystallization age of the sample. The discordant data show evidence of multistage Pb loss of the zircons (Figure 3i). This is shown in Figure 3j as a few peaks at ages younger than the main zircon population.

Sample *93-993* (27°26´S, 119°36´E) is a fine-grained recrystallized granite collected from the eastern side of the greenstone belt (Figure 1). It is almost aplitic and is slightly altered. Zircons separated from this sample are prismatic, transparent to pale pink and have well developed zoning. A few have xenocrystic cores. One of these cores (17.1) was dated at ca. 3670 Ma, and four others at ca. 3.0 Ga (6.1, 4.1 and 5.1 and 16.1; Figure 3k). Although analyses were performed on selected relatively clear sites, many of them still have a high common Pb content (Table 1). However, the 10 concordant and 3 near concordant sites have ^{207}Pb/^{206}Pb ages that are indistinguishable within error, and give a weighted mean age of 2638±10 Ma (Table 1 and Figure 3k). This age is unlikely to be strongly biased by early Pb loss, and by the choice of common Pb composition. As a result, we interpret it as a minimum crystallization age of this granite. The ^{207}Pb/^{206}Pb age spectrum shows three peaks older than ca. 3.0 Ga, and a small peak younger than the crystallization age, which are dated for the disturbed members of the main population (Figure 3l).

Sample *93-1000* (27°08´S, 119°16´E) is a foliated recrystallized biotite granite from the western side of the Gum Creek greenstone belt (Figure 1). The zircons are stubby to elongated, pink in colour, and zoned. Of the 27 analyses, 18 concordant sites give an age of 2682±8 Ma. The main zircon population forms the dominant peak shown in the cumulative ^{207}Pb/^{206}Pb age plot (Figure 3n). We interpret the 2682±8 Ma age as the crystallization age of the rock. An inherited grain from this sample gives ages of 3124±41 Ma (1σ) and 2908±26 Ma (1σ) for two different analysed sites (Table 1 and Figure 3n). Few younger ages were obtained from isotopically disturbed members of the main zircon population. They are shown in Figure 3n as two small peaks to the left of the dominant age peak.

DISCUSSION

Age of the Gum Creek Greenstone Belt and its Stratigraphic Position

Previous geochronological studies of the Southern Cross Province are confined to its central and southern parts, where two greenstone sequences have been recognised [Griffin, 1990; Figure 2a]. In these areas the lower greenstone sequence, which dominates the southern part of the province, consists of a quartzite layer at the bottom, overlain by mafic and ultramafic rocks, with felsic volcano-sedimentary rocks at the top. Griffin [1990] referred to the quartzite as *"basal quartzite"*, although no basement to it has been documented. Froude et al. [1983] reported ca. 3.3 to ca. 3.7 Ga SHRIMP U-Pb zircon ages for detrital zircons from this quartzite.

Nelson [1995] and Savage et al. [1996] obtained SHRIMP U-Pb in zircon ages of 2958±4 Ma and ca. 2.99 Ga for supracrustal rocks from the Ravensthorpe and West River greenstone belt at the southern end of the province. Further north, the felsic to intermediate volcanic rocks in the Lake Johnston greenstone belt have given SHRIMP U-Pb zircon ages of 2921±4 Ma and 2903±5 Ma [Wang et al., 1996b]. These data show that the age of the lower greenstone sequence is between ca. 2.9 and ca. 3.0 Ga. The upper greenstone sequence is a felsic volcano-sedimentary succession of calc-alkaline affinity [Hallberg et al., 1976]. It unconformably overlies the lower sequence and is best developed in the Diemals and Marda areas [Hallberg et al., 1976; Chapman et al., 1981] in the centre of the province to the north of Southern Cross. The most reliable ages for the upper sequence in the Diemals-Marda greenstone belt are the conventional U-Pb zircon ages of 2722±13 Ma by Pidgeon [1986] and 2735±2 Ma by Pidgeon and Wilde [1990].

There have been two estimates of the age of the Gum Creek greenstone belt. Myers [1990] suggested that it is ca. 2750 to ca. 2600 Ma old, and forms part of his Barlee Terrain. On the basis of stratigraphic correlations, Beeson et al. [1993] argued that the age of the greenstone successions is between ca. 2.9 and ca. 3.0 Ga, thus correlating them with the lower greenstone sequence of the Southern Cross Province, and with the Luke Creek Group in the adjacent Murchison Province, which has been dated between ca. 2.9 and ca. 3.0 Ga, e.g. the data summarised by Watkins and Hickman [1990], Pidgeon and Wilde [1990], and Wang et al. [1997].

The data obtained from this study are similar to the published ages for the upper greenstone sequence elsewhere in the Southern Cross Province and the newly obtained 2727±6 and 2702±10 Ma ages for the Mount Farmer Group in the Murchison [Schiøtte and Campbell, 1996], indicating a possible correlation between them. This study shows that at least the upper greenstone units (Unit III to V) in the Gum Creek greenstone belt are ~2.7 Ga old.

These units may unconformably overly Unit I and II. If so, it is possible that the lower mafic and BIF dominated sequences of Unit I and II represent a sequence similar in age to the ca. 2.9 to ca. 3.0 Ga lower greenstone sequences underlying Diemals-Marda complex. Such an interpretation is consistent with the geological make-up of Unit I and II (compare Figure 2a and 2b), and would suggest that the geological setting of Gum Creek is similar to the Diemals area. Unfortunately, no material suitable for dating could be collected from Unit I and II during the course of this study and, as a consequence, no age has been obtained for these units.

Implications from the Xenocrystic Zircon Ages

Xenocrystic zircons were found in a number of granitoid samples during routine analyses. The oldest age of 3671±29 Ma (1σ) is far older than the ca. 3.0 Ga proposed age of the lower greenstone sequence in each of the three granite-greenstone provinces that make up the Yilgarn Craton. Because the xenocrysts were found in granites, they are unlikely to be detrital zircons, and thus give evidence for the existing of old sialic basement beneath the Gum Creek greenstone belt. Detrital zircons up to 3242±33 Ma (1σ) were also found in a pyroclastic volcanic rock (93-998). Detrital zircons, older than ca. 3.0 Ga, have also been dated from within the Southern Cross Province. As previously mentioned, Froude et al. [1983] found two dominant populations of detrital zircons, with ages between ca. 3.7 and ca. 3.6 Ga, and ca. 3.4 and ca. 3.3 Ga respectively, in a quartzite from the Maynard Hill greenstone belt, about 80 km southeast of Sandstone. These observations support the conclusion that there was old sialic crust in the region.

Xenocrystic zircon ages, older than ca. 3.0 Ga, have also been obtained in both the Eastern Goldfields and Murchison Provinces [e.g. Claoué-Long et al., 1988; Compston et al., 1986; Hill et al., 1989; Wang and Campbell, unpublished data]. Claoué-Long et al. [1988] reported ca. 3.1 - 3.2 Ga xenocrysts from a chert layer in the Kapai Slate at Kambalda. Xenocrystic zircon with ages between ca. 3.2 and ca. 3.45 Ga were found in basalts from the same region [Compston et al., 1986]. At Norseman, ca. 3.1 - 3.2 Ga xenocrysts were found in a granite by Hill et al. [1989]. These authors interpreted their observations as providing evidence of old continental crust beneath the southeastern Yilgarn Craton. Furthermore, ca. 3.1 to ca. 3.5 Ga ages were dated for xenocrystic zircons in the Abbotts and Weld Range greenstone belts in the Murchison Province [Wang and Campbell, unpublished data]. Those data and the results presented in this study suggest that the old crust was very extensive and that it underlay most of the Yilgarn Craton.

Timing of Granitoids Emplacement and their Derivation

The 2638 to 2721 Ma ages determined for the four granitoids in the Gum Creek greenstone belt are, within analytical uncertainty, similar to the 2685 to 2737 Ma Pb-Pb whole rock isochron ages obtained from the granitoids near Diemals [Bickle et al., 1983]. In the Eastern Goldfields Province, the ages dated for the granitoid intrusions lie between ca. 2.7 and ca. 2.6 Ga [e.g. Hill et al., 1989, 1992; Hill and Campbell, 1993] with no granitoids older than ca. 2.7 Ga. However, some small to medium sized granitic plutons with ages between ca. 2.7 to ca. 2.75 Ga have been found in both the Southern Cross and Murchison Provinces [Bickle et al., 1983; Wang et al., 1996a; Schiøtte and Campbell, 1996] and this provides a significant difference between these two provinces and the Eastern Goldfields. We suggest that the granitoids in the Gum Creek greenstone belt and elsewhere in the Southern Cross Province were derived from partial melting of older sialic basement that underlay the greenstones [e.g. Oversby, 1975, 1978; Campbell and Hill, 1988] at ~ 2.7 Ga.

CONCLUSIONS

The felsic upper greenstone sequence (Unit III to V) of the Gum Creek greenstone belt has an age of ca. 2.7 Ga, not ca. 2.9 to ca. 3.0 Ga as previously thought. However, the lithological similarity between the mafic and BIF-dominated Unit I and II of the lower Gum Creek greenstone sequence and the lower greenstone sequence observed elsewhere in the Southern Cross Province, implies that they may have similar ages of ca. 2.9 to ca. 3.0 Ga. If this is the case, an unconformity must exist between the Unit II and Unit III at Gum Creek.

As in the Murchison Province, most external granitoids that enclose the greenstone belt in the area, are younger than ca. 2.7 Ga, but some smaller, discrete internal granitoids have been emplaced prior to ca. 2.7 Ga. The xenocrystic zircon ages indicate the existence of old sialic basement beneath the greenstone belt, and partial melting from this pre-existing continental crust is inferred to be the source of the granitoid intrusions.

Acknowledgements. We would like to thank the Renison Goldfields, Western Mining Corporation, BHP and Normandy Poseidon for their financial support. R. Rudowski, K. Massey, and J. Bitmead are thanked for their assistance in sample preparation and analysing respectively. QW acknowledges the receipt of an Australian National University PhD Scholarship. A. Nutman provided comments on early drafts of the manuscript. We thank R. Pidgeon and W. Griffin for their constructive reviews which improved the final version of the article.

REFERENCES

Beeson, J., D. I. Groves, and J. Ridley, Controls on mineralization and tectonic development of the central part of the northern Yilgarn Craton, *Rep. 109*, 92 pp., Mineral and Energy Res. Inst. West. Aust., Perth, WA, 1993.

Bickle, M. J., H. J. Chapman, L. F. Bettenay, and D. I. Groves, Lead ages, reset rubidium-strontium ages and implications for the Archaean crustal evolution of the Diemals area, Central Yilgarn Block, Western Australia, *Geochim. Cosmochim. Acta*, 47, 907-914, 1983.

Campbell, I. H., and R. I. Hill, A two-stage model for the formation of the granite-greenstone terrains of the Kalgoorlie-Norseman area, Western Australia, *Earth Planet. Sci. Lett.*, 90, 11-25, 1988.

Chapman, H. J., M. J. Bickle, J. R. de Laeter, L. F. Bettenay, D. I. Groves, L. S. Anderson, R. A. Binns, and M. P. Gorton, Rb-Sr geochronology of granitic rocks from the Diemals area, central Yilgarn Block, Western Australia, *Spec. Publ.. Geol. Soc. Aust.*, 7, 173-186, 1981.

Claoué-Long, J. C., W. Compston, and A. Cowden, The age of the Kambalda greenstones resolved by ion-microprobe: implications for Archaean dating methods, *Earth Planet. Sci. Lett.*, 89, 239-259, 1988.

Claoué-Long, J. C., W. Compston, J. Roberts, and C. M. Fanning, Two carboniferous ages: a comparison of SHRIMP zircon dating with conventional zircon ages and $^{40}Ar/^{39}Ar$ analysis, *SEPM Special Publ.*, 54, 3-21, 1995.

Compston, W., I. S. Williams, and C. Meyer, U-Pb geochronology of zircon from lunar breccia 73217 using a sensitive high mass-resolution ion microprobe, *J. Geophys. Res.*, 89 Supplement, B525-B534, 1984.

Compston, W., I. S. Williams, I. H. Campbell, and J. J. Gresham, Zircon xenocrysts from Kambalda volcanics: age constraints and direct evidence for older continental crust below the Kambalda-Norseman greenstones, *Earth Planet. Sci. Lett.*, 76, 299-311, 1986.

Cumming, G. L., and J. R. Richards, Ore lead isotope ratios in a continuously changing earth, *Earth Planet. Sci. Lett.*, 28, 155-171, 1975.

Elias, M., J. A. Bunting, and P. H. Wharton, *Explanation notes on Glengarry 1:250000 geological sheet*, Western Australia, 27 pp. Geol. Surv. West. Aust., 1982.

Froude, D. O., W. Compston, and I. S. Williams, Early Archaean zircon analyses from the central Yilgarn Block, *Ann. Rep. Res. Sch. Earth Sci.*, Aust. Nat. Univ., 125-126, 1983.

Gee, R. D., J. L. Baxter, S. A. Wilde, and I. R. Williams, Crustal development in the Archaean Yilgarn Block, Western Australia, *Geol. Soc. Aust. Spec. Publ.*, 7, 43-56, 1981.

Griffin, J. W., Southern Cross Province, in *Geology and Mineral Resources of Western Australia, Western Australia Geological Survey, Memoir 3*, pp. 60-77. Geol. Surv. Western Aust., Perth, 1990.

Hallberg, J. A., C. Johnston, and S. M. Bye, The Archaean Marda igneous complex, Western Australia, *Precambrian Res.*, 3, 111-136, 1976.

Hill, R. I., and I. H. Campbell, Age of granite emplacement in the Norseman region of Western Australia, *Aust. J. Earth Sci.*, 40, 559-574, 1993.

Hill, R. I., I. H. Campbell, and W. Compston, Age and origin of granitic rocks in the Kalgoorlie-Norseman region of Western Australia: implications for the origin of Archean crust, *Geochim. Cosmochim. Acta*, 53, 1259-1275, 1989.

Hill, R. I., B. W. Chappell, and I. H. Campbell, Late Archean granites of the southeastern Yilgarn Block, Western Australia: age, geochemistry, and origin, *Roy. Soc. Edinburgh: Earth Sci.*, 83, 211-226, 1992.

Myers, J. S., Precambrian tectonic evolution of part of Gondwana, southwestern Australia, *Geology*, *18*, 537-540, 1990.

Nelson, D. R., Compilation of SHRIMP U-Pb in zircon geochronology data, *Rec. 1995/3*, 244 pp., Geol. Surv. West. Aust., Perth, WA, 1995.

Otterman, D. W., Gidgee gold deposits, Jonesville, in *Geology of the Mineral Deposits of Australia and Papua New Guinea*, edited by F. E. Hughes, pp. 267-271, Australian Institute of Mining and Metallurgy, Melbourne, Vic., 1990.

Oversby, V. M., Lead isotope systematics and ages of Archaean acid intrusives in the Kalgoorlie-Norseman area, Western Australia, *Geochim. Cosmochim. Acta*, *39*, 1107-1125, 1975.

Oversby, V. M., Lead isotopes in Archaean plutonic rocks., *Earth Planet. Sci. Lett.*, *38*, 237-248, 1978.

Pidgeon, R. T., The correlation of acid volcanics in the Archaean of Western Australia, *Rep. 27*, 102 pp., West. Aust. Mineral and Petroleum Res. Inst., Perth, WA, 1986.

Pidgeon, R. T., and S. A. Wilde, The distribution of 3.0 Ga and 2.7 Ga volcanic episodes in the Yilgarn Craton of Western Australia, *Precambrian Res.*, *48*, 309-325, 1990.

Roddick, J. C., and O. Van Breemen, U-Pb zircon dating: a comparison of ion microprobe and single grain conventional analyses, In *Radiogenic Age and Isotopic Study*, Geol. Surv. Canada Current Res. 1994-F, Rep. 8, pp. 1-9, 1994.

Sambridge, M. S., and W. Compston, Mixture modelling of multi-component datasets with application to ion-probe zircon ages, *Earth Planet. Sci. Lett.*, *128*, 373-390, 1995.

Savage, M. D., M. E. Barley, and N. J. McNaughton, SHRIMP U-Pb dating of 3.0 Ga intermediate to sialic rocks in the Southern Yellowdine Terrain, Yilgarn Craton (abstract), *Geol. Soc. Aust. Abstr.*, *41*, 376, 1996.

Schiøtte, L., and I. H. Campbell, Chronology of the Mount Magnet granite-greenstone terrain, Yilgarn Block, Western Australia: implication for field based predictions of the relative timing of granitoid emplacement, *Precambrian. Res.*, *78*, 237-260, 1996.

Solomon, M., D. I. Groves, and A. L. Jaques, *The Geology and Origin of Australia's Mineral Deposits*, 941 pp., Clarendon Press, Oxford, 1994.

Steiger, R. M., and E. Jäger, Subcommission on geochronology: convention on the use of decay constants in geo- and cosmochronology, *Earth Planet. Sci. Lett.*, *36*, 359-362, 1977.

Tingey, R. J., *Explanation notes on Sandstone 1:250 000 geological sheet, Western Australia*, 37 pp. Geol. Surv. West. Aust., 1985.

Wang, Q., L. Schiøtte, and I. H. Campbell, Geochronological constraints on the age of komatiites and nickel mineralisation in the Lake Johnston greenstone belt, Yilgarn Craton, Western Australia, *Aust. J. Earth Sci.*, *43*, 381-385, 1996a.

Wang, Q., I. H. Campbell, and L. Schiøtte, Age constrain on the supracrustal sequence in the Murchison Province, Yilgarn Block - a new insight from the SHRIMP zircon dating (abstract),*Geol. Soc. Aust. Abstr.*, *41*, 458, 1996b.

Wang, Q., L. Schiøtte, and I. H. Campbell, Geochronology of the supracrustal rocks from the Golden Grove area, Murchison Province, Yilgarn Block, Western Australia, *Aust. J. Earth Sci., in press*, *1997*.

Watkins, K. P., and A. H. Hickman, Geological evolution and mineralization of the Murchison Province, Western Australia, *Bulletin 137*, 267 pp., Geol. Surv. Western Aust., Perth, WA, 1990.

J. Beeson, Goldfields Exploration Pty. Ltd, Post Box 862, Kalgoorlie, WA 6430, Australia

I. H. Campbell, Research School of Earth Sciences, Australia National University, Canberra, ACT 0200, Australia

Q. Wang, Research School of Earth Sciences, Australia National University, Canberra, ACT 0200, Australia